MID-LATITUDE
WEATHER
SYSTEMS

D1230900

MID-LATITUDE WEATHER SYSTEMS

TOBY N. CARLSON

Professor of Meteorology
The Pennsylvania State University

American Meteorological Society
Boston
1998

First published in 1991 by Harper Collins Academic

Reprinted in 1994 by Routledge

Reprinted with corrections and updates in 1998 by the American Meteorological Society

The American Meteorological Society
45 Beacon St.
Boston, MA 02108

Library of Congress Cataloging-in-Publication Data

Mid-Latitude Weather Systems
T. N. Carlson

Includes bibliographic references and index.
ISBN 978-1-878220-30-1
1. Meteorology

CIP 98-073847

Printed in the United States of America
by Braun-Brumfield, Inc., A Sheridan Group Company

The paper used in this book meets the minimum requirements of
American National Standard for Information Sciences—Permanence
of Paper for Printed Library Materials, ANSI Z39.48–1984.

MID-LATITUDE WEATHER SYSTEMS

TOBY N. CARLSON

Professor of Meteorology
The Pennsylvania State University

American Meteorological Society
Boston
1998

First published in 1991 by Harper Collins Academic

Reprinted in 1994 by Routledge

Reprinted with corrections and updates in 1998 by the American Meteorological Society

The American Meteorological Society
45 Beacon St.
Boston, MA 02108

Library of Congress Cataloging-in-Publication Data

Mid-Latitude Weather Systems
T. N. Carlson

Includes bibliographic references and index.
ISBN 978-1-878220-30-1
1. Meteorology

CIP 98-073847

Printed in the United States of America
by Braun-Brumfield, Inc., A Sheridan Group Company

The paper used in this book meets the minimum requirements of
American National Standard for Information Sciences—Permanence
of Paper for Printed Library Materials, ANSI Z39.48–1984.

*This book is dedicated to two former professors,
the late Frank Ludlam and Frederick Sanders,
from whom I learned so much.
It is also dedicated to my wife, AraBelle,
whose patience and understanding were crucial
in maintaining my enthusiasm during many years
preparing the manuscript.*

Preface

Synoptic and dynamic meteorology are both concerned with the motions of the atmosphere and their effect on weather and climate. Synoptic meteorology is primarily concerned with putting together observations in order to understand or predict the weather. Dynamic meteorology is more concerned with quantitative relationships (equations), particularly those equations that govern the motion of the air. Because of the vast wealth of material that must be taught in graduate and undergraduate programs in meteorology, and in order for the student to be presented with at least the rudiments of this science, courses are often divided artificially into separate compartments. Such compartmentalization tends to fragment meteorology into seemingly mutually exclusive branches. On one hand, dynamics tends to introduce the mathematical framework of atmospheric motions without adequately showing their relation to weather maps. On the other hand, synoptic meteorology often is concerned with description, analysis and forecasting of large-scale atmospheric motions, but without detailed physical or mathematical explanations. Much of the material in both disciplines can be bewildering to the novice.

After a year of courses in synoptic and dynamic meteorology, most students are able to grasp individual concepts, although the whole remains to be integrated. This course attempts to provide a fusion in which the behavior of synoptic-scale weather patterns is described in relation to the governing equations. In order to see clearly how the dynamics operate, it is necessary to remain as simple as possible without violating mathematical principles or ignoring important components in the equations.

A proper integration of the material covered by this course means that there must be omissions in the development of the equations and in describing all the interesting aspects of a particular weather situation or case study. To do otherwise infers an integration as extensive as its derivative material. This text, therefore, begins with a set of assumptions concerning the atmosphere, such as the quasi-geostrophic constraint, and proceeds to make further assumptions in order to arrive at certain simple expressions. The aim of this book is to describe these patterns and their evolution using mathematics. The key to understanding is the interpretation of the equations of motion. Because the total system of atmospheric motions implicitly describes everything, and therefore nothing in particular, we must attempt to simplify both what we observe and how we formulate these observations mathematically. We will do that through the use of simplistic models based on the equations set forth in this book.

Although there are some elementary definitions in this course and even some repetition of basic equations throughout the text, this course presumes that the student is well acquainted with the basic dynamics and with elementary map analysis. Thus no effort will be made to derive in detail the fundamental equations (such as the equations of motion and of thermodynamics); they will be merely stated and students will be provided with the assumptions needed to derive all the missing steps in the derivations.

This book was written for the benefit of advanced undergraduate or beginning graduate students in meteorology. Its material has evolved over more than 10 years of teaching at The Pennsylvania State University, in response to a need to fill in some gaps in meteorological pedagogy. While there are a number of very fine books on dynamics and a few concerning synoptic meteorology, there are virtually none that use conventional weather charts and equations to illustrate the behavior and evolution of weather patterns. I hope that this work will redress the balance and place synoptic meteorology squarely within the realm of a quantitative science.

The subject matter is mid-latitude synoptic weather systems, by which I imply a scale of motion that can be treated adequately using quasi-geostrophic theory. The material is subject to some constraints: the subjects would take one or two semesters to cover and the illustrations focus on case studies and topics with which I am intimately familiar. The book does not represent an unbiased survey of the literature but concentrates on selected concepts that I feel are supremely important if one is to gain a deep understanding of mid-latitude, synoptic-scale weather systems.

The text constitutes a suitable reference source for working scientists with some background in physics and mathematics. It assumes some basic knowledge of synoptic meteorology, dynamics and thermodynamics, although the relevant basic equations customarily elaborated upon in dynamics texts are encapsulated in this book. Given the quantitative nature of the material, I must assume that the reader is familiar with the basic equations and relationships (such as the geostrophic wind) outlined in Chapter 1. The purpose of the mathematics is to construct simple models from which one can describe the movement, development and structure of a particular feature or phenomenon.

The presence of so many equations may distress some people and please others. Purists may also fret because complete rigor may be set aside in favor of developing a semi-empirical relationship. Where possible, I have tried to give the reader a sense of what assumptions have been made and where the relationships might differ from a rigorous treatment. Many of the derivations are not absolutely complete in the sense that not all of the steps are given. In some cases, well-known formulae, such as the conservation of potential vorticity, are simply stated where I feel that the mathematical relationships can be found in other texts and the derivation is not essential to the understanding of the discussion that follows. The student is encouraged to recognize the similarities in the underlying physics as expressed in the various models.

If it is an inescapable fact that the language of science is mathematics, then the particular dialect used in meteorology is dynamics. To me, dynamics is happily wedded to observations, and an understanding of how observations relate to equations is an absolute necessity for any modeler or theoretician. I have therefore tried to temper models with observations and mathematical interpretation with empiricism, with just a dash of speculation. A unique aspect of this book is the abundance of weather charts, which feature parameters routinely issued by the United States Weather Service. These charts are used to illustrate the workings of the mathematical models. Students not familiar with weather maps should find this text useful as an introduction to patterns customarily analyzed on weather charts. I hope that those already familiar with the charts will find that this text serves as a means for interpretation of the conventional weather maps within the framework of a simplified dynamics. In the matter of mathematical notation, I have followed normal conventions, but I must point out that there are some superficial differences between this and other texts. For the most part, these differences reflect personal preference of the author. Except where noted by quotation marks, all terms are defined as in the *Glossary of meteorology* (American Meteorological Society, Boston, 1959).

Some readers may express disappointment at not seeing very much surface weather detail, such as precipitation, snowfall amounts or storm destruction statistics. In fact, some of the examples contain rather unspectacular weather but are nevertheless presented to illustrate a particular point. The reader will note that most of the maps center over the eastern two-thirds of the United States. The reason for this parochial choice of maps was not jingoism but convenience. These data are easily available to me and I have become quite familiar, through classroom teaching, with most of the examples, some of which were originally brought to my attention by students.

I have tried to avoid a discussion of local mesoscale effects (mountain breezes, cold air damming, local peculiarities of cyclones, etc.), not simply because mesoscale meteorology is a vast topic in itself, but because the effects are often imposed by local topography. I prefer to generalize the physical arguments. I realize, for example, that lee-side cyclogenesis over the Alps produces different patterns than it does in the lee of the Rocky Mountains in North America, but I am confident that, if the reader comprehends the examples given in this book, it will be a simple matter to apply the concepts to cases of lee-side cyclogenesis in any region. An exception to the exclusion of mesoscale processes is the case of fronts: here I include three chapters devoted to the kinematics and dynamics of fronts and jet streaks, which are byproducts of large-scale processes. The reader will also note that I occasionally use the directions "south" and "north" when I mean to say "equatorward" and "poleward". In most instances I try to avoid hemispheric bias but it is not always feasible to do so; there are just no graceful substitutes for directions such as southwesterly or northwesterly. I hope that Southern Hemisphere readers will overlook this semantic difficulty and make the appropriate coordinate transformations in their minds.

The approach adopted also differs from other texts in that it does not serialize the topics in neat compartments but, instead, attempts to present a picture: subjects are not examined as collections of research results but as elements of a whole, and each topic is related to all others. I wish to explain the behavior of meteorological processes and to give the reader an insight into processes and mechanisms rather than to list facts. It is for this reason that I do not wish to discuss operational forecast models, trends in meteorological research and instrumentation, techniques of computer analysis or current schools of thought, except as part of the historical tapestry. Today's operational models are tomorrow's obsolescence and yesterday's arguments are scarcely memories today. It may ultimately prove to be the case that some arguments are incomplete or partially incorrect and in need of modification. Therefore, I urge the reader, after having considered all the ramifications of the theory, to review the work in the light of his or her own insights and understanding.

Not only do I wish to avoid discussions involving a survey of models or which parameterizations are in favor or which approach yields the best forecast results, but my aim is to avoid the entire subject of weather forecasting, which is an art best left for another book. The contribution of this book to weather forecasting is the body of insights contained herein. Of course, I do use models to illustrate the atmospheric behavior, and I constantly refer to the historical development of different concepts and models. Models are necessary to simplify reality, and without them it would be difficult to isolate the important factors and mechanisms. Their function is simply to explain as many aspects with as much simplicity as possible of a particular set of phenomena. Models, however, are only approximations to the real physics. A perfectly complete model implies a complete mathematical description of the phenomenon and such a description should be left for the computer. Even with the aid of a computer, however, I feel that it is preferable to understand the physics of the atmosphere on simple levels before undertaking to comprehend the difficult. Without an intuitive understanding of the scale and range of the possible answers, one cannot be sure of having performed the calculations correctly with the computer.

The general flow of the discussions is intended to adhere to a simple progression. First, the basic equations are discussed and then reduced to the form of a simplified model, often based on empirical parameters. Next, the model is discussed in reference to schematic illustrations and finally to actual weather charts. Although that sequence of presentation may vary considerably from subject to subject, I hope that a clear sense of this progression remains unbroken in the mind of the reader and that both weather forecasters and mathematical modelers can profit by appreciating the necessary link between empiricism and pure theory. I have tried to minimize meteorological jargon, which abounds in the literature. In some cases the use of popular terms, such as "comma cloud" or "conveyor belt", is inescapable. Such terms, where not fully accepted by the meteorological community or listed in the *Glossary of meteorology*, are surrounded in quotes when first introduced. Terms that have gained more general acceptance are printed in *italics*.

Among some students, the romantic (almost quixotic) notions of weather forecasting have bred resentment toward anything highly mathematical, as if the excitement of reading a weather map will be dissipated by exposing the mechanistic nature of the atmosphere. As a result, the synoptician faces an identity crisis. Once, weather forecasting was performed solely by visual inspection of observations laboriously entered manually on charts. Except in some private enterprises, though, the romance of the independent forecaster has vanished within the lifetime of most of my students. In a way the practitioner is justified in feeling frustrated. First, meteorology differs from other sciences (such as engineering or medicine) in that the theoretician tends to receive more acclaim and reward than the practitioner. Second, weather prediction has become the province of the modeler, who is more often than not a dynamist. Operational weather forecasting is today governed by technology, the high-speed computer and sophisticated measurement systems. Improvement in weather forecasting skill has come about through greater computer capabilities, speed and memory, as well as through better numerical schemes and more efficient automatic data collection. These developments obscure the role of the empiricist and the forecaster, and raise the question whether intuitive understanding of weather processes is of value.

Meteorology has at last arrived as a true branch of physics and today exists more in the realm of fluid mechanics and mathematics. Neither the everyday forecaster nor the modeler can fully comprehend the increasingly complex models upon which routine forecasts are based. This trend toward greater complexity in operational forecasting could lead to a stultification of creative thought. I foresee a day in which operational models, having been developed layer upon layer over many years, will become almost incomprehensible black boxes to all but a very few specialists. There is a possibility that the intuitive scientist in meteorology will become a relic unless a conceptual knowledge can be passed on to the coming generation of meteorologists.

Acknowledgements

I could not have written this book without the advice and active participation of Dan Keyser, who provided me with many ideas and sketches for the material on fronts. I would also like to thank my department, notably its two successive heads during the period of writing, John Dutton and William Frank, for its encouragement and generosity in support of my effort. Thanks are also due to the secretariat, Nancy Warner, Joann Singer and Delores Corman, for their supreme patience with my incessant demands on their time.

I wish also to thank those whose invaluable assistance in reading and criticizing portions of the manuscript was highly beneficial: Peter Bannon, Phillip Smith, Jonathan Merritt, Stan Benjamin, Kevin Trenberth, Paul Kocin, Steve Koch, Louis Uccellini, Bob Burpee, Lloyd Shapiro, Charles Doswell, John Clark, Y.-H. (Bill) Kuo, Nelson Seaman, Alfred Blackadar and Greg Forbes. Figures 10.1 and 10.2 were published by permission of The Norwegian Academy of Science and Letters. The following figures were published by permission of the American Meteorological Society: 10.3, 13.2, 13.10, 15.7, 15.9, 15.11, 15.12, 15.13, 15.19, 16.5, 16.6, 16.7, 16.8, 16.9, 16.11, 16.13, 16.14, 16.15 and 16.17. I am also indebted to those who furnished me figures or data not available in the published literature: Paul Hirschberg, Fred Sanders, Louis Uccellini, Ralph Petersen, Paul Kocin, Jim Belles, Tim Dye, Mike Hemler and Greg Forbes. I would like to thank Mrs Alex Giedroc for her patience and skill in preparing many of the figures and Christopher Schreiner and Mary Young for editorial assistance. Finally, I would like to thank all the students in my advanced synoptic course who inadvertently contributed to the book by allowing me to use the classroom as a forum for my ideas and who provided me with endless material arising from their questions and from their class projects.

Contents

CONTENTS

CONTENTS

1

Introduction and mathematical definitions

1.1 Introduction

This book is about mid-latitude synoptic-scale weather systems and therefore it pertains to scales of motion that are "quasi-geostrophic" but not of planetary size. More concisely, the perturbations have wavelengths significantly below that of the Earth's circumference but with scales having a "Rossby number" below 1.0 (the Rossby number is defined in section 1.1). Generally, this corresponds to waves and cyclones with length scales between 1000 and 6000 km. An exception is made in this text in the case of fronts, which are simultaneously both mesoscale and synoptic-scale phenomena, depending on whether one refers to the width or length of the frontal zone. Indeed, it is shown in later chapters that fronts are forced by synoptic-scale motions and therefore may be treated within a kind of geostrophic system.

Since the subject matter is confined to mid-latitude meteorology, which is weather generally poleward of about 25° latitude, baroclinicity (horizontal temperature gradients) is the major driving mechanism for disturbance development. More precisely, we define mid-latitudes as being the zone within which

(a)   the atmosphere is essentially statistically stable with respect to moist processes (the wet-bulb or equivalent potential temperature increases with height),
(b)   traveling perturbations tend to conform to the Norwegian cyclone model,
(c)   cyclogenesis is primarily associated with baroclinic instability (which we will define and illustrate in subsequent chapters) and
(d)   quasi-geostrophic theory adequately governs the large-scale motions.

These qualifications enable the remaining subject matter to be treated in a manner that is generally beyond the scope of textbooks on dynamic or synoptic meteorology.

1

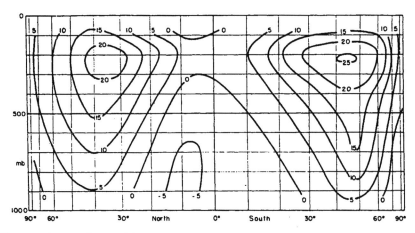

Figure 1.1 The time-and-longitude averaged zonal wind in the northern winter and southern summer (October–March). (From Lorenz, 1967.) Values are in m s^{-1}.

Although the general circulation is a subject that lies outside the scope of this text, it is useful to consider the mean state of the atmosphere over the globe. Mid-latitudes are characterized by westerly winds which increase with height to about 250 mb (Fig. 1.1), the level of the polar front jet. Equatorward of about 25° latitude, the lower tropospheric zonal winds are primarily from the east. That about half of the Earth's surface is covered by easterly winds and the other half by westerly winds is a necessary consequence of the conservation of absolute angular momentum of the Earth–atmosphere system. Another attribute of the mid-latitude winds is that they are inherently unsteady with regard to daily or weekly variations. Traveling waves and cyclones dominate mid-latitude weather, in contrast to the convective nature of tropical weather and the weak perturbations (except for hurricanes) of low latitudes.

Mid-latitude weather systems can be characterized as "pancake eddies", whose lateral dimensions are two or three orders of magnitude larger than in their vertical dimensions. The troposphere, whose depth is about 12–15 km, constitutes the domain over which exchanges of air are carried out by the large-scale eddies.

Though an important contributing factor in mid-latitude cyclogenesis, cumulus convection generally plays a small role in the overall energetics of mid-latitude meteorology. The reason for this is that the statically stable atmosphere at middle latitudes impedes deep convection except where the boundary layer is warm and moist, such as in the warm sector of cyclones. Figure 1.2 shows that dry and moist adiabats are almost horizontal and parallel at high latitudes, but the moist isentropes become increasingly vertical and cross dry isentropes towards the Equator. Equatorward of

2

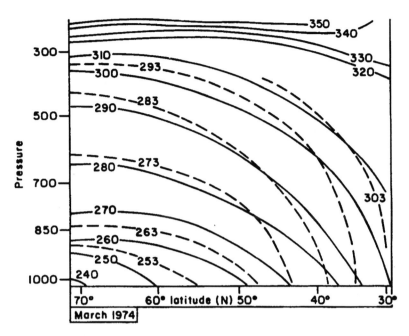

Figure 1.2 Mean vertical isentropic cross section for March 1974 over North America between 30° and 70°N. (From Carlson, 1981.) Full curves are dry isentropes (K); broken curves are moist isentropes (K).

about latitude 25° (according to the season), equivalent or wet-bulb potential temperature decreases with height below about 500 mb, a state that is said to be *potentially* (convectively) unstable. The zone of potential instability and weak horizontal temperature gradients shifts latitudinally with the seasons, resulting in a shift with the seasons in the polar front jet and the zone of extra-tropical cyclones. A hemispheric view of a mid-latitude cyclone can be seen in Figure 1.3. One is struck by the fact that the fractional cover of dense cloud occupies only about 10–20% of the area at mid-latitudes. Moreover, we see that the wave trains of cyclones seem to extend deep into the Tropics. Such cloud patterns are familiar to meteorologists. Clearly, the atmosphere, albeit complex, is selective and not totally chaotic in its structure and organization.

1.2 Basic units and equations

Coordinates referred to in this book are in the x, y, p (or z) and t system, and the units, unless otherwise stated, are in m.k.s. (S.I.), meters for length, kilograms for mass and seconds for time. Abbreviations for units conform to

3

Figure 1.3 Infrared geostationary satellite (GOES) photograph for western sector at 1100 GMT 15 January 1980, showing series of wave/cyclone cloud patterns at middle latitudes.

international usage. A list of symbols is provided in the appendix. Although units generally conform to internationally accepted ones, there are exceptions, such as in the case of pressure. Throughout this text the unit of pressure is the millibar (abbreviated mb here). The internationally accepted unit is the pascal (Pa), 100 Pa being equal to one millibar (1 mb). Use of the pascal is favored in calculations because it is consistent with the m.k.s. system. However, it is awkward to refer to the 50 000 Pa level instead of the 500 mb level. Sometimes, reference may be made to knots or degrees Fahrenheit where figures have been garnered from a relatively old source or where maps have been reproduced from U.S. Weather Service products. Except for demonstrating the magnitude of terms in the equations, there is little need for extensive computations.

Momentum equations

In view of the scale restrictions, it is acceptable to consider an approximate (Cartesian) form of the momentum equations that govern weather systems whose horizontal dimensions are very much less than the circumference of the Earth and whose vertical dimensions and vertical motions are very much less than their horizontal dimensions and horizontal motions. For the x direction the momentum equation is expressed as

$$\frac{du}{dt} = fv - \frac{1}{\rho}\frac{\partial p}{\partial x} + F_x \qquad (1.1a)$$

where u is the component of horizontal motion in the x direction, defined as $u = dx/dt$, f is the Coriolis parameter, ρ is the air density, p is the pressure and F_x is the frictional force per unit mass (whose vector component is F) acting on the particle in the x direction. Here, we refer to the Eulerian expansion of the total derivative (du/dt), which we define below. The Coriolis parameter f is defined as $2\Omega \sin \theta$, where Ω is the Earth's angular velocity $(0.73 \times 10^{-4}\,\text{s}^{-1})$ and θ is the latitude. Similarly, in the y direction the momentum equation is written

$$\frac{dv}{dt} = - fu - \frac{1}{\rho}\frac{\partial p}{\partial y} + F_y \qquad (1.1b)$$

where v is the y component of wind velocity $(v = dy/dt)$ and F_y is the y component of the frictional force.

Equations (1.1a) and (1.1b) express a Newtonian balance between the acceleration acting on an air parcel (dV/dt) and the Coriolis, pressure gradient and frictional forces per unit mass. The Coriolis force $(-2\Omega \times V)$ (the first terms on the right-hand side of (1.1a) and (1.1b)) acts to the right of

5

the motion in the Northern Hemisphere, the pressure gradient force $(-(1/\rho)\,\nabla_p p)$ acts at right angles to the isobars toward low pressure, and the frictional force (F) acts against the movement of the parcel.

A useful scaling length (L) and scaling velocity (U) for the ratio of the inertial force (U^2/L), which represents the acceleration term in (1.1a) or (1.1b), and the Coriolis force (fU) is the Rossby number (Ro, after the Swedish meteorologist C. G. Rossby)

$$Ro \equiv (U^2/L)/fU = U/fL.$$

Where the Rossby number is much less than 1.0, the inertial terms take on values much smaller than the pressure gradient and Coriolis terms, and the atmosphere is said to be in approximate geostrophic balance. Scale analysis shows that the Rossby number is much less than 1.0 for mid-latitude synoptic-scale weather systems. This is not to imply that accelerations are unimportant on these scales but that the atmosphere observes an approximate geostrophic balance.

The vertical equation of motion is written as

$$\frac{dw}{dt} = -\frac{1}{\rho}\frac{\partial p}{\partial z} - fu\cot\theta - g + F_z \tag{1.2}$$

where $w\,(= dz/dt)$ is the vertical motion, g is the acceleration due to gravity (the gravitational constant) and F_z is the vertical component of the frictional force. Synoptic-scale observations show that the change of pressure with height is almost exactly equal to $-\rho g$; the vertical acceleration is about two to three orders of magnitude smaller than g, which is almost exactly balanced by the vertical pressure gradient force. Equation (1.2) can be approximated for synoptic-scale motions as

$$\partial p/\partial z = -\rho g \tag{1.3}$$

which is the "*hydrostatic approximation*". The hydrostatic approximation follows from the idea that the horizontal size of synoptic-scale eddies (L), which are the prime focus of this book, is two or three orders of magnitude larger than the vertical depth of the eddies (H), which is approximately that of the troposphere. Thus, if $H \ll L$, the motions are almost horizontal ($V \gg w$), implying that the vertical accelerations are small compared to gravity, which must almost exactly balance the vertical pressure gradient force. (A rigorous treatment shows that the horizontally varying pressure component is in hydrostatic equilibrium with the horizontally varying density field.) The above equation states that the force of gravity is *exactly* balanced by the vertical pressure gradient force.

Thermodynamics

The thermodynamic state of the atmosphere (the pressure, temperature and density) is approximated by the equation of state, also known as the ideal gas law. For dry air

$$p = \rho R_d T \tag{1.4}$$

where p is the pressure, T is the temperature of air and R_d is the ideal gas constant for dry air ($R_d = 287$ J kg^{-1} K^{-1}). For the case of a moist atmosphere, the mixture of dry air and water vapor is less dense than the same amount of dry air. The gas constant for this mixture (R_m) is

$$R_m = R_d(1 + 0.61q) \tag{1.5a}$$

where q is the specific humidity, defined as the ratio of the mass of water vapor to a unit mass of air. The mixing ratio (r) is numerically almost identical to the specific humidity, being the ratio of the mass of water per unit mass of dry air. It is convenient to express the adjustment to R_d in the temperature, defining the "*virtual temperature*" T_v as

$$T_v = T(1 + 0.61q) \tag{1.5b}$$

so that the equation of state becomes

$$p = \rho R_d T_v. \tag{1.5c}$$

Another equation governing the thermodynamic state is the first law of thermodynamics, which, for dry air, is

$$c_p \frac{dT}{dt} - \frac{1}{\rho} \frac{dp}{dt} = \dot{Q} \tag{1.6}$$

where $\dot{Q}$ is the state of addition of heat to the parcel and c_p is the specific heat of dry air at constant pressure ($c_p = 1004$ J kg^{-1} K^{-1}), equal to $R_d + c_v$ where c_v is the specific heat of dry air at constant volume. For an ideal gas undergoing an *adiabatic* process (one in which there is no heat exchange with the environment), $\dot{Q}$ is zero and the first law can be written (after dividing (1.6) by T and applying the gas law (1.5c)) as

$$c_p \frac{d \ln T}{dt} - R_d \frac{d \ln p}{dt} = 0 \tag{1.7a}$$

7

Integrating this expression from a reference pressure p_0 and temperature θ to a pressure p and temperature T and taking the antilogarithm, we obtain

$$\theta = T(p_0/p)^\kappa \qquad\qquad (1.7b)$$

where $\kappa = R_d/c_p$. This relationship is known as *Poisson's equation* and θ is called the *potential temperature*, which is the temperature that a parcel of air at pressure p would have if compressed or expanded adiabatically from a standard pressure p_0, customarily taken to be 1000 mb. This process of dry adiabatic ascent, illustrated with regard to the schematic skew T–$\log p$ diagram (Fig. 1.4), conserves its value of θ as it ascends from a level p (denoted by the filled circle). Moist adiabatic ascent occurs once the parcel (p) reaches saturation. Its saturation level, or lifting condensation level (*LCL*), is determined by its initial values of specific humidity (q) and potential temperature (θ). Above the *LCL* further ascent brings the parcel along a moist adiabat (θ_w). The equivalent potential temperature (θ_e) is calculated by a dry adiabatic descent after all of the parcel's water vapor is

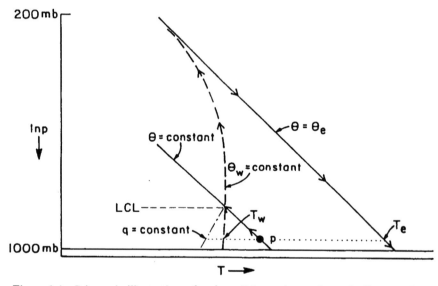

Figure 1.4 Schematic illustration of a skew T–$\log p$ thermodynamic diagram. The figure shows the process lines (full lines are constant potential temperature (θ) and broken curve is constant wet-bulb potential temperature (θ_w)) for a parcel. The parcel ascends from a reference level p (with specific humidity q) along constant potential temperature to its lifting condensation level (*LCL*) at the intersection of q and θ. It then ascends to the upper troposphere along the appropriate moist adiabat and finally down to 1000 mb, where its temperature (T_e) is the equivalent potential temperature (θ_e).

condensed following moist adiabatic ascent, as shown in Figure 1.4. We will continue to make use of this type of thermodynamic diagram throughout the text.

It can be shown easily (from (1.3), (1.4) and (1.7b)) that the adiabatic "lapse rate" $(\gamma_d) = -(dT/dz)_a = g/c_p$, which is approximately a constant over the troposphere and has the value of about 9.8 °C $(1000 \text{ m})^{-1}$. In general the actual lapse rate differs from the dry adiabatic one by an amount

$$\gamma_d - \gamma = \frac{\partial \theta}{\partial z}\left(\frac{T}{\theta}\right).$$

Subsequent discussions will refer to the stratification of the atmosphere. For a stably stratified atmosphere a parcel displacement upward or downward results in a return of the parcel to the original level. A dry atmosphere is said to be stably stratified if θ increases with height $(\partial\theta/\partial z > 0)$ and hence $\gamma < \gamma_d$.

Moist thermodynamics also are considered in this text. Two variables of importance are the equivalent potential temperature and the wet-bulb potential temperature, θ_e and θ_w. More properly, these quantities refer to pseudo-adiabatic potential temperatures, and the form given here can be found in standard dynamic or thermodynamic texts. In addition, derivative parameters, such as the "saturation wet-bulb potential temperature" $(\theta_{sw}$; the value of θ_w for a parcel with the temperature of the environment at the saturation specific humidity) are included. Qualitatively, one need only be acquainted with the evaluation of these thermodynamic variables with the aid of a conventional thermodynamic diagram, i.e. the skew T–log$p$ chart.

Continuity

The mass of air (dm) between two given pressure surfaces is a constant and equal to dp/g. This implies that vertical stretching of a column is exactly balanced by horizontal divergence of mass. Consequently, the conservation of mass in a volume bounded by pressure surfaces can be expressed simply and exactly as

$$\frac{\partial u}{\partial x} + \frac{\partial v}{\partial y} + \frac{\partial \omega}{\partial p} = 0 \tag{1.8a}$$

where $\omega \; (= dp/dt)$ is the vertical motion in pressure coordinates. Use of pressure rather than height as the vertical coordinate is particularly useful because it eliminates the time derivatives and density terms that occur when height is used. Equation (1.8a) is equivalent to saying that the three-dimensional mass divergence is zero. The two-dimensional divergence on a quasi-horizontal pressure surface (D_p) is defined as

$$D_p \equiv \left(\frac{\partial u}{\partial x} + \frac{\partial v}{\partial y}\right)_p = -\frac{\partial \omega}{\partial p}. \tag{1.8b}$$

These forms of the continuity equation (rather than that in x, y, z coordinates) are used almost exclusively throughout this text. In x, y, z coordinates, the form of the continuity equation is

$$-\frac{1}{\rho}\frac{d\rho}{dt} = \frac{\partial u}{\partial x} + \frac{\partial v}{\partial y} + \frac{\partial w}{\partial z} \tag{1.8c}$$

where the right-hand side closely resembles (1.8a). Indeed, the magnitude of the left-hand side of the equation is, for synoptic-scale weather systems, two orders of magnitude smaller than individual derivatives on the right-hand side, resulting in the incompressibility approximation that

$$\frac{\partial u}{\partial x} + \frac{\partial v}{\partial y} + \frac{\partial w}{\partial z} \approx 0. \tag{1.8d}$$

We can express the divergence as a fractional change of area (A) on a horizontal surface. Imagine a ring of air parcels enclosing area A. The equation governing the change with time of this area is

$$\frac{1}{A}\frac{dA}{dt} = \nabla_p \cdot V \equiv D_p = -\frac{\partial \omega}{\partial p}. \tag{1.8e}$$

Thus, the horizontal contraction or expansion of an area must be compensated by a vertical expansion or contraction of a column.

Total differentiation

Expansion of the total derivative d/dt for a quantity (say T) varying in x, y, p and t is expressed as

$$\frac{dT}{dt} = \frac{\partial T}{\partial t} + \frac{dx}{dt}\frac{\partial T}{\partial x} + \frac{dy}{dt}\frac{\partial T}{\partial y} + \frac{dp}{dt}\frac{\partial T}{\partial p}$$

or $\tag{1.9a}$

$$\frac{dT}{dt} = \frac{\partial T}{\partial t} + u\frac{\partial T}{\partial x} + v\frac{\partial T}{\partial y} + \omega\frac{\partial T}{\partial p}.$$

Similarly,

$$\frac{du}{dt} = \frac{\partial u}{\partial t} + u\frac{\partial u}{\partial x} + v\frac{\partial u}{\partial y} + \omega\frac{\partial u}{\partial p} \tag{1.9b}$$

and

$$\frac{\mathrm{d}v}{\mathrm{d}t} = \frac{\partial v}{\partial t} + u\,\frac{\partial v}{\partial x} + v\,\frac{\partial v}{\partial y} + \omega\,\frac{\partial v}{\partial p}\,. \tag{1.9c}$$

Derivatives have a physical significance in meteorology. *Total* differentials ($\mathrm{d}/\mathrm{d}t$) describe motion following a parcel, whereas *local* (partial) derivatives (with respect to x, y, p and t) describe changes occurring at a point in space. Total differentiation is sometimes called Lagrangian, as opposed to the Eulerian derivatives, which represent local changes in time at a fixed point. The local derivative is also called a "tendency", a practical example of which is the pressure tendency, which is customarily reported in conventional weather observations as a three-hour change in pressure at a station over the time interval immediately preceding the time of observation. Equations written in terms of the tendency of atmospheric variables are said to be "prognostic" because, in principle, integration of such equations with respect to time yields the state of the atmosphere in the future. Equations in which the total derivative with respect to time is equal to zero are said to be conservation equations. In a conservation equation, the scalar property remains unchanged (or *conserved*) following an air parcel.

The chain rule

The Eulerian expansion of a scalar variable (e.g. temperature) is written

$$\frac{\mathrm{d}T}{\mathrm{d}t} = \left(\frac{\partial T}{\partial t}\right)_{x,y,p} + \left(\frac{\partial T}{\partial x}\right)_{y,p,t}\frac{\mathrm{d}x}{\mathrm{d}t} + \left(\frac{\partial T}{\partial y}\right)_{x,p,t}\frac{\mathrm{d}y}{\mathrm{d}t} + \left(\frac{\partial T}{\partial p}\right)_{x,y,t}\frac{\mathrm{d}p}{\mathrm{d}t} \tag{1.10}$$

where the subscripts x, y, z and t refer to the coordinate axis along which the local derivative of temperature is held constant. Coordinate velocity components u, v and ω are defined, respectively, as $\mathrm{d}x/\mathrm{d}t$, $\mathrm{d}y/\mathrm{d}t$ and $\mathrm{d}p/\mathrm{d}t$. Thus, omitting subscripts, we arrive at the form given in (1.9a).

It is often useful to transform horizontal derivatives on a constant-height surface to quasi-horizontal derivatives on a constant-pressure surface. Consider the pressure distribution on a horizontal surface varying in a direction x and in the vertical. Expanding $\mathrm{d}p$ with the aid of the chain rule in terms of x and z only yields

$$\mathrm{d}p = \left(\frac{\partial p}{\partial x}\right)_z \mathrm{d}x + \left(\frac{\partial p}{\partial z}\right)_x \mathrm{d}z. \tag{1.11a}$$

If the variation of pressure ($\mathrm{d}p$) is along a pressure surface then $\mathrm{d}p = 0$ and (1.11a) expresses the transformation of coordinates from a horizontal variation of pressure on a surface of constant height to the variation of geo-potential height Z on a constant-pressure surface. Thus, invoking (1.3),

$$\left(\frac{\partial p}{\partial x}\right)_z = -\left(\frac{\partial p}{\partial z}\right)_x \left(\frac{\partial Z}{\partial x}\right)_p = \rho g \left(\frac{\partial Z}{\partial x}\right)_p \qquad (1.11b)$$

which is a very useful relationship in that it permits expression of the pressure gradient terms in the momentum equations as one variable, the height of a pressure surface (Z). Thus,

$$\frac{1}{\rho}\left(\frac{\partial p}{\partial x}\right)_z = g\left(\frac{\partial Z}{\partial x}\right)_p ; \qquad \frac{1}{\rho}\left(\frac{\partial p}{\partial y}\right)_z = g\left(\frac{\partial Z}{\partial y}\right)_p \qquad (1.11c)$$

can be substituted in the pressure gradient terms in the momentum equations (1.9b, c). Note that (1.11) resembles (1.3) but without the minus sign.

For movement on a constant-pressure surface, the u and v momentum equations take the form

$$\frac{du}{dt} = \frac{\partial u}{\partial t} + u\frac{\partial u}{\partial x} + v\frac{\partial u}{\partial y} + \omega\frac{\partial u}{\partial p} = -g\frac{\partial Z}{\partial x} + fv + F_x \qquad (1.12a)$$

and

$$\frac{dv}{dt} = \frac{\partial v}{\partial t} + u\frac{\partial v}{\partial x} + v\frac{\partial v}{\partial y} + \omega\frac{\partial v}{\partial p} = -g\frac{\partial Z}{\partial y} - fu + F_y. \qquad (1.12b)$$

In this form of the equations, the horizontal derivatives are taken on the constant-pressure surface, although the u and v components of the wind remain defined with respect to motion on a constant-height (horizontal) surface). Unless otherwise indicated, all other horizontal derivatives pertain to p surfaces.

The temperature equation

The thermodynamic (temperature) equation can be written after expanding the total derivative in terms of dry adiabatic and non-dry adiabatic processes, and employing (1.6) and the definition of ω ($= dp/dt$), as

$$\frac{dT}{dt} = \frac{\partial T}{\partial t} + u\frac{\partial T}{\partial x} + v\frac{\partial T}{\partial y} + \omega\frac{\partial T}{\partial p} = \frac{\dot{Q}_{nd}}{c_p} + \frac{1}{\rho c_p}\omega. \qquad (1.13a)$$

The right-hand side of (1.13a) expresses the total contribution to the rate of change of temperature following a parcel due to non-dry adiabatic heating plus dry adiabatic expansion or compression. We can solve for the local time derivative $(\partial T/\partial t)$ and consolidate the vertical motion term into the expression for the dry adiabatic change of temperature following an air parcel

$$(dT/dt)_{ad} = \omega(R_d T_v/c_p p)$$

which is the last term on the right-hand side of (1.13a). Thus,

$$\frac{\partial T}{\partial t} = -\left(u \frac{\partial T}{\partial x} + v \frac{\partial T}{\partial y}\right) + \omega\left(\frac{R_d T_v}{c_p p} - \frac{\partial T}{\partial p}\right) + \frac{\dot{Q}_{nd}}{c_p} \qquad (1.13b)$$

is the temperature tendency equation. This is treated in further detail in Chapter 3. Letting the virtual temperature $T_v = T$ and multiplying (1.13b) by $(p_0/p)^{\kappa} = \theta/T$ yields the analogous equation for potential temperature, which is

$$\frac{\partial \theta}{\partial t} = -\left(u \frac{\partial \theta}{\partial x} + v \frac{\partial \theta}{\partial y}\right) - \omega \frac{\partial \theta}{\partial p} + \frac{\dot{Q}_{nd}}{c_p}\left(\frac{\theta}{T}\right). \qquad (1.13c)$$

Vector notation

In vector notation the momentum equations can be written as

$$dV/dt + fk \times V = -g\nabla_p Z + F \qquad (1.14)$$

where the horizontal velocity vector (V) is $ui + vj$, i and j being the unit vectors along x and y, respectively. (In some instances it will be necessary to write the three-dimensional velocity vector (V_3).) The cross product of two vectors $A \times B$ is defined in x, y, z space as

$$A \times B = (A_x i + A_y j + A_z k) \times (B_x i + B_y j + B_z k)$$
$$= i(A_y B_z - A_z B_y) + j(A_z B_x - A_x B_z) + k(A_x B_y - A_y B_x). \qquad (1.15a)$$

Defining the three-dimensional gradient operator (∇_z) in x, y, z space as

$$\nabla_z = i \frac{\partial}{\partial x} + j \frac{\partial}{\partial y} + k \frac{\partial}{\partial z} \qquad (1.15b)$$

equation (1.15a) is expressed as a matrix

$$A \times B = \begin{vmatrix} i & j & k \\ A_x & A_y & A_z \\ B_x & B_y & B_z \end{vmatrix}. \qquad (1.16a)$$

13

Thus, the frequently used expression $\nabla_z \times A$ is

$$\nabla_z \times A = \begin{vmatrix} i & j & k \\ \dfrac{\partial}{\partial x} & \dfrac{\partial}{\partial y} & \dfrac{\partial}{\partial z} \\ A_x & A_y & A_z \end{vmatrix}. \tag{1.16b}$$

The dot product of the vectors $(A \cdot B)$ is a scalar frequently defined as

$$A \cdot B = |A| \, |B| \cos \alpha \tag{1.17a}$$

where α is the angle between the two vectors A and B. (Let us briefly note, however, that these mathematical coordinates are defined differently than meteorological coordinates, in which angles increase clockwise.) Use of the dot product in meteorology enters in the expansion for the total derivative, e.g. dT/dt, where

$$\frac{dT}{dt} = \frac{\partial T}{\partial t} + V_p \cdot \nabla_p T + \omega \frac{\partial T}{\partial p} \tag{1.17b}$$

where $V_p \cdot \nabla_p T$ is the dot product between the horizontal velocity vector on a constant-pressure surface (V_p) and the horizontal gradient vector of temperature $(\nabla_p T)$, defined as

$$\nabla_p T = i \frac{\partial T}{\partial x} + j \frac{\partial T}{\partial y} \tag{1.17c}$$

where the horizontal gradient vector on a constant-pressure surface $\nabla_p T$ is directed toward higher values. The dot product of the horizontal wind vector and the gradient (of temperature) is

$$V_p \cdot \nabla_p T = u \frac{\partial T}{\partial x} + v \frac{\partial T}{\partial y}. \tag{1.17d}$$

Thus, the thermodynamic and momentum equations can be written in vector notation, employing the dot product for the advection terms. Here, horizontal advection is defined on quasi-horizontal pressure surfaces as the rate of change of temperature due only to advection:

$$-V \cdot \nabla_p T = \left(\frac{\partial T}{\partial t} \right)_{\text{advection}}. \tag{1.18}$$

Note that the definition of advection involves a negative sign in front of the advection terms.

Continuity may also be expressed in terms of vector notation, employing the dot product; thus, (1.8) becomes

$$\nabla_p \cdot \mathbf{V}_p = \left(\frac{\partial u}{\partial x} + \frac{\partial v}{\partial y} \right)_p \equiv D_p. \tag{1.19}$$

Another differential operator is called the Laplacian (∇_z^2 or ∇_p^2), which here refers to the horizontal second derivative operator in x, y space, on either a constant-height or a constant-pressure surface. Thus, for temperature

$$\nabla_p^2 T = \left(\frac{\partial^2 T}{\partial x^2} + \frac{\partial^2 T}{\partial y^2} \right)_p. \tag{1.20}$$

Finally, a useful operator is the Jacobian. The Jacobian ($J_{xy}(a, b)$) of two scalars a and b is defined as

$$J_{xy}(a, b) = \frac{\partial a}{\partial x} \frac{\partial b}{\partial y} - \frac{\partial b}{\partial x} \frac{\partial a}{\partial y} \tag{1.21a}$$

where the subscripts x and y of the Jacobian refer to the direction of the derivatives of a and b, in this case along the x and y axes. An identity is that

$$J_{xy}(a, b) = - J_{xy}(b, a). \tag{1.21b}$$

The Jacobian can also be written as a determinant

$$J_{xy}(a, b) = \begin{vmatrix} \partial a/\partial x & \partial b/\partial x \\ \partial a/\partial y & \partial b/\partial y \end{vmatrix} \tag{1.22}$$

and therefore as a cross product (e.g. on a horizontal surface)

$$J_{xy}(a, b) = \mathbf{k} \cdot (\nabla a \times \nabla b) \tag{1.23}$$

which has the useful property of being inversely proportional to the area of the solenoids formed by intersecting contours of a and b on the horizontal surface.

The geostrophic approximation

In mid-latitude synoptic-scale weather systems the Coriolis force (fu; fv) and pressure gradient forces ($- g(\partial Z/\partial x)_p$; $- g(\partial Z/\partial y)_p$) are approximately an order of magnitude larger than the acceleration and frictional forces. The resulting approximate balance between the Coriolis and pressure gradient forces is termed geostrophic balance. As we have said, this balance is maintained in regions where the Rossby number is much less than 1.0, i.e. for synoptic-scale systems at mid-latitudes. The equation of Coriolis to pressure gradient forces is valid to within about 10–20%; this degree of balance seems to hold to fairly low latitudes. Therefore, (1.14) becomes, when transformed by (1.11), the definitions of u_g and v_g. The geostrophic wind balance is

$$-fv_g = -g\left(\frac{\partial Z}{\partial x}\right)_p \tag{1.24a}$$

and

$$fu_g = -g\left(\frac{\partial Z}{\partial y}\right)_p \tag{1.24b}$$

where the subscript g denotes a geostrophic value. Equations (1.24) relate the geopotential height field at a given time to the geostrophic wind components. In rapidly changing weather patterns (e.g. cyclogenesis), the non-geostrophic component may become locally rather large. Moreover, in the planetary boundary layer (to be defined later) and in regions of strong turbulent dissipation of kinetic energy (for example, near jets), the frictional component may become as large as the Coriolis and pressure gradient forces.

That the geostrophic wind closely approximates the actual wind is an expression of the essential balance of the atmosphere. Weather patterns do change with time relatively slowly, allowing the meteorologist to recognize continuity in weather patterns from day to day. In regions of strong curvature, the winds may be better approximated by gradient wind balance, which expresses a force balance between Coriolis force, pressure gradient force and centrifugal force. The latter represents the effects of curvature on the flow and requires knowledge of the radius of curvature of the trajectories when evaluated in cylindrical coordinates. It is likely that the atmosphere maintains a gradient wind balance to within 5–10% on the synoptic scale. Practical use of the gradient wind equations, however, is a bit cumbersome, and the evaluation of the terms involving curvature is difficult and imprecise without the aid of a computer. Gradient wind balance is not treated in this text because it is not a sufficient improvement over geostrophic balance to warrant the added complexity.

In vector notation (1.24) can be expressed as

$$V_g = (g/f)k \times \nabla_p Z. \tag{1.25}$$

16

The momentum equations (1.12) can be written using (1.24) as

$$\frac{du}{dt} = \frac{\partial u}{\partial t} + u\frac{\partial u}{\partial x} + v\frac{\partial u}{\partial y} + \omega\frac{\partial u}{\partial p} = f(v - v_g) + F_x = fv_{ag} + F_x \quad (1.26a)$$

and

$$\frac{dv}{dt} = \frac{\partial v}{\partial t} + u\frac{\partial v}{\partial x} + v\frac{\partial v}{\partial y} + \omega\frac{\partial v}{\partial p} = -f(u - u_g) + F_y = -fu_{ag} + F_y \quad (1.26b)$$

where u_{ag} and v_{ag} are the ageostrophic wind components, defined as $u_{ag} = u - u_g$ and $v_{ag} = v - v_g$. Equations (1.26) suffer no loss of generality from (1.12), except that they allow the momentum equations to be expressed in terms of the deviation of the wind velocity components from a base (geostrophic) reference state. Since it is possible, without a great loss in accuracy, to assume that the acceleration or momentum terms in (1.26a, b) are small compared to the pressure gradient and Coriolis terms, we could define the inertial (acceleration) terms as being subject to geostrophic constraint. We can expand the total geostrophic derivative for the momentum equation to obtain

$$\frac{D_g u}{dt} = \frac{\partial u_g}{\partial t} + u_g\frac{\partial u_g}{\partial x} + v_g\frac{\partial u_g}{\partial y} + \omega\frac{\partial u_g}{\partial p}$$

and

$$\frac{D_g v}{dt} = \frac{\partial v_g}{\partial t} + u_g\frac{\partial v_g}{\partial x} + v_g\frac{\partial v_g}{\partial y} + \omega\frac{\partial v_g}{\partial p}.$$

(Later in this text, we will define a set of equations in which the advecting winds are the total winds, but the gradients of the wind components are geostrophic. This system of equations (Ch. 14) is known as the geostrophic momentum approximations.) In terms of (1.26) the above expression can be written as

$$D_g u/dt = fv_{ag} + F_x \quad (1.27a)$$

and

$$D_g v/dt = -fu_{ag} + F_y. \quad (1.27b)$$

If the vertical derivative terms in the momentum components of (1.27) $(\omega\partial u_g/\partial p;\ \omega\partial v_g/\partial p)$ are set equal to zero and friction is neglected, the resulting form of the equations constitutes the *quasi-geostrophic approxi-*

17

mation, which is justified in Chapter 3. This important modification to (1.27) is expressed thus for the quasi-geostrophic momentum equations with friction:

$$\frac{d_g u}{dt} = \frac{\partial u_g}{\partial t} + u_g \frac{\partial u_g}{\partial x} + v_g \frac{\partial u_g}{\partial y} = f v_{ag} \tag{1.28a}$$

$$\frac{d_g v}{dt} = \frac{\partial v_g}{\partial t} + u_g \frac{\partial v_g}{\partial x} + v_g \frac{\partial v_g}{\partial y} = -f u_{ag}. \tag{1.28b}$$

Equation (1.28) serves as a foundation for quasi-geostrophic theory and for many important derivations pertaining to cyclogenesis that we present in subsequent chapters. Note that quasi-geostrophic theory allows for the fact that there is non-geostrophic flow via inclusion of the ageostrophic components in the momentum equations. We will see that it is impossible to describe cyclogenesis without retention of these ageostrophic terms.

The temperature equation, compatible with the quasi-geostrophic approximation in (1.13), is expressed

$$\frac{d_g T}{dt} = \frac{\partial T}{\partial t} + u_g \frac{\partial T}{\partial x} + v_g \frac{\partial T}{\partial y} + \omega \frac{\partial T}{\partial p} \tag{1.29a}$$

or

$$\frac{d_g \theta}{dt} = \frac{\partial \theta}{\partial t} + u_g \frac{\partial \theta}{\partial x} + v_g \frac{\partial \theta}{\partial y} + \omega \frac{\partial \theta}{\partial p}. \tag{1.29b}$$

Geostrophic advections can be expressed as a Jacobian; for example, in the case of geostrophic temperature advection

$$-V_g \cdot \nabla_p T = -\left(u_g \frac{\partial T}{\partial x} + v_g \frac{\partial T}{\partial y} \right) = \frac{g}{f} J_{xy}(T, Z). \tag{1.30}$$

Thus, a Jacobian of a scalar variable and geopotential height can be interpreted physically as an advection. Considerable use is made of the Jacobian in the discussion of fronts and transverse (ageostrophic)/vertical motion in Chapters 14 through 16. In Chapter 14 it will be shown that the sign and magnitude of a Jacobian give the direction and strength of the transverse/vertical circulations in two-dimensional planes.

Equations (1.28) and (1.29) constitute the basis for the development theory contained in Chapters 4 through 11. The geostrophic constraint can be relaxed, if necessary, to include components of the advection of momentum that are not geostrophic, such as in the case of fronts where ageostrophic components normal to the front (y direction) are retained in both the momentum and temperature equations. These more general geostrophic momentum equations are discussed in some detail in Chapter 14.

Hypsometric equation and thermal wind relationships

Substitution of the hypsometric equation of state (1.4) in the hydrostatic approximation (1.3) yields the hypsometric relationship

$$\partial Z/\partial p = -R_d T_v/gp \tag{1.31}$$

which, when integrated between two pressure surfaces, gives

$$\Delta Z = -\frac{R_d}{g} \int_{p_b}^{p_t} T_v \, d\ln p \tag{1.32a}$$

where p_t and p_b are the pressures at the upper and lower levels, respectively, and ΔZ is the thickness of the layer between the two pressure surfaces. If one defines the mean virtual temperature in the layer as

$$\bar{T}_v \equiv \frac{\displaystyle\int_{p_t}^{p_b} T_v \, d\ln p}{\displaystyle\int_{p_t}^{p_b} d\ln p} \tag{1.32b}$$

the integration yields the hypsometric equation

$$\Delta Z = \frac{R_d}{g} \bar{T}_v \ln\left(\frac{p_b}{p_t}\right). \tag{1.32c}$$

For deep layers (e.g. 1000–500 mb), it is sufficiently accurate to ignore the virtual temperature correction and to let $T_v = T$.

Now, taking the geostrophic wind equations (1.24) at two pressure surfaces and subtracting yields, upon substitution of (1.32c),

$$u_{gt} - u_{gb} = -\frac{g}{f}\frac{\partial}{\partial y}\Delta Z = -\frac{R_d}{f}\ln\left(\frac{p_b}{p_t}\right)\frac{\partial \bar{T}_v}{\partial y} \tag{1.33a}$$

and

$$v_{gt} - v_{gb} = \frac{g}{f}\frac{\partial}{\partial x}\Delta Z = \frac{R_d}{f}\ln\left(\frac{p_b}{p_t}\right)\frac{\partial \bar{T}_v}{\partial x} \tag{1.33b}$$

This vertical shear of the geostrophic wind, which is a function of the mean horizontal virtual temperature gradient, is called the thermal wind. In vector notation the thermal wind equation can be written

$$V_{gt} - V_{gb} = \frac{R_d}{f} \ln\left(\frac{p_b}{p_t}\right) k \times \nabla_p \bar{T}_v \equiv V_T \qquad (1.34a)$$

or

$$V_{gt} - V_{gb} = \left(\frac{g}{f}\right) k \times \nabla_p(\varDelta Z) \equiv V_T \qquad (1.34b)$$

or

$$V_{gt} - V_{gb} = \left(\frac{g}{f}\right) k \times \nabla_p(Z_t - Z_b). \qquad (1.34c)$$

Expressed as components of the geostrophic wind shear, we have

$$\frac{\partial u_g}{\partial p} = \frac{R_d}{pf}\frac{\partial T}{\partial y} = \gamma\frac{\partial \theta}{\partial y} \qquad (1.34d)$$

and

$$\frac{\partial v_g}{\partial p} = -\frac{R_d}{pf}\frac{\partial T}{\partial x} = -\gamma\frac{\partial \theta}{\partial x} \qquad (1.34e)$$

where

$$\gamma \equiv \left(\frac{R_d}{pf}\right)\left(\frac{p}{p_0}\right)^{\kappa}. \qquad (1.34f)$$

The hodograph and static stability

According to (1.34a–f), the thermal wind vector is the vector difference between the geostrophic wind at two levels. The thermal wind vector is parallel to the mean layer isotherms or thickness lines, and its magnitude is proportional to the mean horizontal temperature or thickness gradient. Graphically this relationship can be illustrated with reference to a polar diagram or hodograph, such as shown in Figure 1.5. The two vectors (V_1 and V_2) are assumed to be the geostrophic wind velocities at the pressure levels p_1 and p_2. The thermal wind vector (V_T) extends from the head of the lower vector to the head of the upper one. The layer-mean isotherm ($\bar{T}_v$) is the same direction as V_T and the gradient of $\bar{T}_v$ is proportional to the magnitude of the

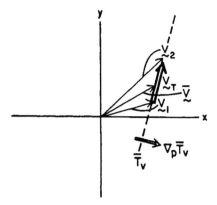

Figure 1.5 Hodograph of two geostrophic wind vectors at two levels in the atmosphere, the lower one labeled V_1 and the upper one V_2. The mean virtual temperature isotherm in the layer is labeled $\bar{T}_v$ and the direction and magnitude of the thermal wind (V_T) is indicated by the bold-faced vector arrow. The gradient of mean virtual temperature is labeled accordingly. $\bar{V}$ represents the mean wind velocity over the layer.

thermal wind. Just as there are higher geopotential heights on the right side (facing downwind) of the geostrophic wind vector, there is warmer air (higher thickness) on the right side of the thermal wind vector. Thus, a "right-hand rule" is that mean (layer-averaged) isotherms bear the same relationship to the magnitude and direction of the mean temperature gradient as the geostrophic wind velocity bears to the field of geopotential height Z.

Now, it can also be shown by referring to the hodograph that, if the wind varies continuously between V_1 and V_2 in the layer ΔZ, the mean wind in the layer ($\bar{V}$) must lie between V_1 and V_2 and the advection of temperature can be expressed by the component of the mean wind normal to the isotherm. Thus, the example of Figure 1.5 is one of cold advection. In general, cold advection occurs if the geostrophic wind direction "backs" (counterclockwise rotation) with height, and warm advection occurs if the wind "veers" (clockwise rotation) with height. (There may be some ambiguity if the wind varies by an amount close to 180°, however.)

An important factor is the static stability. Static stability is defined in terms of temperature (T) or potential temperature as

$$\gamma \equiv -\frac{\partial T}{\partial z} \qquad \Gamma \equiv \frac{\partial \theta}{\partial z}$$

$$\left[\text{or} -\frac{\partial \theta}{\partial p} \right]$$

$$\gamma_d = -\left(\frac{\partial T}{\partial z}\right)_d \qquad \gamma_m = -\left(\frac{\partial T}{\partial z}\right)_m \qquad \Gamma_{m,d} = 0$$

21

where the Greek letter gamma (γ, Γ) refers to the vertical derivative of temperature (lower case) or potential temperature (capital), and subscript d refers to the dry adiabatic and m to the moist adiabatic *lapse rate*. Mean static stability over a given atmospheric depth can be defined with respect to the mean isotherms in two sublayers

$$\bar{\gamma} = - \Delta T / \Delta Z$$

where the overbar signifies the vertical mean stability over the layer ΔZ.

A diagnosis of static stability and static stability advection can be made using the hodograph by plotting three geostrophic wind velocities for three levels on the hodograph (p_1, p_2, p_3) and considering vertical differences of mean layer temperature with height between the two layers. If the respective pairs of pressure levels have the same ratios and therefore the same values of the logarithm (e.g. $\log(1000/500) = \log(500/250)$), the vectors connecting the heads of the two thermal wind vectors (when the latter are displaced to a common origin on the hodograph) give the direction of the mean layer static stability isopleths (the "stability wind"). The magnitude of the gradient of static stability is exactly analogous to the mean temperature gradient, as given by two geostrophic wind velocities, which determine the direction of the thermal wind and the magnitude of the temperature gradient. Given the mean wind in the layer, the advection of static stability can be diagnosed in a fashion analogous to that of the advection of mean layer temperature using the thermal wind vectors and the right-hand rule.

Friction

Friction constitutes an important force in the planetary boundary layer and causes the winds to deviate substantially from geostrophic balance near the ground. Balance is no longer approximately maintained between pressure gradient and Coriolis forces near the surface as it is at mid-levels, but is approximately between these two forces and friction, which acts in a direction *opposite* to the motion of the air parcel. The effect of friction on the velocity of the air parcel is thus to introduce a substantial ageostrophic component near the surface. Inspection of any surface weather map shows that winds tend to blow across isobars from high to low pressure. This cross-isobaric component is largely due to friction.

Treatment of friction must therefore lie outside the realm of quasi-geostrophic theory, although we will return to a discussion of the role of friction in cyclogenesis in regard to the vorticity equation (Ch. 3) and the omega equation (Ch. 8).

Complicating the treatment of friction is the fact that there is more than one process in the atmosphere that constitutes friction. Friction causes a

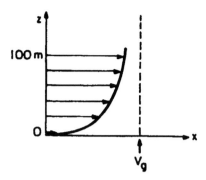

Figure 1.6 Schematic vertical profile of wind speed near the surface. Arrows are proportional to the speed, and the surface geostrophic wind speed is indicated by the vertical broken line labeled V_g.

dissipation of kinetic energy by transferring momentum from the air to the ground. Friction also causes internal eddy dissipation above the surface due to horizontal and vertical mixing, but this process merely transfers the kinetic energy from larger to smaller scales of motion. We say that this motion energy is "lost", whereas the energy is actually transferred to chaotic motion on scales that are not resolvable by conventional observations. The ultimate sink for this energy is turbulence and heat. Turbulent dissipation is produced by large values of horizontal or vertical wind shear, and also by buoyant forces. Thus friction, as defined in the broader sense, can become a significant term in the energetics, not only near the surface, but at upper levels near strong jets and in regions of deep cumulus convection.

The symbols F_x and F_y in (1.12) represent the net effect of different contribution by surface drag and internal mixing. Only surface drag is considered here because it is easier to treat analytically and constitutes an important kinetic-energy sink for cyclones. We normally think of friction in terms of solid bodies, objects rubbing against each other. Friction also occurs in fluids as a result of the transfer of momentum by eddies, rather than by molecular processes, along a gradient in wind speed. Figure 1.6 depicts the typical surface-layer wind speed profile as a function of height. The vertical wind shear, $\partial V/\partial z$, increases toward the surface as a consequence of the decreasing efficiency and size of the eddies with decreasing distance from the lower boundary.

According to Newton's law of viscosity, the shearing *stress* (the drag force per unit area, τ) is proportional to the vertical gradient of the wind speed in the direction of the wind. That constant of proportionality is called the dynamic coefficient of viscosity (μ). Thus, the law is written

$$\tau_{xz} = \mu \frac{\partial u}{\partial z} \approx - \rho g \mu \frac{\partial u}{\partial p} \qquad (1.35a)$$

23

for the stress component in the x direction. The viscous force per unit mass on a thin horizontal slab embedded within the air is written in the y, z plane as

$$\tau_{yz} = \mu \frac{\partial v}{\partial z} \approx - \rho g \mu \frac{\partial v}{\partial p} . \qquad (1.35b)$$

It is also possible to write these stress components in terms of the eddy flux of momentum or eddy stresses,

$$\tau_{xz} = - \rho \overline{u' w'} \qquad (1.36a)$$

$$\tau_{yz} = - \rho \overline{v' w'} \qquad (1.36b)$$

where the primed quantities refer to the departures of the speed components from a time average, which is symbolized by the overbar. This equation illustrates the fact that momentum (as well as heat, water vapour and other constituent) fluxes in the atmosphere are accomplished almost exclusively by eddy motion rather than by molecular transport. The coefficient (μ) for momentum transport in (1.35) represents the eddy viscosity and a measure of the eddy sizes in the planetary surface layer. Since eddy size decreases toward the lower boundary, while the shear increases, the eddy momentum stresses tend to remain relatively constant with height over the lowest 30–100 m and only slowly decrease with height up to the top of the planetary boundary layer, which we define as the hypothetical level at which $\tau = 0$.

Between the top of the planetary surface layer and the top of the planetary boundary layer (about 0.5–1.0 km), the vertical profile of wind speed gradually approaches geostrophic balance, following a somewhat complicated route known as the Eckman spiral. Most of the change in wind speed with height occurs in the planetary surface layer, as suggested by Figure 1.6. For all practical purposes (at least for synoptic-scale processes) the wind stress approaches zero at the top of the planetary boundary layer. The average value of the frictional force components (F_x and F_y) in the planetary boundary layer are therefore equal to

$$F_x = - g \, \tau_{x0} / \Delta p_f \qquad (1.37a)$$

and

$$F_y = - g \, \tau_{y0} / \Delta p_f \qquad (1.37b)$$

where τ_{x0} and τ_{y0} are the stress components at the surface and Δp_f is the depth of the friction layer in pressure coordinates. Equations (1.37a, b) can be expressed in terms of the *friction velocity* (U_*), which is defined as

$$\tau = \rho U_*^2.$$

The *surface drag coefficient* C_d can be expressed in terms of the surface wind speed V_s as

$$C_d = U_*^2/V_s^2.$$

This allows us to express the surface stress components in terms of the surface wind speed components (u_s; v_s) as

$$\tau_{xz} = \rho_s C_d |V_s| u_s \tag{1.38a}$$

$$\tau_{yz} = \rho_s C_d |V_s| v_s. \tag{1.38b}$$

Henceforth, we will assume that the quantity ($\rho_s C_d$) in (1.38) can be treated as a constant. We will see that these simplifications allow frictional dissipation of momentum and vorticity to be expressed in relatively compact forms.

Thus, frictional stresses, which represent forces acting opposite to the velocity vector, are expressed solely in terms of a surface drag and are assumed to vanish at the top of the planetary boundary layer. Surface drag is assumed to apply at anemometer level (nominally 10 m). The surface drag coefficient (C_d) tends to vary with wind speed and static stability and roughness. It has the value of about 1×10^{-3} over water, but may be several times that magnitude over land. In general, the surface wind speed (V_s) at 10 m is about one-third the magnitude of the surface geostrophic wind speed over land and about two-thirds that of the surface geostrophic wind speed over the ocean. We will make use of this scaling idea in expressing frictional effects in terms of the 1000 mb geostrophic wind speed.

Problems

1.1 If the mean virtual temperature in a vertical column between 1000 and 100 mb has a measurement error of 1 °C, what is the resulting error implied in the derived height of the 100 mb surface, assuming that the 1000 mb height is correct? How much of a vertically averaged error must there be in the mixing ratio (or specific humidity) in order for this error in thickness to be due entirely to the measurement of specific humidity?

1.2 What is the corresponding change in height of the 1000 mb surface in a hydrostatic atmosphere for a 1 mb change in sea-level pressure, assuming the temperature is 300 K? How about the change in the 500 mb height for a change in 1 mb of the 5.8 km surface? (Assume a reasonable temperature at 500 mb.)

1.3 Draw to scale on the axis three vectors representing, respectively, the 1000 mb geostrophic wind (at 10 m s^{-1} toward the east; wind direction 270°), the 500 mb geostrophic wind (30 m s^{-1} toward the northeast; wind direction 225°) and the 250 mb geostrophic wind (30 m s^{-1} toward the north; wind direction 180°). Determine the direction of the thermal wind and the orientation of the mean isotherms in each of the two layers, and label the cold and warm sides of the isotherm. Determine the sense (cold or warm) of the mean temperature advection in each of the two layers. Finally, translate one of the thermal wind vectors to the origin of the other and construct a vector from the head of the lower to that of the upper vector. Determine the direction of the isopleth of mean static stability over the 1000–250 mb column and the direction of the gradient of static stability in that layer; determine the sense of the static stability advection.

Further reading

Dutton, J. A. 1976. *The ceaseless wind.* New York: McGraw-Hill.

Holton, J. R. 1979. *An introduction to dynamic meteorology,* 2nd edn (Int. Geophys. Ser., Vol. 23). New York: Academic Press.

Lorenz, E. N. 1967. *The nature and theory of the general circulation of the atmosphere.* WMO Publ. no. 218. TP. 115.

McIlveen, R. 1988. *Basic meteorology. A physical outline.* London: Van Nostrand Reinhold.

Petterssen, S. 1956. *Weather analysis and forecasting II.* New York: McGraw-Hill.

2

Vorticity and vertical motion

A fundamental goal of synoptic meteorology is formulating quantitative techniques for evaluating the vertical motion field. Vertical motions are necessarily small for the synoptic scale, which is approximately hydrostatic, and hence cannot be measured accurately using the continuity equation. The vertical motion field is important not only as a result of its effect in producing clouds and precipitation, but also because of its relationship to the vertical profiles of horizontal divergence and convergence associated with the cyclogenesis process, which is quantified by a localized temporal increase in the vertical component of vorticity. Even if vertical motion could be measured accurately using the continuity equation, it is nevertheless useful to be able to relate the vertical motions and the surface or sea-level pressure tendency $(\partial p/\partial t)_s$ to the equations of motion in order to understand the physical processes that govern the behavior of disturbances. Since we live at the surface, we are most aware of changes in pressure, temperature and weather at the surface. Accordingly, the subsequent discussion is aimed at understanding these changes.

Vorticity is probably one of the most overworked and misunderstood quantities used in meteorology. Although not a primary quantity such as momentum, it describes a very important aspect of a fluid, which is its spin. Inspection of daily weather maps shows that cyclones and anticyclones rotate in circular eddies or move within wave-like undulations in the basic current. Growth and decay of cyclones and anticyclones is very much related to changes in the fluid's vorticity. The vertical component of vorticity is just a transformation of the horizontal velocity field and its tendency captures the essentials of the horizontal equations of motion (1.12). In the quasi-geostrophic system, which is the underlying basis of this text, changes in surface vorticity are related directly to changes in surface pressure, while advections of vorticity (and temperature) at upper levels are used to diagnose these surface vorticity changes. Cyclone development, the process by which cyclones or anticyclones become more intense with time, is expressed using the vorticity tendency equation. Generally, we will refer to the positive tendency of vorticity as "spin up" and the negative vorticity tendency as "spin down". In special cases, we may also refer to the specific development of cyclones spin up (or spin down) to denote a general positive (or negative) tendency of vorticity at the center. For highs, spin up refers to a negative vorticity tendency and spin down to a positive vorticity tendency.

In this chapter the concept of vorticity is examined, and the vertical component of vorticity is related to features on conventional weather maps with an emphasis on geostrophic balance. We also examine a related kinematic property of a fluid, the divergence, which is a transformation of the vertical motion profile. In the next chapter we show how vertical motion and vorticity are intimately related, setting the stage for the development of simple models for diagnosing vertical motion and surface pressure tendency in Chapter 4.

2.1 Vorticity

Vorticity is a kinematic property quantifying the local spin or rotation of parcels within a fluid. Unlike temperature and momentum, it is not needed in current weather prediction models to make a forecast, although it is a familiar output product. As a conceptual tool, vorticity is clearly of great importance because it describes the rotational motion of the air and therefore the perturbations on weather maps. Because vorticity is difficult to visualize, we will first derive it in various mathematical forms and later show how it relates to conventional weather maps. Mathematically, vorticity is a three-dimensional vector expressed as the curl of the velocity ($\nabla_\phi \times V_3$), which is customarily expressed in the Cartesian reference frame with vertical coordinate z as

$$\nabla_z \times V_3 = i\left(\frac{\partial w}{\partial y} - \frac{\partial v}{\partial z}\right) + j\left(\frac{\partial u}{\partial z} - \frac{\partial w}{\partial x}\right) + k\left(\frac{\partial v}{\partial x} - \frac{\partial u}{\partial y}\right). \tag{2.1}$$

Each of the components in (2.1) describes the fluid spin of parcels in the plane orthogonal to its respective unit vector denoted by the independent variables in the parenthesized expression (e.g. the y, z plane for the i component).

The vertical component of vorticity, denoted as

$$\zeta \equiv k \cdot \nabla_z \times V_3 = \frac{\partial v}{\partial x} - \frac{\partial u}{\partial y} \tag{2.2}$$

represents the local spin of the horizontal winds about a vertical axis. It is of considerable interest to synoptic meteorologists because it is this component of the vorticity that is represented on horizontal surfaces and therefore on conventional weather charts. Moreover, it is the vertical component of the vorticity that can be related to development of cyclones and to vertical motion through quasi-geostrophic theory. Historically, the development of modern weather forecasting began with the development of the equivalent barotropic and quasi-geostrophic vorticity equations during the 1940s and 1950s. The advantage in replacing the vertical component of the vorticity by two fundamental momentum variables, u and v, is that ζ combines both momentum components into a single quantity which can be expressed in terms of products that are more easily diagnosed from conventional weather maps.

One can visualize the vertical component of vorticity by the following mental exercise. Imagine that a pencil is placed upright within a moving fluid, which has a horizontal variation in velocity. Attached to the pencil beneath the surface of the fluid are small vanes that project into the sur-rounding fluid, causing the pencil to rotate. Since the vanes are absolutely flat they do not rotate like anemometer cups but only when the force of the fluid against them is greater on one side of the pencil than it is in that same direction on the opposite side. Such an arrangement measures the twisting

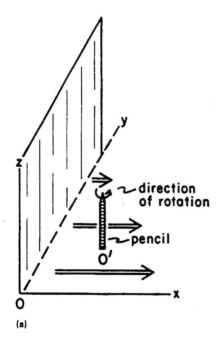

(a)

Figure 2.1 (a) Pencil rotated by horizontal shear of fluid (speed of fluid proportional to length of arrows in x, y plane).

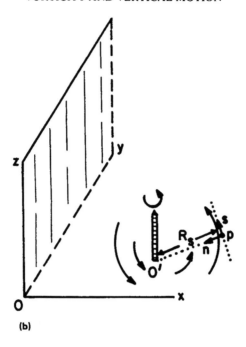

(b)

Figure 2.1 (b) Pencil rotated by curved flow in x, y plane. The natural (s, n) coordinate system is centered on the point p, which represents a fluid parcel moving in the direction along the s axis a distance R_s from the center of rotation.

effect of the fluid on the pencil rather than the speed of the current. Given a lateral shear as depicted in Figure 2.1a, or a rotation as in Figure 2.1b, the pencil will rotate because of the horizontal twist imparted to it. The twist can be imparted either by a gradient of wind speed, due to rotation about the point, or by a lateral shear. In Figures 2.1a and b the center of rotation for either pure rotation or pure shear is denoted by the letter O' (to distinguish it from the coordinate system origin at O).

If the arrow lengths are taken to represent the flow speed, the lateral shear would rotate the pencil counterclockwise about its axis at a rate equal to $-\partial u/\partial y > 0$, where $\partial v/\partial x = 0$ for the laterally sheared flow in Figure 2.1a. The same sense of spin occurs if the sheared flow in Figure 2.1a is replaced by a curved flow shown in Figure 2.1b. In the curved flow, both $\partial v/\partial x$ and $\partial u/\partial y$ are non-zero in general. Positive spin or vorticity, referred to as *cyclonic*, is in the same sense as the counterclockwise rotation of the Earth viewed from above the pole in the Northern Hemisphere (Fig. 2.1d); conversely, negative vorticity is referred to as anticyclonic. The terminology is reversed in the Southern Hemisphere; viewed from above the South Pole, the Earth appears to possess a clockwise spin.

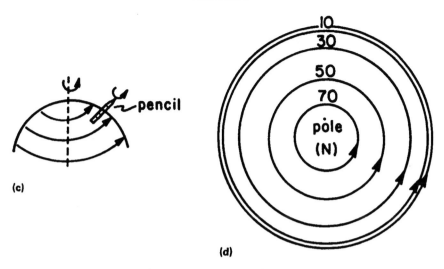

Figure 2.1 (c) Side view of the Northern Hemisphere showing the solid movement of points on latitude circles about the polar axis, and the sense of the rotation of a vertically oriented pencil due to the rotation of the Earth. (d) Schematic top (polar) view of (c) showing movement and spacing of Earth points on equispaced latitude circles.

The distinction between shear and curvature vorticity can be quantified in so-called *natural coordinates*, in which orthogonal axes are defined with respect to the local flow pattern. In the natural coordinate system the x axis is taken along the direction of the wind vector; the y axis points to the left of the wind vector. (To avoid confusion, however, we will assign different symbols to these rotated axes, which we will call natural coordinates.) As illustrated in Figure 2.1b, the s axis (replacing the x axis) lies along the direction of motion of the horizontal flow, and the n axis (replacing the y axis) is orthogonal to the s axis with positive n to its left. In natural coordinates the vertical component of the vorticity is expressed as

$$\zeta = -\frac{\partial V}{\partial n} + \frac{V}{R_s} \qquad (2.3)$$

where V is the wind speed and R_s is the radius of streamline curvature. The first and second terms on the right of (2.3), respectively, represents the contributions of shear and curvature to the vorticity. Evaluating (2.3) for the case of Figure 2.1b shows that $-\partial V/\partial n$ and V/R_s are both positive. (Note that R_s is defined as positive for cyclonically curved flow; the center of curvature lies on the positive n axis.) In fact, both shear and curvature are aspects of the same thing.

31

To illustrate further, let us imagine that Figure 2.1b depicts pure solid rotation (as in the case of a turntable) about an axis that is not necessarily located at the origin of the x, y plane. Now, $V = \Omega r$, where r is the distance from the axis of rotation between points O' and p, and Ω is the angular rotation rate. Since, for cyclonic rotation, $\partial r/\partial n = -1$ (r and n are oppositely directed), $\partial V/\partial n = \Omega$ and $R_s = r$, the vorticity (ζ), which is defined in (2.3), is equal to 2Ω. (If (2.3) is applied to an Earth-based coordinate system, the above sign convention for R_s applies in the Northern Hemisphere and is reversed in the Southern Hemisphere.) Note that equal amounts of vorticity (Ω) are contributed by the shear and curvature terms (i.e. $\zeta_E = \Omega + \Omega = 2\Omega$, where ζ_E is the vorticity of the turntable).

This type of rotation, called solid rotation, is not simply confined to points on the Earth or to rotating disks. Intense cyclones, such as hurricanes, exhibit solid rotation in the wind field near their centers. The above example considers the pure solid rotation of a plane about a vertical axis. Figure 2.1d treats the rotation of a solid sphere representing the Earth about the North Pole. For an observer viewing the Earth from above the North Pole, the latitude circles appear to spin in circular motion about the polar axis; however, with decreasing latitude, the local vertical (represented by the pencil in Figure 2.1c) increasingly tilts out of the plane normal to the polar axis and experiences less rotation about that axis. At the Equator there is no rotation about the polar axis, although points on the Earth's surface are moving faster than those at higher latitudes by virtue of their distance from the polar axis (Fig. 2.1d). Because of the sphericity of the Earth, the vorticity, f, of a local vertical axis (Fig. 2.1c) is not 2Ω but

$$f \equiv 2\Omega \sin \theta \qquad (2.4)$$

where θ is the latitude and the $\sin \theta$ term gives the component of rotation of the Earth about the local vertical (the pencil) and may be viewed as a correction due to sphericity. The Earth's vorticity, f, is called the *Coriolis parameter* and it represents the Earth's rotation rate with respect to the fixed stars ($\Omega = 7.292 \times 10^{-5} \text{ s}^{-1}$).

The vorticity of air parcels moving relative to the rotating Earth (the quantity denoted by ζ, computed from wind observations using (2.2) or (2.3)), can be added to the Earth's vorticity, $\zeta_E = f$, to obtain the vorticity of the air parcels with respect to the fixed stars, which is referred to as the absolute vorticity,

$$\zeta_a = \zeta + f. \qquad (2.5)$$

In the Southern Hemisphere f is negative (θ is taken to be less than zero in (2.4)), and cyclonic (anticyclonic) vorticity is negative (positive) and refers to clockwise (counterclockwise) rotation. As in the Northern Hemisphere, cyclonic vorticity has the same sign as that of the local rotation of the Earth, given by f.

One of the tenets of quasi-geostrophic theory is that, for synoptic-scale motions, the relative vorticity can be closely approximated by its geostrophic value. Thus, knowledge of the observed wind field is not required because the relative vorticity field can be inferred directly from the geopotential height patterns. The geostrophic relationships, as defined in (1.24), refer to horizontal derivatives of the height Z on a constant-pressure surface, a subscript p usually being understood. It turns out that, for meteorological applications, simplifications occur if the relative vorticity is evaluated on constant-pressure surfaces rather than constant-height surfaces as indicated in (2.2) and (2.3). Consequently, we consider

$$\zeta_p = \left(\frac{\partial v}{\partial x}\right)_p - \left(\frac{\partial u}{\partial y}\right)_p. \tag{2.6}$$

Evaluating the relative geostrophic vorticity on a pressure surface, ζ_{pg}, using (1.24) and (2.6), gives

$$\zeta_{pg} = \frac{g}{f} \nabla_p^2 Z + \frac{\beta}{f} u_g \tag{2.7}$$

where the Laplacian operator is

$$\nabla_p^2 = \left(\frac{\partial^2}{\partial x^2}\right)_p + \left(\frac{\partial^2}{\partial y^2}\right)_p.$$

Now, $\beta = df/dy$, which is approximately $1.6 \times 10^{-11} \, \text{s}^{-1} \, \text{m}^{-1}$ at 45° latitude. The second term on the right-hand side of (2.7) usually is small compared to the first; its magnitude rarely exceeds $1 \times 10^{-6} \, \text{s}^{-1}$ while the first term is usually on the order of $1 \times 10^{-5} \, \text{s}^{-1}$. Moreover, in keeping with the assumption that weather systems are small compared to the radius of the Earth, a reasonable first approximation is that f is a local constant. Consequently, to a reasonable approximation, the relative geostrophic vorticity is

$$\zeta_g = (g/f_0) \nabla_p^2 Z \tag{2.8}$$

where the subscript p in the wind components is dropped for notational convenience. Note that f is replaced by a spatially constant value, f_0, representing the local region of interest.

Equation (2.8) establishes a relationship between the geopotential height

33

field and the horizontal wind field. This relationship can be appreciated by approximating the partial derivatives in (2.8) with centered finite differences. This particular finite difference form of the Laplacian, evaluated at point o in Figure 2.2, is

$$\nabla_p^2 Z = (1/d^2)[Z(a) + Z(b) + Z(g) + Z(h) - 4Z(0)] \qquad (2.9)$$

where a, b, g and h are points on the x and y coordinate axes passing through o and are at a distance d from o. Defining

$$\tilde{Z} \equiv [Z(a) + Z(b) + Z(g) + Z(h)]/4 \qquad (2.10)$$

permits us to rewrite (2.8) in finite difference form:

$$\zeta_g = (4g/f_0 d^2)[\tilde{Z} - Z(0)]. \qquad (2.11)$$

Equation (2.11) expresses the relative geostrophic vorticity in terms of the difference between the average geopotential height, $\tilde{Z}$, surrounding a central point with geopotential height $Z(0)$ at the center. Where the average geopotential height around the center point is higher (lower) than the geopotential height at the center point, the relative geostrophic vorticity is cyclonic (anticyclonic) and positive (negative) in the Northern Hemisphere. Thus, relative minima in the geopotential height field (valleys or troughs) possess positive relative geostrophic vorticity, and relative maxima (hills or ridges) possess negative relative geostrophic vorticity. A fluid parcel has negative or positive relative geostrophic vorticity only in regard to the average geopotential height of the surroundings. In other words, if the average geopotential height around the circumference of a circle of radius d is higher than that of the center, the Laplacian of the height field is positive at that

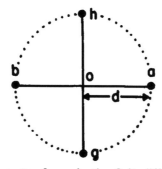

Figure 2.2 Coordinate system for evaluating finite differences about point o. Points a, b, g and h lie on a circle of radius d.

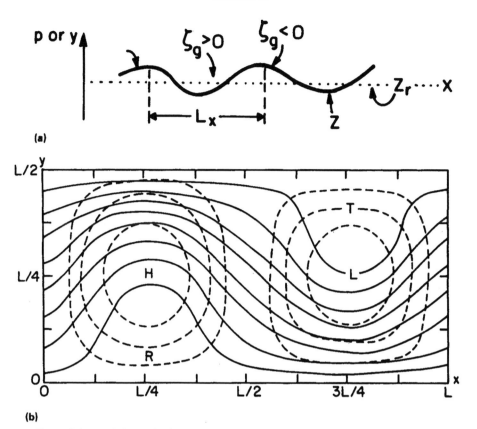

Figure 2.3 (a) Schematic sinusoidal geopotential contour with wavelength L about reference height Z_r in the x, y or x, p plane. Relative geostrophic vorticity is indicated. (b) Idealized square-wave sinusoidal perturbation in x and y with wavelength L. Broken contour lines are isopleths of geopotential height for a given amplitude with no zonal current (representing the geopotential height field at 1000 mb); full contour lines are for 500 mb with same amplitude and a superimposed uniform zonal current of strength $\partial \check{Z}/\partial y$. Symbols R and T, respectively, refer to the ridge and trough positions of the 500 mb wave and H and L denote the location of the 1000 mb high and low centers.

point. Note, however, that the precise value of the relative vorticity is dependent on the choice of d.

Relative geostrophic vorticity can be viewed as a measure of the local trough-like or ridge-like character of the geopotential height field. It also quantifies the relative intensity and scale of perturbations. Relative intensity of a disturbance in the geopotential height field can thus be defined as the Laplacian of Z, which is proportional to $[\check{Z} - Z(0)]/d^2$. Computationally, the exact value of vorticity should be dependent on the choice of d.

Let us consider an idealized pattern of Z in Figure 2.3a, which depicts a one-dimensional sine wave in x, y or x, p coordinates. For a simple sinusoidal wave

$$Z = Z_r + \hat{Z}\sin(2\pi x/L_x) \qquad (2.12)$$

where Z_r is a domain-averaged reference value, $\hat{Z}$ is the amplitude and L_x is the wavelength. Substituting (2.12) into (2.8) and ignoring y derivatives, we find that

$$\zeta_g = -\frac{4\pi^2 g}{f_0 L_x^2}\hat{Z}\sin\left(\frac{2\pi x}{L_x}\right) = \frac{4\pi^2 g}{f_0 L_x^2}(Z_r - Z) \qquad (2.13)$$

where the negative sign results from the second differentiation of the sine function. The equation says that relative geostrophic vorticity is inversely proportional to the square of the wavelength of the perturbation and proportional to the difference between the local geopotential height and a reference value of Z (Z_r), and which we can imagine to be a latitudinal or area-averaged mean; $Z_r - Z$ is thus a measure of the relative geopotential height perturbation in x. Note the similarity of (2.13) to (2.11).

The assumption that $L_x = L_y = L$ implies the wave is "square", which is to say that its meridional and zonal dimensions are identical. Regions of positive and negative relative geostrophic vorticity are, respectively, regions of negative and positive height departures from the reference value, Z_r. Consequently, for a given value of $\hat{Z}$, the closer together the troughs and ridges are situated, the larger the relative geostrophic vorticity. The Laplacian of Z is therefore a measure of the *intensity* of a perturbation in the Z field.

The pattern illustrated in Figure 2.3a can be extended to both x and y. Figure 2.3b represents a sinusoidally varying wave in both the x and y directions, except that the latter varies only one-half a wavelength from zero along the x axis at $y = 0$ to a maximum at $y = L/4$ to zero again at $y = L/2$. The figure depicts both a wave and a cellular pattern with identical distributions of relative vorticity and differing only by a uniform zonal current with zero relative geostrophic vorticity.

In both wave and cellular pattern, the perturbation intensities are identical, since the amplitudes are equal. Let us imagine that the cellular and wave patterns in Figure 2.3b resemble the geopotential height patterns in the lower and middle troposphere, say 1000 and 500 mb, respectively. The configuration of the geopotential height in Figure 2.3b does resemble flow patterns at these levels, except that in reality the vorticity pattern tends to be much more intense at 500 than at 1000 mb. Here the relative geostrophic vorticity pattern is identical for both the wave and cellular geopotential fields, and *its isopleths* are essentially parallel to the geopotential height contours of the cellular field. The sinusoidal pattern in Figure 2.3b is expressed mathematically as

$$Z = Z_r + a_z y + \hat{Z} \sin\left(\frac{2\pi x}{L_x}\right) \sin\left(\frac{2\pi y}{L_y}\right) \qquad (2.14a)$$

where a_z expresses the mean meridional geopotential height gradient ($\partial Z / \partial y$). Note that a_z is also a measure of the mean zonal current, which we will call U.

If the flow is geostrophic,

$$U = -\frac{g}{f_0} a_z \equiv -\frac{g}{f_0}\left(\frac{\partial \hat{Z}}{\partial y}\right). \qquad (2.15)$$

Z_r is now interpreted as the zonally averaged geopotential height at $y = 0$. Differentiating (2.14) using (2.8) yields

$$\zeta_g = -\frac{8\pi^2 g}{f_0 L^2} \hat{Z} \sin\left(\frac{2\pi x}{L_x}\right) \sin\left(\frac{2\pi y}{L_y}\right) \equiv \hat{\zeta} \sin\left(\frac{2\pi x}{L_x}\right) \sin\left(\frac{2\pi y}{L_y}\right)$$

$$= \frac{8\pi^2 g}{f_0 L^2} (Z_r + a_z y - Z) \qquad (2.16)$$

where $\hat{\zeta}$ is the amplitude of the vorticity perturbation.

We assume that $L_x = L_y = L$ (the special case of the square-wave pattern, shown in Figure 2.3b). The square-wave assumption allows us to express the amplitude of the relative geostrophic vorticity (which is $8\pi^2 g \hat{Z} / f_0 L^2$) in terms of a single wavelength. The term in parentheses on the right-hand side of (2.16) represents the local value of the geopotential height perturbation. In particular, lows ($Z < a_z y + Z_r$) and highs ($Z > a_z y + Z_r$) correspond, respectively, to regions of positive and negative relative geostrophic vorticity. Equation (2.16), which will be cited in subsequent discussions, is analogous to (2.13) and may be interpreted in the same manner except that Z_r represents a zonally averaged geopotential height at $y = 0$. Then $Z_r + a_z y$ represents a zonally averaged value of Z at the latitude y. Here $Z_r + a_z y - Z$ is a measure of the relative geopotential height perturbation in x and y.

The important points regarding vorticity are brought out by (2.16). This expression shows that the magnitude of the relative geostrophic vorticity is inversely proportional to the square of the wavelength. The difference between the minimum or maximum geopotential height and that in the surroundings is a rough measure of the depth or amplitude of the geopotential height perturbation on an isobaric surface. Because of the inverse-square relationship in (2.16), the intensity of the geopotential height perturbation

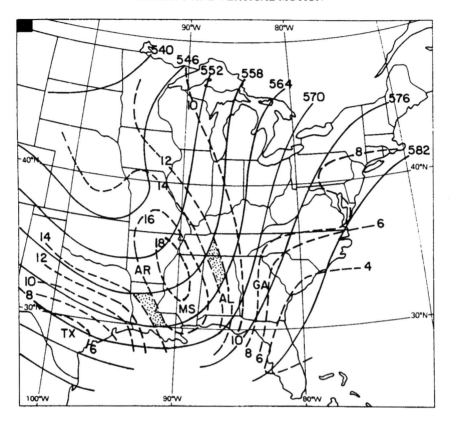

Figure 2.4 The 500 mb geopotential height contours (full curves labeled in dam; 1 dam = 10 m) and absolute vorticity (broken curves at intervals of $2 \times 10^{-5}\,s^{-1}$) for 0000 GMT 26 November 1979. The shaded regions over the southern United States denote vorticity advection solenoids. The blackened area in the upper left-hand corner of the map is the area of a unit advection solenoid. (Geographical locations of states are given by conventional two-letter code or otherwise indicated in text.)

and the relative geostrophic vorticity rapidly increase with decreasing wavelength.

The equation also demonstrates that height perturbation and the relative geostrophic vorticity evince opposite signs. Thus, highs tend to possess negative relative geostrophic vorticity and lows positive relative geostrophic vorticity. More generally, the Laplacian of a variable X, such as geopotential height, vertical motion, geopotential height tendency and streamfunction, is proportional to the negative of that variable. We will make use of this relationship in later discussions.

Finally, this equation shows that, for a uniform zonal current (a_z constant), the relative geostrophic vorticity is the same as for a cellular pattern with no zonal current ($a_z = 0$) provided that the amplitudes of the patterns are the same. In this case the relative geostrophic vorticity is proportional to the geopotential height. The vorticity of the cellular and wave patterns is identical in Figure 2.3b. Because the geopotential height and vorticity contours are parallel in the cellular pattern, the horizontal vorticity advection is zero. Thus, the $a_z y$ term, which describes the basic zonal current speed, accounts for the horizontal advection of relative vorticity. Although we may think of a wave as consisting of a vortex superimposed on a zonal current, the two patterns differ profoundly with respect to the advection of vorticity. Geopotential height lines intersect the vorticity isopleths in the wave pattern but not in the cellular one, i.e. advection of relative vorticity is accomplished by the zonal current. The only difference between the smoothed 500 mb geopotential height contours and the dashed 1000 mb height contours in Figure 2.3b is the absence of a uniform westerly current in the latter. Both sets of height contours possess identical values of the vertical component of the vorticity at each point in the diagram, except that the 1000 mb height field has a cellu-

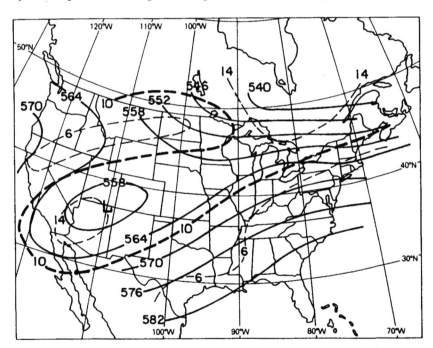

Figure 2.5 The 500 mb geopotential height contours (full curves labeled in dam) and absolute vorticity isopleths (broken curves at intervals of $4 \times 10^{-5}\,\text{s}^{-1}$) for 0000 GMT 9 December 1980, showing the relationship between positive relative vorticity, which is enclosed by the $10 \times 10^{-5}\,\text{s}^{-1}$ vorticity isopleth (heavy broken curve), and negative height departures from a uniform geopotential height distribution.

lar shape and the other field is more wave-like. Horizontal vorticity advection in the cellular pattern is zero, however, while that for the wave would consist of a dipole pattern of positive and negative vorticity advection, as in Figure 2.4.

An example of an observed geopotential height pattern for the 500 mb level is shown in Figure 2.4. A deep trough is located over the southern Mississippi Valley of the United States. Absolute vorticity isopleths (of the actual wind; see definition in (2.5)) depict a maximum slightly in excess of $18 \times 10^{-5}\,s^{-1}$ centered on the trough axis at the border of Arkansas (AR) and Mississippi (MS). Vorticity minima (less than $4 \times 10^{-5}\,s^{-1}$) are located on either side of the trough axis; one over southern Texas (TX) and the other about 1200 km due east of the vorticity maximum, along the coast (GA). Consequently, the wavelength of this system is about 2400 km and the vorticity amplitude ζ is about $7 \times 10^{-5}\,s^{-1}$ (we will henceforth refer to units of $10^{-5}\,s^{-1}$ as vorticity units or simply *units* for the sake of convenience).

It is customary to display the absolute vorticity field on 500 mb charts. The relative vorticity field closely resembles that of absolute vorticity, provided that the former is not so weak that the latitudinal variation of f dominates. That the relative vorticity is a measure of the perturbation in the geopotential height field can be seen in Figure 2.5. The 10 unit absolute vorticity isopleth (corresponding to f at 43 °N) tends to separate negative height anomalies (positive relative vorticity) from positive height anomalies (negative relative vorticity). The cutoff low centered over the southwestern United States is characterized by low heights and positive relative vorticity, and the ridge over the eastern part of the country has positive heights and negative relative vorticities.

Vorticity advection

The horizontal advection of absolute vorticity can be written in vector form or in the natural coordinate system as

$$- V \cdot \nabla_p (\zeta + f) = - V \frac{\partial}{\partial s}(\zeta + f). \tag{2.17}$$

Equation (2.17) states that if the absolute vorticity decreases (increases) downstream the advection is positive (negative). We will see that the advection of absolute vorticity is a crucial concept in understanding cyclogenesis. Note that the negative sign is necessary in the definition of advection in order to obtain the correct sign of the tendency.

Now, let us examine a real vorticity advection pattern, shown in Figure 2.4. Qualitative inspection of this figure shows that there is strong positive (cyclonic) vorticity advection (PVA) over western Alabama (AL), downstream of the absolute vorticity maximum situated over the southern Mississippi Valley. We can evaluate such a pattern visually in a quantitative manner using the *method of solenoids*. For advection of absolute geostrophic vorticity by the geostrophic wind, i.e. rather than by the actual wind as in (2.17), the natural coordinate expression leads to a simple graphical interpretation of vorticity advection. Using the expression for the geostrophic wind (2.6) rewritten in natural coordinates ($V_g = -(g/f_0)\partial Z/\partial n$), the expression (2.17) results in

$$-V_g \cdot \nabla_p(\zeta_g + f) = \frac{g}{f_0} \frac{\partial Z}{\partial n} \frac{\partial}{\partial s}(\zeta_g + f). \qquad (2.18)$$

Equation (2.18) can be approximated in finite difference form as

$$-V_g \cdot \nabla_p(\zeta_g + f) \approx \frac{g}{f_0} \frac{\Delta Z \Delta(\zeta_g + f)}{\Delta n \Delta s} \qquad (2.19)$$

where Δn is the distance to the left of the geostrophic wind vector between contours of constant Z, and Δs is the distance along the geostrophic wind vector between isopleths of constant absolute vorticity. The increments ΔZ and $\Delta(\zeta_g + f)$ correspond to the contour intervals of the geopotential height and absolute vorticity fields. As (2.19) is written, ΔZ is always negative (Z decreases along the n axis, which points to the left of the wind vector) and $\Delta(\zeta_g + f)$ changes sign depending on the sense of the downstream variation of the absolute vorticity. The overlay pattern of geopotential height and absolute vorticity in Figure 2.4 forms a network of solenoids, two of which are shaded (also, see the schematic in Figure 2.6). The stronger the geostrophic wind or the larger the absolute vorticity gradient, the smaller the area of the solenoid. Given that the contour intervals for the height and vorticity fields are fixed by convention on weather charts (e.g. for the U.S. Weather Service, 60 m and 2×10^{-5} s^{-1} at 500 mb), expression (2.19) can be rewritten as

$$-V_g \cdot \nabla_p(\zeta_g + f) = \frac{g}{f_0} \frac{\Delta Z \Delta(\zeta_g + f)}{A_s} \qquad (2.20)$$

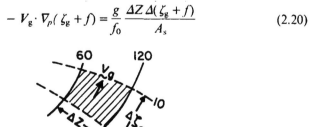

Figure 2.6 Schematic illustration of an advection solenoid (shaded). The broken isopleths are absolute geostrophic vorticity $\zeta + f$ (labeled in units of 1×10^{-5} s^{-1}) and the full isopleths are the geopotential height (Z; labeled in m). The geostrophic wind velocity at the center of the solenoid is V_g.

where $A_s(= \Delta n \Delta s)$ is the approximate area of an advection solenoid (the shaded area in Figure 2.6) formed by adjacent contours of geopotential height and absolute vorticity. Because the Coriolis parameter (f_0) is only weakly dependent on latitude at 40° from the Equator, it may be considered a constant at mid-latitudes for a given choice of contour intervals on a weather chart. Equation (2.20) becomes exact for infinitesimally small solenoids, as suggested in (2.18). In practice, solenoids are not perfect parallelograms. If the magnitude of the advection is either small or varying rapidly in space, solenoids may not "close off". For example, the advection is clearly positive over western Georgia (GA) in Figure 2.4, despite the absence of solenoids.

The solenoid concept is useful in that it allows one to visualize the advection pattern, including magnitude, upon inspection of conventional geopotential height and vorticity analyses. As evident from (2.20), the magnitude of the vorticity advection is inversely proportional to the area of an advection solenoid or, stated alternatively, directly proportional to the number density of solenoids per unit area. Because of the latitudinal variation of the Coriolis parameter, advection solenoids carry greater weight for a given size at lower latitudes; but as pointed out above, this dependence is quite weak. Numerical values can be assigned to (2.20); so that it is applicable to the 500 mb analysis in Figure 2.4. Choosing $f_0 = 1 \times 10^{-4}\,\mathrm{s}^{-1}$ and using the standard contour spacing referred to earlier for the geopotential height and absolute vorticity, and $A_s = (1°\ \text{latitude squared}) = (111\ \text{km})^2$, one finds that (2.20) yields values of $95.4 \times 10^{-10}\,\mathrm{s}^{-2}$, which is on the order of 100 vorticity units per 24 hours. The size of this "unit" solenoid is given by the blackened square on the upper-left corner of Figure 2.4. Inspection of Figure 2.4 reveals that the positive and negative advections are largest in magnitude northeast and southwest, respectively, of the vorticity center. Because the shaded positive vorticity advection solenoid is about three times as large as a unit solenoid, the maximum value in Figure 2.4 is roughly $30 \times 10^{-10}\,\mathrm{s}^{-2}$, or about 30 vorticity units per day. It is evident that the magnitude of the advection of absolute vorticity by the geostrophic wind exceeds 10 vorticity units per day over a large region of Figure 2.4.

The quantity estimated in (2.18)–(2.20) is the advection of the absolute geostrophic vorticity of the total wind field by the geostrophic wind, while the corresponding quantity customarily evaluated on these weather charts is the advection of the absolute vorticity of the total horizontal wind by the total horizontal wind field, which is expressed in (2.17). The geostrophic advection of the absolute vorticity of the total wind field is emphasized in this text because it approximates the vorticity advection by the total horizontal wind field and because it can be easily estimated from conventional weather charts.

The approximation given by (2.20) to the absolute vorticity advection by the total wind is sufficiently accurate for qualitative diagnostic purposes, but will be in error quantitatively when the geostrophic wind direction or speed differ substantially from that of the total wind. A common example of the latter discrepancy occurs in slowly moving troughs and ridges, where the total wind speeds are subgeostrophic ($|V| < |V_g|$) and supergeostrophic ($|V| > |V_g|$), respectively, on account of the prevalence of cyclonic and anticyclonic parcel trajectory curvature in these regions. Consequently, the solenoid method (2.20) tends to overestimate (underestimate) the vorticity advection by the total wind in troughs (ridges). The more important vorticity advections, however, tend to occur in straight flow between troughs and ridges.

An additional estimate for the magnitude of the vorticity advection by the horizontal wind for mid-latitude synoptic-scale waves can be determined analytically for a typical sinusoidal wave pattern shown in Figure 2.3b. Here and in subsequent discussions we make the assumption, for the purpose of calculating the magnitude and sign of various mathematical terms, that the isopleths of height, temperature and vertical motion are sinusoidal. This artifice allows us to determine maximum and minimum values of the quantities without recourse to extensive calculations. Although real patterns are seldom exactly sinusoidal in shape, the deviations from this imposed periodicity will seldom be of first-order importance. Accordingly, we can use expressions such as (2.14) in order to investigate the magnitude and scale dependence of mathematical quantities in physical models of the atmosphere, provided that we can specify the amplitude and wavelength of the pattern.

Short waves

Meteorologists sometimes refer to so-called *short waves* to distinguish a particular type of feature on upper-level wind charts. Short waves are generally less than a few thousand kilometers between successive troughs or ridges. They are felt to be important because of their association with surface weather and with cyclone development. Short waves usually move rapidly, with speeds approaching that of the mean current in which they are embedded. Long waves, on the other hand, tend to move slowly or not at all.

We can also distinguish short and long waves on the basis of vorticity advection. Let us examine the properties of the geostrophic vorticity advection using some simple sinusoidal patterns of geopotential height. According to (2.20), geostrophic vorticity advection is determined solely by the geopotential height field and latitude. We can demonstrate the scale dependence of vorticity advection on wavelength using (2.14) and the definitions of the geostrophic wind (1.24) and the geostrophic vorticity (2.8). Neglecting the latitudinal variation of f (the β effect), (2.14) yields an expression for *relative* geostrophic vorticity advection:

$$-\left(\frac{g}{f_0}\right)^2 \frac{16\pi^3}{L^3}\, a_z \hat{Z} \cos\left(\frac{2\pi x}{L}\right) \sin\left(\frac{2\pi x}{L}\right).$$

The average *magnitude* of the above expression over one wavelength is determined by multiplying by $2/\pi$, which gives

$$\left(\frac{g}{f_0}\right)^2 \left(\frac{32\pi^2}{L^3}\, |a_z \hat{Z}|\right).$$

For a zonal geostrophic wind component of 20 m s^{-1}, $a_z = -2.0 \times 10^{-4}$ (approximately 100 m change in geopotential height per 5° of latitude). For $L = 2400$ km and $\hat{Z} = 120$ m (estimates compatible with the pattern in Figure 2.4), one obtains an estimate of relative vorticity amplitude (through (2.14)) of about 8 vorticity units and a magnitude of geostrophic vorticity advection (from (2.32)) of about 50×10^{-10} s^{-2} (50 units of vorticity change per day), which is of comparable value to that determined from other examples. Another interesting implication of the above scaling expression is that the vorticity advection decreases very rapidly with increasing wavelength (for a given amplitude) because of the inverse-cube dependence. Consequently, vorticity advection is likely to be small in very long waves and large in short ones. Although we will show later how vorticity advection and wavelength affect the surface weather, the growth of cyclones and the motion of waves at mid-levels, let us now adopt the qualitative definition that *a short wave is one that is associated with a number of vorticity advection solenoids*. We should also note that vorticity advection solenoids tend to be concentrated in a relatively small region not far from the vorticity maximum (e.g. Fig. 2.4), with relatively weak advections elsewhere.

2.2 Vertical motion and continuity

Equation (1.8a) states that, as a column of air is squeezed horizontally (convergence) or pulled apart (divergence) on a pressure surface, it must stretch or compact vertically; therefore, for mass to be conserved, vertical stretching or shrinking of an air column is exactly balanced by convergence or divergence. Horizontal divergence may be visualized by considering the kinematic expression

$$\left(\frac{\partial u}{\partial x} + \frac{\partial v}{\partial y}\right)_p = \nabla_p \cdot V = -\frac{\partial \omega}{\partial p} = \frac{1}{A}\frac{dA}{dt} \equiv D_p \qquad (2.21)$$

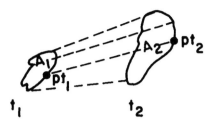

Figure 2.7 Horizontal divergence of a ring of fluid on a pressure surface, enclosing area A_1 at time t_1 and area A_2 at time t_2; the dot labeled p is an arbitrary element of fluid on the boundary of the area.

which states that *horizontal divergence is equal to the fractional rate of increase of an area (A) on a pressure surface enclosed within an identifiable ring of air parcels*, as shown in Figure 2.7. Horizontal divergence and convergence differ fundamentally from diffluence and confluence (defined in Ch. 13); the latter produces a change in shape of an area enclosed by a ring of air parcels through deformational processes, but does not necessarily result in a change in the size of the area enclosed by the ring. Consequently, confluence and diffluence are not directly associated with vertical motions through (1.8). The horizontal divergence may be expressed in terms of natural coordinates as

$$D_p = \frac{\partial V}{\partial s} - V \frac{\partial \gamma}{\partial n} \tag{2.22}$$

where γ is the angle of the meteorological wind direction, along the n axis (as defined in the discussion of the natural coordinate form of the vorticity (2.3)); γ increases as the angle is rotated *clockwise*. Note that (2.22) applies to any quasi-horizontal coordinate surface, as does (2.3).

Use of the geostrophic relationships (1.24) in the definition of D_p results in the expression for geostrophic divergence, which is

$$D_{pg} = - (\beta/f) v_g. \tag{2.23a}$$

For light to moderate values of the geostrophic wind speed ($V_g = 20$ m s^{-1}), D_{pg} is on the order of 10^{-6} s^{-1} at mid-latitudes, which is roughly an order of magnitude less than the divergence values characteristic of synoptic-scale systems (10^{-5} s^{-1}). Consequently, the contribution to the geostrophic divergence due to the curvature of the Earth (the beta (β) effect) is usually neglected. Therefore, we will henceforth assume that

$$\frac{\partial u_g}{\partial x} + \frac{\partial v_g}{\partial y} \approx 0. \tag{2.23b}$$

The assumption that the geostrophic wind is non-divergent is analogous to the neglect of the β effect in the geostrophic relative vorticity (2.7). Use of this fact in (2.22) results in an expression relating the meteorological angle of the wind direction, and the geostrophic wind speed, V_g:

$$\left(\frac{\partial \gamma}{\partial n}\right)_p = \frac{1}{V_g}\left(\frac{\partial V_g}{\partial s}\right)_p. \tag{2.24}$$

Equation (2.24) states that, if the geostrophic wind speed increases downstream, the geostrophic streamlines, which correspond to contours of the geopotential height of a pressure surface, are merging. This is called confluence. The reverse, when air decelerates, corresponds to streamlines widening along the direction of flow. Merging or widening of streamlines, called confluence or diffluence, are often mistaken for convergence or divergence, whereas angular variations of the streamlines are compensated by wind speed changes along the direction of flow. Thus, the winds can be confluent but have no convergence.

Because the divergence tends to be relatively small (except perhaps at the surface), it is subject to large errors in measurement. Consequently, the computation of vertical motion using continuity (2.21) tends to produce very large errors in ω. Construction of isotachs must therefore take account of the inability of the analyst to capture the correct divergence pattern kinematically, and thus the isotachs should elongate in the direction of the streamlines in order to minimize spurious divergence.

Vertical stretching or shrinking of a column of air tends to be maximized at the lower and upper boundaries of the atmosphere, whereas the vertical motions tend to vanish because air cannot penetrate the Earth's surface or escape from the upper limits of the atmosphere. The simplest vertical motion profile (excluding the trivial case of ω identically zero) satisfying mass continuity (2.21) and the above boundary constraints is sketched in Figure 2.8. The vertical motion, ω, which is of uniform sign and is maximized in the middle atmosphere (roughly 500 mb), vanishes at the top of the atmosphere ($p = 0$) and also approaches zero at the surface ($p_s = 1000$ mb). For the profile illustrating sinking motion in Figure 2.8, D_p is positive below the level of maximum ω and negative above.

Calculated vertical motion profiles (for synoptic-scale weather systems) tend to resemble Figure 2.8, provided that complicating factors such as fronts or gravity waves, processes such as deep cumulus convection and vertical motions induced by sloping topography and frictional effects in the planetary boundary layer are small. The simple "bowstring" profile is expected to appear in a composite or averaged sense for a collection or ensemble of synoptic-scale weather systems, where the effects of random observational errors and small-scale features are removed. For flat, level

46

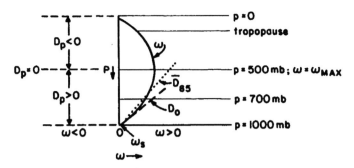

Figure 2.8 Schematic vertical motion profile (the bowstring model). The vertical motion (ω) obeys a sinusoidal distribution in the vertical but is slightly non-zero at the surface (1000 mb). In this case of descending motion ($\omega > 0$), there is convergence aloft and divergence below the level of non-divergence, where the vertical motion is a maximum (ω_{max}). The sloping broken and dotted lines are approximations to $\partial\omega/\partial p$ at the lower boundary (1000 mb) and at 850 mb using the mean slope between 1000 and 700 mb.

terrain the vertical motion in z coordinates ($w = dz/dt$) must vanish at a horizontal surface, since air cannot penetrate vertically through the ground. In pressure coordinates the vertical velocity at the surface, ω_s, may be non-zero. Non-zero ω_s implies the vertically averaged divergence over a column extending from the surface to the top of the atmosphere is also non-zero. We can prove this by integrating (2.21) from $p = 0$ to $p = p_s$ with

$$\bar{D}_p = \frac{1}{p_s}\int_0^{p_s}\left(\frac{\partial u}{\partial x} + \frac{\partial v}{\partial y}\right)dp = -\frac{1}{p_s}\int_0^{p_s}\frac{\partial\omega}{\partial p}\,dp = -\frac{\omega_s}{p_s} \qquad (2.25)$$

since $\omega(p = 0) = 0$. The definition of the vertical average in (2.25) is

$$(^-) \equiv \frac{1}{p_s}\left(\int_0^{p_s}(\ )dp\right).$$

In this instance if the vertical motion in pressure coordinates is exactly equal to zero at the surface, divergence and convergence in the air column exactly cancel, requiring that there be at least one level of non-divergence. Of course, there would be no surface pressure changes either and the weather map would not be very interesting. One can show, however, that ω_s is small compared to the usual values of vertical motion found in the atmosphere but large enough to account for the observed surface pressure tendencies. Thus, approximately compensating upper- and lower-level divergence and convergence are still compatible with a net surface pressure change. Let us pursue this idea by deriving an expression for ω_s, which may be obtained by expanding dp/dt and evaluating it at the surface:

$$\omega_s = \left(\frac{\partial p}{\partial t}\right)_s + (V \cdot \nabla_z p)_s + w_s \left(\frac{\partial p}{\partial z}\right)_s. \tag{2.26a}$$

The second term on the right-hand side of (2.26a) vanishes over flat terrain, since $V_s = 0$ at the surface (a no-slip boundary condition); alternatively, if the wind were geostrophic at the surface, the term would also vanish because V_g would be parallel to the isobars. The third term on the right vanishes by virtue of the assumption of flat terrain ($w_s = 0$). Therefore we have

$$\left(\frac{dp}{dt}\right)_s \equiv \omega_s \approx \left(\frac{\partial p}{\partial t}\right)_s. \tag{2.26b}$$

The sea-level pressure tendency is usually reported on surface weather charts as a three-hour change for the period ending at the observation time.

It is useful to compare ω_s with typical values of vertical motion found at upper levels. Typical values of $(\partial p/\partial t)_s$ range from 1 to 5 mb (3 h^{-1}) in active synoptic-scale weather systems (although values up to 10 mb (3 h^{-1}) are occasionally observed in intense mid-latitude storms). According to (2.26b), these values of pressure tendency correspond roughly to ω_s of 0.1–0.5 μb s^{-1} and a vertically averaged divergence ($\bar{D}_p$) of (1 to 5) $\times 10^{-7}$ s^{-1}. Typical values of ω in the mid-troposphere are an order of magnitude greater than ω_s as can be seen from Figure 2.9, which shows a 700 mb ω field. Ascending vertical motion ($\omega < 0$) in excess of 2 μb s^{-1} covers much of the southeastern United States and maximum ascent exceeds 4 μb s^{-1}. For a 700 mb ω of 3 μb s^{-1}, the vertically averaged divergence between the surface and 700 mb determined from (2.27) is about 1×10^{-5} s^{-1}. This value is two orders of magnitude greater than the vertically averaged divergence for the entire depth of the atmosphere (1×10^{-7} s^{-1}). This seemingly contradictory situation of such a difference in magnitudes between the vertically averaged divergences for the lower troposphere and for the entire atmosphere can be reconciled by postulating a divergence profile in the vertical that changes sign at least once. For the *bowstring profile* in Figure 2.8, the divergence changes sign once, at the level of maximum ω.

The configuration of high-level convergence overlying low-level divergence (or vice versa), of nearly equal magnitude, is referred to as "Dines' compensation". Viewed alternatively, if Dines' compensation were not the rule, i.e. if lower-level convergence (or divergence) and upper-level divergence (or convergence) did not nearly balance each other, surface pressure tendencies would be much larger than are now observed. Since the surface pressure tendency is thus almost equal to ω_s,

$$\left(\frac{\partial p}{\partial t}\right)_s \approx - \int_0^{p_s} \nabla_p \cdot V \mathrm{d}p \qquad (2.27)$$

which states that the surface pressure tendency must be equal to minus the integral of column divergence. If this divergence is exactly zero, the surface pressure does not change with time. Therefore, for the surface pressure to

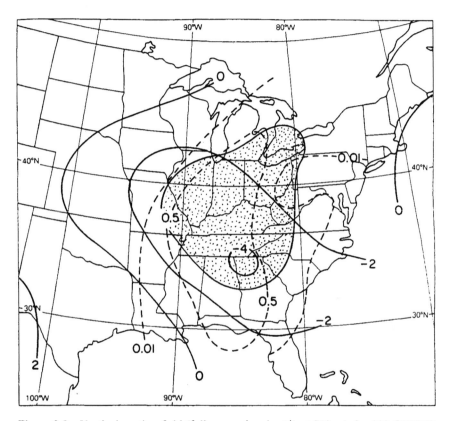

Figure 2.9 Vertical motion field (full curves in $\mu b\ s^{-1}$) at 700 mb for 000 GMT 26 November 1979, from the 12 h forecast by the National Meteorological Center's Limited Area Fine Mesh (LFM) model. Broken curves are predicted 12 h precipitation amounts in inches. (In this and some subsequent figures, the 12 h vertical motion forecast for the particular map time is used to represent the analysis of vertical motion as it would be obtained from a set of initial conditions at that same time. This assumption is necessitated by the inability of the model to produce a meaningful vertical motion pattern at the initial forecast time. The 12 h forecast is felt to be a fairly close representation of the actual vertical motions at the map time.) Shading denotes the region in which the 700 mb relative humidity is greater than 70%.

change there must be a net divergence or convergence in the column, the former corresponding to pressure falls and the latter to pressure rises at the bottom of the column. Since ω_s tends to be small, for surface pressures to decrease, ω_s need be only slightly negative, approximately equal to the value of the surface pressure tendency.

A useful approximation that simplifies interpretation of the quasi-geostrophic equations is that the surface divergence is approximately equal to the 850 mb divergence. The latter is estimated by evaluating the vertical derivative of ω between 700 and 1000 mb. We can demonstrate the validity of these approximations by assuming a sinusoidal variation of ω in the vertical, which is

$$\omega = \hat{\omega}(x, y)\sin\left[\pi\left(\frac{p - p_t}{p_*}\right)\right] \tag{2.28}$$

where p_t is the pressure at the top of the profile ($p_t = 0$ in Figure 2.8), $p_* = p_s - p_t$ and $\hat{\omega}(x, y)$ is the amplitude of the profile and a function of x and y. Differentiating (2.28) with respect to pressure yields an expression for $\partial\omega/\partial p$ proportional to $\cos[(p - p_t)/p_*]$, the magnitude of which is maximized at p_s and p_t. We will make use of (2.28) in subsequent discussions.

The linear approximation to the 1000 mb divergence is expressed as

$$-D_0 \approx \left(\frac{\omega_0 - \omega_7}{p_0 - p_7}\right) = \left(\frac{\omega_0 - \omega_7}{300 \text{ mb}}\right) \tag{2.29a}$$

where the subscript 7 denotes 700 mb and 0 for 1000 mb. Now let us consider the 1000 mb divergence, D_0, which is the slope of the broken line in Figure 2.8. This "surface-level" divergence is approximated by the mean between the surface (1000 mb) and 700 mb, which is called $(\bar{D}_{85})$ in Figure 2.8. The dotted straight line drawn between ω_0 ($= \omega_s$, which is approximately zero) at 1000 mb and ω_7 at 700 mb is a close approximation to the slope of the line at 1000 mb and hence to the exact divergence at the surface; therefore $\bar{D}_{85}$ can be used in place of D_0 without much loss of accuracy. One can show with (2.28) that the value of D_0 (at the surface) for an idealized sinusoidal profile of ω is 14% greater than the vertically averaged divergence between 1000 and 700 mb, given by the differentiated form of the sinusoidal profile. As pointed out previously, ω_0 (or ω_s) can be neglected relative to ω_7, provided that the terrain slope is sufficiently small (it is assumed that the Earth's surface and the 1000 mb level nearly coincide). Consequently, the 700 mb vertical motion pattern in Figure 2.9 can be considered a map of the 1000 mb (surface) divergence patterns, i.e.

$$D_0 \approx \omega_7/(300 \text{ mb}). \tag{2.29b}$$

This equation figures importantly in our formulation of cyclone development theory, which is expanded in subsequent chapters. It states that surface divergence and 700 mb vertical motion are related.

Problems

2.1 Realizing that

$$\left(\frac{\mathrm{d}p}{\mathrm{d}t}\right)_s = \omega_s = \left(\frac{\partial p}{\partial t}\right)_s + V_s \cdot \nabla_z p_s + w_s \left(\frac{\partial p}{\partial z}\right)_s$$

(a) Verify or contest the assumption that neglect of the surface horizontal pressure advection compared with the surface pressure tendency in the equation is valid by making some spot calculations of surface isobaric advection ($- V_s \cdot \nabla_z p_s$) near the area of rapid surface pressure falls in Figure 3.2e. Using Figure 3.1a, make your calculation in natural coordinates and assume 3 dam (30 m) equals 4 mb for the contour spacing.

(b) Show that the magnitude of ω_s in that same region is much less than ω at 700 mb.

2.2 (a) For a *unit* advection solenoid ($\Delta s = \Delta n = 1°$ latitude $= 111$ km), calculate the strength of the geostrophic vorticity advection ($\Delta Z_s = 60$ m; $\Delta(\zeta_{gs} + f) = 2 \times 10^{-5} \mathrm{s}^{-1}$) and thickness advection ($\Delta Z_0 = 30$ m; $\Delta h = 60$ m) at 55°, 45° and 30° latitude. Discuss the error implicit in assuming that the geostrophic advections within this latitude interval (30–55°) is computed at the latitude 45°.

(b) Find the *smallest* vorticity advection solenoid in Figure 3.1c within the continental United States and estimate the vorticity advection at that location.

Further reading

Astling, E. G. 1976. Some aspects of cloud and precipitation features associated with mid-latitude cyclones. *Mon. Wea. Rev.* **104**, 1466–73.

Haltiner, G. J. 1971. *Numerical weather prediction.* New York: Wiley.

Holton, J. R. 1979. *An introduction to dynamic meteorology*, 2nd edn (Int. Geophys. Ser., Vol. 23). New York: Academic Press.

McIlveen, R. 1988. *Basic meteorology. A physical outline.* London: Van Nostrand Reinhold.

Petterssen, S. 1956. *Weather analysis and forecasting II.* New York: McGraw-Hill.

Saucier, W. J. 1955. *Principles and practice of synoptic analysis.* Chicago, IL: University of Chicago Press.

3

The vorticity and thermodynamic equations

Two equations that have proven useful in relating the vertical motion field to observable quantities are the prognostic equations for the vertical component of the vorticity and that for temperature. We will begin this chapter by evaluating terms in the vorticity equation in order to show the relative importance of each term at different levels in the atmosphere. This magnitude analysis will lead to simplifications in the vorticity equation that will be consistent with the quasi-geostrophic equations introduced in Chapter 1. Combined with the quasi-geostrophic version of the temperature equation, versions of the quasi-geostrophic vorticity equation allow us to develop simplified equations describing the vertical motion and pressure (geopotential height) tendency equations. In order to determine representative magnitudes of terms in the vorticity equation, we refer to Figures 3.1 and 3.2, which constitute a typical (but not very intense) mid-latitude weather pattern.

3.1   The vorticity equation

The vorticity equation is formed by partially differentiating the u momentum equation (1.12a) with respect to y, the v momentum equation (1.12b) with respect to x and subtracting the former result from the latter. After rearranging the order of time and space differentiation, one obtains an expression for the time rate of change of absolute vorticity $(\zeta + f)$, which is expressed as

$$
\begin{aligned}
\frac{\partial(\zeta + f)}{\partial t} = & -V \cdot \nabla_p(\zeta + f) - \omega \frac{\partial \zeta}{\partial p} - (\zeta + f)\left(\frac{\partial u}{\partial x} + \frac{\partial v}{\partial y}\right) \\
& - \left(\frac{\partial \omega}{\partial x}\frac{\partial v}{\partial p} - \frac{\partial u}{\partial p}\frac{\partial \omega}{\partial y}\right) + \left(\frac{\partial F_y}{\partial x} - \frac{\partial F_x}{\partial y}\right).
\end{aligned} \tag{3.1}
$$

This equation expresses the local rate of change of the spin of the fluid. Note that the prognostic expression for absolute vorticity is given in pressure coordinates and therefore the solenoid terms, which appear when the equation is derived with respect to height, vanish.

The remainder of this section is concerned with evaluating the magnitude of and physically interpreting each term in the vorticity equation for the case of a representative mid-latitude cyclone. The magnitudes of the terms in (3.1) are summarized in Table 3.1. Note also that the use of the subscript p in (3.1) to denote vorticity or divergence on isobaric surfaces will be dropped when its meaning is clear from the context of the discussion. The term on the left-hand side of (3.1) is the local tendency of the relative vorticity (or the absolute vorticity, since the local tendency of f is independent of time). Approximating $\partial \zeta / \partial t$ by $\partial \zeta_g / \partial t$ allows use of (2.16) to determine that

$$\frac{\partial \zeta_g}{\partial t} \approx - \frac{8 \pi^2 g}{f_0 L^2} \frac{\partial Z}{\partial t} \tag{3.2}$$

which applies to the "square" sinusoidal wave pattern as illustrated in Figure 2.3b. Note that f_0 is a local constant.

If (3.2) is applied near the surface, for example at 1000 mb, $(\partial \zeta / \partial t)_0$ can be equated with $(\partial p / \partial t)_s$ (by imposing the hypsometric equation (1.31)), yielding

$$\left(\frac{\partial Z}{\partial t} \right)_0 \approx \left(\frac{R_d T_v}{p_s g} \right) \left(\frac{\partial p}{\partial t} \right)_s \equiv K_0 \left(\frac{\partial p}{\partial t} \right)_s . \tag{3.3}$$

The coefficient multiplying $(\partial p / \partial t)_s$, which is the hypsometric constant K_0, is approximately constant at mid-latitudes. (For $T_v = 280$ K, $p = 1000$ mb, $R_d = 287$ J kg^{-1} K^{-1} and $g = 9.8$ m s^{-2}, $K_0 \approx 8$ m mb^{-1}.) Use of the constant K_0 allows one to equate sea-level (or surface) pressure changes with changes in the 1000 mb geopotential height. Similarly, one could equate changes in geostrophic vorticity at 500 mb to changes in geopotential height at an appropriate level, say 5.8 km.

This result given by (3.2) demonstrates that the geostrophic constraint requires that a local change in vorticity involves a corresponding change in geopotential height of a pressure surface (or a change in pressure on a constant-height surface).

For a wavelength of 2400 km, a vorticity tendency of 1×10^{-10} s^{-2} (approximately one unit of vorticity change per day) and $f_0 = 1 \times 10^{-4}$ s^{-1}, equations (3.2) and (3.3) imply that $(\partial p / \partial t)_s$ is approximately 0.1 mb (3 h)$^{-1}$. Thus, a small change of p_s equal to 0.1 mb (3 h)$^{-1}$ is approximately equivalent to a 1000 mb vorticity tendency of 1×10^{-10} s^{-2}. This latter tendency, which will be used as a standard for comparing subsequent terms in the vorticity equation, is considered to be insignificant at any level of the atmosphere.

In Chapter 2 we showed that a typical value for the 500 mb absolute horizontal vorticity advection for active weather patterns is at least 30 vorticity units per day. Since wind speeds tend to increase with altitude in the troposphere (Fig. 1.1), the magnitude of the vorticity increases with increasing height below the height of the tropopause (250 mb), in accordance with the poleward decrease in temperature up to this level. Stronger wind speeds at higher levels also contribute to larger horizontal vorticity advections (in (2.20) the spacing of the height contours is smaller). Inspecting the advection solenoids in Figure 3.1 reveals an increase of vorticity advection with height through the troposphere. It is important to observe, by comparing the pattern of advection solenoids in Figure 3.1a with that in Figure 3.1d, that the 1000 mb horizontal vorticity advection is significantly less than that in the mid- and upper troposphere.

Besides being weaker at low levels than at middle levels, the wind and height patterns are much less wave-like and more cellular near the surface;

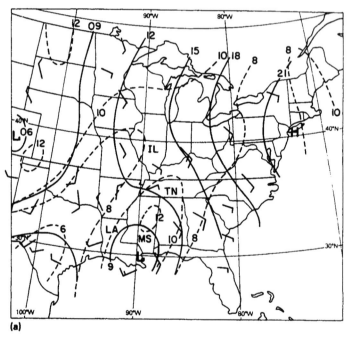

(a)

Figure 3.1 (a) The 1000 mb height (full contours labeled in dam) and geostrophic absolute vorticity (broken curves labeled at intervals of 2×10^{-5} s^{-1}) for 1200 GMT 25 November 1979. Plotted winds are for the surface (1 full barb = 10 kt = 5 m s^{-1}).

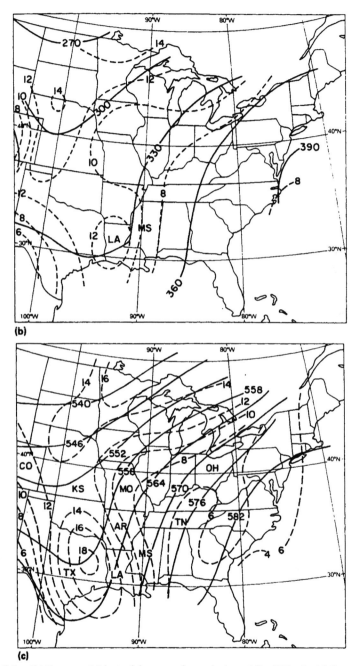

Figure 3.1 (b) Same as (a) but without surface winds and for 700 mb. (c) Same as (b) but for 500 mb.

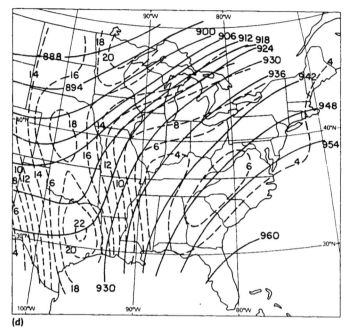

(d)

Figure 3.1 (d) Same as (b) but for 300 mb.

thus vorticity and height contours typically intersect at small angles and there are fewer vorticity/geopotential height solenoids at 1000 or 850 mb than at 500 mb. For this reason subsequent discussions will often presuppose the fact that 1000 mb vorticity advections are negligible compared to those at 500 mb. Exceptions are possible in the vicinity of intense surface cyclones or in areas where the 500 mb vorticity advections are very weak or vanishing, such as along the axes of vorticity maxima or minima.

The magnitude of the vertical advection term in (3.1), $- \omega \, \partial \zeta / \partial p$, can be estimated from the vertical motion field in Figure 3.2b and the vorticity patterns in Figure 3.1. An ascent rate of $\omega_7 = - 4 \, \mu b \, s^{-1}$ over southern Mississippi (MS) implies that the maximum value of vertical motion $(\omega_{max} = \hat{\omega})$ is about $- 5 \, \mu b \, s^{-1}$ at 500 mb, which we obtain using (2.28) with $p_* = 1000$ mb and $p_t = 0$ mb. At the same point, the absolute vorticity varies from roughly 10 units at 700 mb (Fig. 3.1b) to 6 units at 300 mb (Fig. 3.1d), yielding a vertical advection term with a magnitude of $5 \times 10^{-10} \, s^{-2}$. This value is roughly half an order of magnitude weaker than the values for horizontal advection ($30 \times 10^{-10} \, s^{-2}$ in the region of vorticity advection solenoids at 500 mb (Fig. 3.1c)). Moreover, the vertical advection term is small in the lower troposphere and negligible at the surface, where ω_s effectively vanishes in the absence of significant topographical effects.

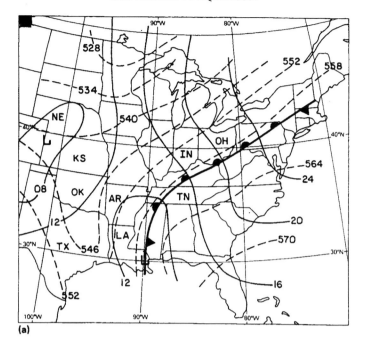

(a)

Figure 3.2 (a) Sea-level pressure isobars (full curves labeled in mb above 1000 mb) for 1200 GMT 25 November 1979, and isopleths of 1000–500 mb thickness (broken curves labeled in dam). The blackened square at the upper left represents the area of a unit advection solenoid. Surface frontal positions are also indicated.

A characteristic of developing mid-latitude baroclinic systems is the westward tilt of the lows and highs with altitude. This tilt is associated with the vertical variation in horizontal divergence. More properly, the tilt is really a lag, since the 500 mb trough is not necessarily continuous in the vertical with the surface low. The same is true for highs. This lag in weather systems is an essential aspect of the dynamics and energetics of mid-latitude weather systems and is a consequence of the baroclinic nature of the atmosphere, as is discussed in more detail in subsequent chapters.

It can be seen from Figure 3.1 that the trough and associated vorticity maxima at 700 and 500 mb are situated to the west of the 1000 mb low and vorticity maximum. Because of this lag, anticyclonic relative vorticity in the upper troposphere tends to overlie weak cyclonic relative vorticity associated with the 1000 mb low. The reverse can be said to be true for upper cyclonic vorticity overlying weak relative anticyclonic vorticity near the 1000 mb high. This change of vorticity with height is important in considering the vertical advection term. The vertical advection of vorticity transports positive vorticity upward (positive vorticity downward) in the rising air

57

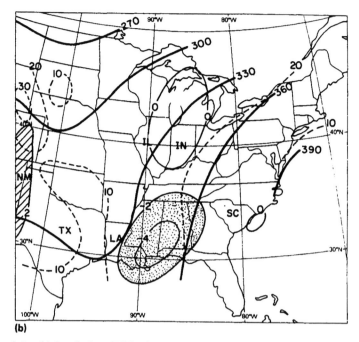

(b)

Figure 3.2 (b) Isopleths of 700 mb geopotential height (thick full curves labeled in dam), geostrophic wind speed (broken curves in m s^{-1}) and the 12 h predicted 700 mb vertical motions (thin full curves in μb s^{-1}, shaded or hatched inside of -2 or $+2$ contours) for 1200 GMT 25 November 1979.

ahead of surface cyclones (in sinking air near the surface high). For example, vertical advection of positive vorticity occurs ahead of and behind the surface cyclone near northwestern Mississippi (MS) for the case represented in Figures 3.1 and 3.2 (see specifically Figures 3.1a, 3.1c and 3.2b).

Vertical and horizontal advections neither create nor destroy an existing scalar field but merely move it about. The divergence term, however,

$$-(\zeta+f)\left(\frac{\partial u}{\partial x}+\frac{\partial v}{\partial y}\right)_p = -(\zeta+f)D_p = (\zeta+f)\frac{\partial \omega}{\partial p}$$

represents a *source or sink of vorticity*, the sign of which is virtually governed by the sign of the divergence, since hydrodynamic stability demands that absolute vorticity be positive on the synoptic scale. The presence of negative absolute vorticity initiates a form of inertial instability, which rapidly restores positive absolute vorticity by generating gravity waves and turbulence. In the divergence (or *development*) term, convergence corresponds to an increase of cyclonic vorticity with time and divergence corresponds to a

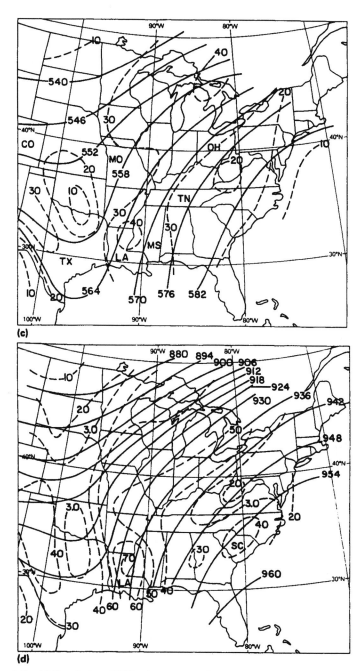

Figure 3.2 (c) Isopleths of 500 mb geopotential height (full curves labeled in dam) and geostrophic wind speed (broken curves labeled in m s^{-1}) for 1200 GMT 25 November 1979. (d) Same as (c) but for 300 mb.

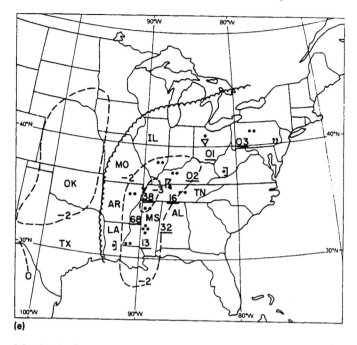

Figure 3.2 (e) Surface weather depiction chart using conventional meteorological symbols for 1200 GMT 25 November 1979. Six-hour rainfall totals in hundredths of an inch are underlined and isopleths of 3 h sea-level pressure tendency are shown as broken curves. Scalloped border denotes western edge of overcast cloud cover.

decrease of positive vorticity with time. Thus, convergence is associated with the "spin up" and divergence with "spin down" of an area for cyclonic vorticity. The idea that the horizontal contraction of a ring of fluid (convergence; Fig. 2.7) is associated with spin up is consistent with the conservation of angular momentum and is well known to figure skaters and acrobatic dancers, though their rotation is not derived significantly from the Earth's spin but from conservation of local (relative) angular momentum.

As has been stated, a typical vertical motion at 700 mb in active mid-latitude weather patterns is a few microbars per second, as can be seen in Figures 2.9 and 3.2b. For a vertical motion of 4 μb s^{-1} at 700 mb, the surface divergence is about 1.3×10^{-5} s^{-1} and, therefore, for an absolute vorticity of 10×10^{-5} s^{-1}, the development term possesses a magnitude of 13×10^{-10} s^{-2} near the surface and diminishes toward zero at middle levels. The bowstring model of Figure 2.8 suggests that the development term exerts its greatest influence in the lower (and upper) troposphere and is near zero near mid-troposphere. Obviously, development does occur at all levels in the atmosphere, but should take place more rapidly at the surface than at

500 mb, where systems maintain themselves in a more conservative state.

The development term $(-(\zeta + f)D_p)$ is a product of two kinematic properties: absolute vorticity and divergence. It is clear that, for a given divergence, development must occur more rapidly where there is already an existing absolute vorticity maximum. Thus, one should expect that existing local absolute vorticity maxima at the surface, such as occur at fronts or in cyclones, would constitute pre-existing sites favorable for surface cyclogenesis. Once a disturbance has formed, moreover, it should remain a favorable site for further development. On the other hand, existing sites of anticyclonic vorticity may correspond to relatively small values of absolute vorticity. Therefore highs would be unfavorable sites for cyclogenesis or anticyclogenesis. This is one explanation for why lows grow more rapidly than highs.

Examination of the 1000 mb absolute vorticity pattern in Figure 3.1a reveals that ζ_0 is fairly uniform, differing on the average by only $2 \times 10^{-5}\,\text{s}^{-1}$ from the value of f, at least in the lower troposphere. Although there are small areas in the vicinity of the cyclone where the value of the relative vorticity at 1000 and 500 mb exceeds that of the Coriolis parameter (as in Figs 3.1a & c), the spatially averaged relative vorticity is generally (but not always) much smaller than f at middle latitudes. Accordingly, it is not only convenient but consistent with scale analyses (and with quasi-geostrophic theory) to consider $(\zeta + f)$ as a spatial constant (f_0) in the development term. While we may refer to the complete version of the development term in discussions, the neglect of ζ with respect to f allows one to simplify mathematical treatment of the vorticity and derivative equations, such as the quasi-geostrophic omega equation, without great loss of accuracy.

Another term, representing a source or sink of vorticity, constitutes the tilting effect

$$\left(\frac{\partial \omega}{\partial y} \frac{\partial u}{\partial p} - \frac{\partial \omega}{\partial x} \frac{\partial v}{\partial p} \right).$$

In the three-dimensional sense, absolute vorticity is conserved with respect to tilting. The tilting terms describe the effect of rotating vorticity from one plane to another. In vorticity equation (3.1) the tilting terms produce or destroy relative vorticity by rotating vertical wind shear into and out of the horizontal plane by means of horizontal gradients of vertical motion. In other words vorticity about a horizontal axis becomes vorticity about a vertical axis or vice versa. The tilting terms constitute a minor "leak" of vorticity from one plane to another. In Figure 3.3, the horizontal winds create a westerly vertical wind shear along the x axis. There is ascent south of the origin $(\omega < 0)$ and descent north of the origin, $\partial \omega / \partial y > 0$. This effect is to rotate the vertical shear $(\partial u / \partial p < 0)$ out of the vertical plane and into the

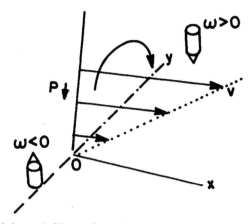

Figure 3.3 Schematic illustration of vertical wind shear along x axis in presence of vertical motions (tubular arrows). Ascent to south ($-y$) and descent to the north are causing the vertical shear to tilt the winds into the x, y plane in the direction of the thin curved arrow, creating anticyclonic relative vorticity.

horizontal, in the sense described by the curved arrow, exchanging the vorticity of vertical shear for that in the horizontal. One can easily picture the rotation of such a shear increasing clockwise (anticyclonic) wind shear in the horizontal plane $(d(-\partial u/\partial y)/dt < 0)$ and increasing the anticyclonic vorticity with time.

At the surface the tilting effect is negligible because ω approaches zero. At upper levels, however, the vertical motions and vertical wind shears may be sufficiently large to assume a first-order importance. Consider that the vertical motions at 700 mb for the case presented in Figures 3.1 and 3.2 are approximately the same at 500 mb as at 700 mb (Fig. 3.2b), and that the vertical wind shears at 500 mb are given by the vertical difference in the wind speeds at 300 and 700 mb (Figs. 3.2b & d). The strongest geostrophic wind speeds at 300 mb occur over LA and SC (maximum values of 72 and 46 m s^{-1}, respectively). At the location of these wind maxima, the vertical wind shears are approximately 55 and 30 m s^{-1} over the 400 mb layer. Using the 700 mb vertical motions to approximate those at 500 mb, the horizontal gradient of ω over these areas (Fig. 3.2b) is about 4 μb s^{-1} per 600 km. Taking an average of the vertical wind shears, one arrives at an estimate for the magnitude of the tilting term of about 7×10^{-10} s^{-2}, which is certainly not negligible but is much smaller than the horizontal advection term at mid-levels in the region of strongest vorticity advections. Similar reasoning leads to slightly larger estimates of the tilting term at 300 mb. In an active weather pattern, therefore, the tilting terms may be important and even dominate at upper levels in the vicinity of fronts or jet streaks where strong vertical wind shears normally accompany strong baroclinicity; however, the tilting effects are small compared to the vorticity advections at mid-levels and they tend to vanish near the surface. In future mathematical developments, the tilting terms are neglected in discussions of quasi-geostrophic theory at either 500 or 1000 mb, an assumption that is consistent with quasi-geostrophic scaling and thus with the other simplifications discussed in this text.

In some respects friction is very difficult to describe because its effects feed back into each term in the vorticity equation and indirectly affect the atmosphere above the layer actively retarded by friction. The net effect of friction undoubtedly diminishes cyclonic vorticity in regions of positive cyclonic vorticity and anticyclonic vorticity in regions of anticyclonic vorticity. We illustrate this degradation using some simple mathematical relationships introduced in Chapter 1.

Proceeding from (1.35)–(1.38), we obtain

$$F_x = -g \frac{\partial \tau_{xp}}{\partial p} \qquad F_y = -g \frac{\partial \tau_{yp}}{\partial p}$$

$$F_0 \equiv k \cdot (\nabla_p \times F) = g \left(\frac{\partial}{\partial y} \frac{\partial \tau_{xp}}{\partial p} - \frac{\partial}{\partial x} \frac{\partial \tau_{yp}}{\partial p} \right). \tag{3.4a}$$

Given (1.38) and assuming that $\rho g C_d$ is a spatial constant, (3.4a) can be written as

$$k \cdot (\nabla_p \times F) = F_0 = -g \rho_s C_d \zeta_s |V_s| / \Delta p_f. \tag{3.4b}$$

Note that (3.4) shows that the vorticity tendency is negative if the local surface-layer relative vorticity is positive, i.e. the cyclonic (anticyclonic) vorticity becomes less cyclonic (anticyclonic) with time at a rate proportional to the relative vorticity. This degradation with time of the relative vorticity perturbations is greater in regions of stronger winds and higher drag coefficients. This would include regions of rough terrain or strong surface heating. It is important to note that the above equations refer to bulk properties of the planetary boundary layer (PBL) and not to individual parcels.

Although it is beyond the scope of this book to treat turbulence in detail, the parametric expression for F_0 (equation (3.4)) is meant to provide a physical basis for discussion of the effects exerted by friction on large-scale vorticity patterns. Therefore, the friction term is considered, for the present, only as a lower boundary effect that is confined to a shallow friction layer of depth Δp_f, above which the frictional drag is considered to be zero. (This text treats Δp_f as the depth of the planetary boundary layer.)

A representative magnitude for F_0, the frictional dissipation rate of vorticity, may be obtained from (3.4). Over land the anemometer wind speed is roughly about one-third the magnitude of the surface geostrophic wind speed; over the ocean that fraction is probably closer to two-thirds. Let us treat the surface wind speed (V_s) as some fraction (r) of the 1000 mb geostrophic wind speed (V_0), $V_s = rV_0$, $\zeta_s = r_0\zeta_0$. The expression for F_0 becomes

$$F_0 = -\left(\frac{r^2 g \rho_s C_d |V_0|}{\Delta p_f}\right)\zeta_0 = -C_f\zeta_0$$

where C_f represents a frictional decay coefficient and $r^2 \sim 0.1$ for land surfaces. For a drag coefficient $C_d = 3 \times 10^{-3}$, $g = 9.8$ m s^{-2}, $\rho_s = 1$ kg m^{-3}, $\zeta_0 = 10 \times 10^{-5}$ s^{-1}, $V_0 = 10$ m s^{-1} and $\Delta p_f = 10\,000$ Pa (100 mb), F_0 is roughly 3×10^{-10} s^{-2}, which is much smaller than the development term near the surface.

Estimates of F_0 suggest that this term is occasionally important. F_0, however, is partly offset by frictional forced convergence or divergence in the development term.

An important effect of friction is to retard the wind speed. In Chapter 9 we will show that frictional balance requires that the winds are subgeostrophic, have a cross-isobaric flow toward lower pressure and experience convergence in regions of positive relative vorticity; the reverse is true for areas of negative vorticity with regard to divergence. Thus friction opposes its own dissipative effect on vorticity in the planetary boundary layer by means of feedback to the divergence term. This feedback does not cancel frictional dissipation when the total atmospheric column is considered because the frictionally driven divergence changes sign with height at the level p_f and produces the reverse tendency of vorticity above the friction layer. This reverse tendency, which affects the vorticity above the friction layer, is called "secondary spin down". Normally the frictionally induced vertical motion is

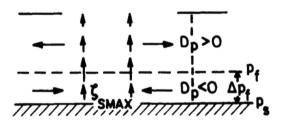

Figure 3.4 Schematic representation of frictional convergence and frictionally induced vertical motion (arrow proportional to strength of the irrotational wind or vertical motion) in the vicinity of a relative surface vorticity maximum. Friction layer depth is Δp_f. Secondary divergence, occurring above the friction layer and induced by the primary convergence in the friction layer, is shown above p_f in the layer $D_p > 0$.

very much smaller than that imposed by the large-scale dynamics, except intermittently near the top of the friction layer.

In Figure 3.4 the arrows denote the divergent part of the wind, rather than the total wind velocity. There is a local maximum of relative cyclonic vorticity at the surface. Convergence is occurring in the boundary layer, resulting in ascent over the column; above p_f, there is divergence of the winds due to frictionally induced ascending motion.

Below p_f, the top of the friction layer, there is an approximate balance of forces between friction, Coriolis and pressure gradient (see Fig. 9.1) and between the friction, divergence and advection of vorticity terms in the vorticity equation. Even near the surface these three terms do not precisely cancel, but rather leave a residual vorticity tendency that will act to degrade the existing vorticity pattern. The effective rate at which the vorticity pattern degrades due to friction is likely to be less than the term F_0 because of compensation of F_0 with the development term.

Nevertheless, we can set the tendency exactly equal to the friction term in the vorticity equation to obtain a lower limit to the time constant for spin down of the entire atmosphere were it allowed to run down purely through frictional dissipation. In this case,

$$\partial \zeta_0 / \partial t = F_0 = - C_f \zeta_0$$
$$\zeta_0 = \zeta_0|_{t=0}(e^{-C_f t}).$$

The time constant for frictional decay (C_f^{-1}) in the solution of this equation is on the order of 10 days. This is a justifiable result for a vorticity equation that has been integrated over the entire atmosphere and globe (over which the net divergence, as we have seen, is nearly zero and the advection terms vanish). This result indicates that baroclinic disturbances at middle latitudes can lose much of their kinetic energy to friction over the length of their lifetimes.

In Figure 3.1a the effects of friction on the wind direction can be seen in the frictionally induced cross-isobaric flow toward the cyclone along the Gulf of Mexico coastline. Frictional convergence increases the development term, compensating for the negative effect of F_0. The cross-isobaric flow advects air with lower values of positive vorticity toward the center of positive vorticity, helping to fill the cyclone in the absence of other influences. On the other hand, frictional divergence causes an export of negative relative vorticity near the high, as can be seen in Figure 3.1a.

Use can be made of the quasi-geostrophic approximations in the presence of friction without contradicting the principle of geostrophic balance. This is done in some numerical treatments by considering ω_f (the frictionally induced vertical motion at the top of the friction layer) to be the vertical motion at the lower boundary, e.g. 1000 mb. Recall that in the vorticity equation the divergence term is $-(\zeta+f)D_p$, or approximately $f_0\,\partial\omega/\partial p$ if ζ is neglected with respect to f, a justifiable assumption according to scale considerations. If advection is neglected, and we confine our arguments only to the friction effect, there is an approximate balance between the divergence term and friction in the vorticity equation within the friction layer. Therefore, if we let $F_0 = -f_0\,\partial\omega/\partial p$, then

$$F_0 = -\frac{r^2 g\rho_s C_d |V_0|\zeta_0}{\Delta p_f} = -f_0\,\frac{\partial\omega}{\partial p}. \qquad (3.5a)$$

Integration of the above expression from the bottom to the top of the friction layer (whose depth is Δp_f) yields

$$\omega_f = -r^2 g\rho_s C_d |V_0|\,\zeta_0/f_0. \qquad (3.5b)$$

Thus, by neglecting the other terms in the vorticity equation we can gain an approximate expression for ω_f in terms of the 1000 mb relative geostrophic vorticity and wind speed. This expression allows us to examine the physical processes associated with frictional forcing. Vertical motion due to friction induced at the top of the friction layer is upward in regions of cyclonic relative vorticity in the boundary layer, and downward where there is anticyclonic relative vorticity in the boundary layer.

The magnitude of this frictional forcing is a maximum (ω_f) at the top of the friction layer ($p = p_f$), where the frictional effect vanishes. Frictionally forced ascent at the top of the planetary boundary layer is sometimes referred to as "Ekman pumping", after the Swedish oceanographer, Ekman. Above p_f the sign of the divergence is opposite to that in the friction layer (as can be inferred from Fig. 3.4). The average value for the rate of change of vorticity with time over the entire atmospheric column above Δp_f due to friction is approximately $f_0\omega_f/p_f$. For $p_f = 900$ mb, $\omega_f = 1\ \mu\text{b s}^{-1}$ and $f_0 = 1 \times 10^{-4}\,\text{s}^{-1}$, the vorticity tendency averaged over the atmosphere above p_f is approximately $1 \times 10^{-10}\,\text{s}^{-2}$. It seems likely, however, that the influence of frictional divergence or convergence in the column above p_f diminishes rapidly with height so that its effect is probably negligible at 500 mb.

Table 3.1 reviews the typical magnitudes of the terms in the vorticity equation (3.1) for three levels in the atmosphere, 1000, 500 and 300 mb. It is readily apparent that the vertical advection term and (in most cases) the friction and tilting terms are small with respect to the other terms at all three levels. Hence, the quasi-geostrophic vorticity equation may be written as

$$\frac{d(\zeta+f)}{dt} = \frac{\partial\zeta_g}{\partial t} + V_g \cdot \nabla_p(\zeta_g+f) = (\zeta_g+f)\frac{\partial\omega}{\partial p} \approx f_0\frac{\partial\omega}{\partial p}. \qquad (3.6a)$$

Table 3.1 Typical magnitudes of terms in vorticity equation at middle latitudes for three levels of the atmosphere, 1000, 500 and 300 mb. (Units are 10^{-10} s^{-2}; one unit advection solenoid is about equal to 100×10^{-10} s^{-2}.)

	Horizontal advection	Vertical advection	Development	Tilting	Friction, internal/surface	Local change
Low levels (~1000 mb)	<5	<5	10	<5	<5 Occasionally large	10
Middle levels (~500 mb)	30	5	5	<5	<5 Occasionally large	30
High levels (~300 mb)	50	5	15	10 Occasionally large	<5 Occasionally large	50

The same result as (3.6a) can be obtained directly by deriving the vorticity tendency equation using the quasi-geostrophic momentum equations (1.28a, b) discussed in Chapter 1. Justification for use of these equations is based on scale analysis of the terms in the momentum equations, a rigorous analysis of which was first performed by J. Charney in the 1940s. Table 3.1 indicates that there is an approximate balance for mid-latitude baroclinic cyclones between the local tendency and development terms at 1000 mb. Hence, at 1000 mb (3.6a) reduces to

$$\left(\frac{\partial \zeta_g}{\partial t}\right)_0 \approx f_0 \left(\frac{\partial \omega}{\partial p}\right)_0. \tag{3.6b}$$

On the other hand, at the level of non-divergence, near the middle troposphere, horizontal advection is a first-order effect and the divergence term is relatively small. In this case, the magnitude of $f_0 \, \partial \omega / \partial p$ is much less than $- V_{gs} \cdot \nabla_p(\zeta_{gs} + f)$, and equation (3.6a) reduces to

$$\left(\frac{\mathrm{d}(\zeta + f)}{\mathrm{d}t}\right)_s \approx \frac{\partial \zeta_{gs}}{\partial t} + V_{gs} \cdot \nabla_p(\zeta_{gs} + f) \approx 0 \tag{3.6c}$$

which is identical to saying that the latter equation expresses the conservation of absolute vorticity following an air parcel. Equation (3.6c) is sometimes referred to as the "*barotropic*" vorticity equation, which is treated in more detail in Chapter 6. For most of this text, we will make frequent references to the quasi-geostrophic vorticity equations (3.6a–c).

3.2 The thermodynamic equation

Equations (3.6a–c) are dynamic equations that govern the winds but do not explicitly account for the effect of temperature (mass). The latter is governed by thermodynamic equations. The total rate of change of temperature (T) for an air parcel can be expanded to arrive at (1.13), which expresses the total rate of temperature change due to dry adiabatic compression or expansion and due to all other (diabatic) causes, respectively expressed as $(\mathrm{d}T/\mathrm{d}t)_{ad}$ and $(\mathrm{d}T/\mathrm{d}t)_{nd}$. Restating (1.29) as

$$\frac{\mathrm{d}T}{\mathrm{d}t} = \frac{\partial T}{\partial t} + V \cdot \nabla_p T + \omega \frac{\partial T}{\partial p} = \left(\frac{\mathrm{d}T}{\mathrm{d}t}\right)_{ad} + \left(\frac{\mathrm{d}T}{\mathrm{d}t}\right)_{nd} \tag{3.7}$$

we make the following substitutions involved in obtaining (1.13), which are

$$\left(\frac{\mathrm{d}T}{\mathrm{d}t}\right)_{ad} = \frac{R_d T_v}{c_p p} \omega$$

$$\left(\frac{\mathrm{d}T}{\mathrm{d}t}\right)_{\mathrm{nd}} = \frac{\dot{Q}_{\mathrm{nd}}}{c_p} \qquad (3.8)$$

where $\dot{Q}_{\mathrm{nd}}$ represents the total diabatic heating rate. This heating rate is associated with temperature changes of the parcel due to all processes except for dry adiabatic ascent and descent. For dry adiabatic motion ($\dot{Q}_{\mathrm{db}} = 0$), the first law of thermodynamics (1.6) can be expressed in terms of the time rate of adiabatic temperature change $(\mathrm{d}T/\mathrm{d}t)_{\mathrm{ad}}$. It should be noted here that the temperature rather than the virtual temperature has been used in order to simplify further derivations using the temperature equation. Actual and virtual temperatures are very nearly equal except in the low troposphere when the humidity is high. Henceforth, most treatments of equations ignore the difference between temperature and virtual temperature except in equations involving thicknesses.

Substituting (3.8) into (3.7) yields

$$\frac{\partial T}{\partial t} = -\,V \cdot \nabla_p T + \omega \left(\frac{R_\mathrm{d} T}{c_p p} - \frac{\partial T}{\partial p}\right) + \frac{\dot{Q}_{\mathrm{nd}}}{c_p}. \qquad (3.9)$$

The terms in parentheses in (3.9) can be expressed with the aid of the hydrostatic relationship (1.3), the ideal gas law (1.5) and the relationship $\gamma_\mathrm{d} = g/c_p$, as

$$\frac{R_\mathrm{d} T}{gp} (\gamma_\mathrm{d} - \gamma).$$

Alternatively, this term can be expressed in terms of potential temperature (θ) by taking the logarithm of Poisson's equation (1.7b) and differentiating with respect to pressure, yielding

$$\left(\frac{R_\mathrm{d} T}{c_p p} - \frac{\partial T}{\partial p}\right) \equiv s_\mathrm{d} = -\frac{T}{\theta}\frac{\partial \theta}{\partial p}$$

which is approximately $-\partial\theta/\partial p$ in the lower troposphere. (Note that temperature is henceforth used for virtual temperature.) Both of these expressions are a measure of the dry static stability (s_d). Thus, (3.9) becomes

$$\frac{\partial T}{\partial t} = -\,V_\mathrm{g} \cdot \nabla_p T + s_\mathrm{d}\omega + \frac{\dot{Q}_{\mathrm{nd}}}{c_p}. \qquad (3.10)$$

The first term on the right-hand side of (3.10) is the advection of temperature and the second term is the vertical motion term. The term $\dot{Q}_{\mathrm{nd}}/c_p$ contains two components of heating: (1) that due to diabatic processes (which we shall call $\dot{Q}_{db}$), such as radiation, turbulent mixing, surface sensible heating and cumulus convection; and (2) a component ($\dot{Q}_{ma}$) due to moist adiabatic ascent. Convection is generally identified with diabatic temperature changes due to processes occurring in or around cumulus clouds. Since buoyant ascent within cumulus clouds is also a form of adiabatic vertical motion, we must agree to define $\dot{Q}_{db}$ as being due to all forms of heating (or cooling) not resulting from dry or moist adiabatic ascent or descent in synoptic-scale systems; thus, $\dot{Q}_{nd} = \dot{Q}_{ma} + \dot{Q}_{db}$.

It can be shown (see standard texts on dynamics for details) that for *pseudo-adiabatic* ascent or descent (following a moist adiabat)

$$\dot{Q}_{ma} = - \frac{\delta L_e F}{p} \omega \qquad (3.11)$$

where L_e is the latent heat of vaporization for water (2.5×10^6 J kg^{-1}). Now F is defined in (3.11) as

$$F = qT \left(\frac{L_e R_d - c_p R_v T}{c_p R_v T^2 + q_s L_e^2} \right)$$

where q_s is the specific humidity at saturation for temperature T and R_v is the universal gas constant for water vapor. Here, δ is a discrete operator, which is equal to 1 for saturated ascent ($q = q_s$ and $\omega < 0$) and 0 for descent or where $q < q_s$. For moist (pseudo-adiabatic) ascent (3.10) becomes

$$\frac{\partial T}{\partial t} = - V_s \cdot \nabla_p T + s_m \omega + \frac{\dot{Q}_{db}}{c_p} \qquad (3.12)$$

where s_m is the *moist* static stability, equal to

$$s_m = \left(\frac{R_d T}{c_p p} - \frac{\partial T}{\partial p} \right) - \left(\frac{\delta L_e F}{c_p p} \right)$$

$$= \qquad s_d \qquad - \qquad s_m'$$

$$= \frac{R_d T}{g p} (\gamma_m - \gamma) = - \frac{T}{\theta_e} \frac{\partial \theta_e}{\partial p}. \qquad (3.13)$$

In (3.13) the first term in the parentheses is recognizable as the dry adiabatic stability (s_d). Since the term to its right (s_m') is always positive, the value of s_m is less than that of s_d except when the atmosphere is perfectly dry, where the two are equal. The difference between the dry and moist lapse rates, greatest upon inspection of thermodynamic diagrams, is evident below mid-troposphere. A measure of the dry adiabatic lapse rate (γ_d) is given by (3.13) for the case where the correction to s_d within the parentheses is equal to zero. For non-zero s'_m, (3.13) expresses the moist adiabatic lapse rate (γ_m). (Freezing of water under pseudo-adiabatic ascent along ice adiabat θ_f contributes to a small positive correction to (3.13); this effect will be neglected in further discussions.)

Equation (3.13) also defines s_m in terms of the *equivalent potential temperature* (θ_e). The *wet-bulb potential temperature* (θ_w) is similar to the equivalent potential (θ_e). Both are conserved with respect to moist adiabatic processes; the two have very different numerical values, however. Since the numerical value of θ_w usually does not change by more than several degrees over the troposphere at mid-latitudes, it is roughly comparable to the surface temperature at any level below the tropopause. For this reason, we favor the use of θ_w over θ_e in this text.

For synoptic-scale ascent or descent in unsaturated air, the governing equation is (3.10). Above the planetary boundary layer $\dot{Q}_{db}$ is primarily the result of condensation in cumulus clouds, vertical turbulent mixing and long-wave radiation losses to space. The latter is normally small, the equivalent of a cooling rate of about 1–1.5 °C day^{-1} in the middle troposphere. For ascent in synoptic-scale stratiform cloud systems, the static stability for the layer of condensation is probably closely approximated by s_m. That s_m tends to be smaller than s_d can be seen from inspection of typical profiles of θ and θ_e, as shown in Figure 3.5. In general, the profiles of θ and θ_e resemble each other in winter, at high latitudes or in the upper troposphere, where the absolute humidity of the air is relatively small.

During summer, or at lower latitudes, there are large differences in the dry and moist static stabilities in the lower troposphere, notably in the planetary boundary layer where the lapse rate of θ_e tends to be negative. In such

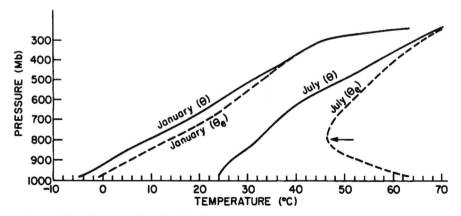

Figure 3.5 Mean vertical distribution of potential temperature (θ) and equivalent potential temperature (θ_e) for Dayton, Ohio, January and July 1979. The horizontal arrow marks the top of the potentially unstable layer for July.

71

instances the atmosphere is *convectively* unstable with respect to convection and parcels of air can rise vertically following a moist adiabat from near the surface to upper levels. (Convective instability is sometimes referred to as *potential instability*, which is the favored usage.) Indeed, in the case of the θ_e profile for July, shown in Figure 3.5, there is potential instability since $\Delta\theta_e/\partial z$ is less than zero below about 850 mb (the level marked with a horizontal arrow). This figure is consistent with Figure 1.2 and with the introductory remarks made in Chapter 1. Note that with decreasing latitude the wet-bulb potential temperature profile becomes increasingly vertical. Eventually θ_w decreases with height from the surface to mid levels equatorward of about 25–30° latitude.

Convection, however, is a topic largely outside the scope of this book, because the inclusion of negative static stability presents mathematically ill-posed conditions for the solution of the quasi-geostrophic equations. Consequently, the dry static stability is used except when discussing moist adiabatic ascent in stratiform clouds. Although such cloud layers tend to possess lapse rates approaching that of the moist adiabat, they occupy a relatively small fraction of the atmosphere (see Ch. 12). In highly baroclinic weather patterns, the lapse rates throughout the middle and upper troposphere are generally between the dry and moist adiabatic. This situation is known as *conditional instability*. Hence, most mid-latitude cyclones contain embedded convective elements in which both evaporation and mixing may be occurring. The percentage of area occupied by convective elements, though never very large even in tropical disturbances, increases toward lower latitudes as the θ_e profile becomes less stable. At mid-latitude, however, it is unlikely that s_d or s_m can become negative over large-scale regions or over deep layers.

Thickness and thermal wind

In keeping with the quasi-geostrophic, quasi-adiabatic premise of this text, the assumption is made that the advecting wind in the thermodynamic equation is geostrophic ($V = V_g$) and that the static stability, whether dry or moist, represents a local spatial average ($\tilde{s}_d$ or $\tilde{s}_m$). These assumptions are to be used in subsequent equations in which the symbol s refers to either the dry or moist adiabatic stability and the tilde overbar ($\tilde{\ }$) symbolizes the fact that s_d or s_m do not vary locally in the horizontal. Thus, the thermodynamic equation can be expressed as

$$\frac{\partial T}{\partial t} = -V_g \cdot \nabla_p T + \tilde{s}\omega + \frac{\dot{Q}_{db}}{c_p}. \tag{3.14}$$

Henceforth, the subscript for $\tilde{s}$ is dropped, but it is understood that $\tilde{s}_d$ or $\tilde{s}_m$

belongs in this equation, depending on whether dry adiabatic or moist adiabatic processes are occurring. The thermodynamic equation can also be written in terms of the thickness Z between two pressure surfaces.

Expressing (1.31) in finite difference notation allows us to solve directly for the thickness (ΔZ) and the thickness tendency $\partial(\Delta Z)/\partial t$. Now, from (1.31)

$$\frac{\partial Z}{\partial p} = - \frac{R_d \bar{T}_v}{gp} .$$

Here $\bar{T}_v$ is the virtual temperature averaged with respect to the logarithm of pressure over a column of depth Δp centered on pressure $\bar{p}$ and bounded by the pressure surfaces p_b below and p_t above. The vertical averaging operator is defined in (1.32) as

$$(\bar{})_{\ln p} \equiv \frac{1}{\ln(p_b/p_t)} \int_{p_t}^{p_b} (\ \ ) \, d\ln p \approx \left(\frac{1}{p_b - p_t} \right) \int_{p_t}^{p_b} (\ \ ) \, dp. \qquad (3.15)$$

At this point it is instructive to derive a tendency equation for the 1000–500 mb thickness ($h = (Z_5 - Z_0)$). To do this, all variables in (3.14) must represent vertical averages with respect to pressure as defined in (3.15). Let us assume that the thermal wind does not change direction with height between 1000 and 500 mb, i.e. that

$$\frac{\partial T}{\partial x} = C \frac{\partial T}{\partial y}$$

(where C is a scalar constant) and therefore

$$\frac{\partial v_g}{\partial \ln p} = - C \frac{\partial u_g}{\partial \ln p} .$$

(The justification for this restriction on the vertical variation of the geostrophic wind is presented below.) With these approximations (3.14) becomes

$$\frac{\partial}{\partial t} \left(\frac{\partial Z}{\partial \ln p} \right) = - V_g \cdot \nabla_p \frac{\partial Z}{\partial \ln p} - \frac{R_d \bar{s}\omega}{g} - \frac{R_d}{gc_p} \dot{Q}_{db} \qquad (3.16a)$$

which for the 1000–500 mb layer can be written

$$\frac{\partial}{\partial t} (\Delta Z) \equiv \frac{\partial h}{\partial t} = - V_g \cdot \nabla_p h + \left(\frac{R_d \bar{s}\Delta p_s}{g\bar{p}} \right) \bar{\omega} + \left(\frac{R_d \Delta p_s}{gc_p \bar{p}} \right) \dot{Q}_{db} \qquad (3.16b)$$

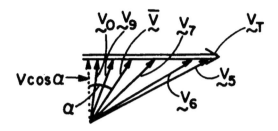

Figure 3.6 Schematic hodograph showing geostrophic winds, represented by the arrows, at various levels between 1000 and 500 mb including the logarithmically averaged wind velocity in the layer ($\bar{V}$). The figure illustrates the case where the isotherms do not change direction with height and the geostrophic wind speed varies linearly with respect to the logarithm of pressure between 1000 and 500 mb. All arrows terminate along the shaft of the thermal wind vector V_T. The dotted vector is antiparallel to the mean temperature gradient in the layer and points toward colder air: its magnitude is given as $V \cos \alpha$, where α is the angle between any wind vector in the layer and the dotted gradient vector.

where $\bar{\omega}$ represents a logarithmically averaged vertical motion between 1000 and 500 mb. The vertically averaged pressure between 1000 and 500 mb ($\bar{p}$) is very close to 700 mb and therefore the assumption is made that $\bar{p} = p_7$. Similarly, $\bar{s}$ is assumed to be that averaged between 1000 and 500 mb (and averaged in the horizontal) and $\bar{V}_g$ is the advecting geostrophic wind velocity over the layer $\Delta Z = h$, where h is the 1000–500 mb thickness. Equation (3.16b) is analogous to the temperature tendency (3.14) and is referred to in subsequent discussions (henceforth, the logarithmic average is understood in this equation).

Thickness advection $(- \bar{V}_g \cdot \nabla_p h)$ can be evaluated by the solenoid method, as described for the vorticity advection in Chapter 2. To do this, it is necessary to determine a wind that can suitably represent $\bar{V}_g$. Consider a hodograph as pictured in Figure 3.6, in which the vector geostrophic winds V_{g0}, V_{g1}, V_{g2}, etc., are drawn from a common origin but the heads of the vectors lie along the thermal wind vector (bold-faced arrow, $V_T \, (= V_{g5} - V_{g0})$), which pertains to the layer between 500 and 1000 mb. In this special case the *direction* of the horizontal temperature gradient ($\nabla_p T$) is constant with p, an assumption that is used to derive (3.16). The dotted arrow is drawn perpendicular to the thermal wind vector so that it points down the thickness gradient. This arrow represents the projection of the vertically averaged wind velocity ($\bar{V}_g$) and *all other wind velocities in the layer* on the thickness gradient vector. Now, if the angle between a wind vector and the normal to V_T is called α, the advection can be written

$$- \bar{V}_g \cdot \nabla_p h = - |\bar{V}_g| |\nabla_p h| \cos \alpha = - |V_{g0}| |\nabla_p h| \cos \alpha_0$$

$$= -|V_{s9}||\nabla_p h|\cos\alpha_9 = \ldots = -|V_{g5}||\nabla_p h|\cos\alpha_5. \quad (3.17)$$

As long as the winds are geostrophic and vary in such a way that the direction of the thermal wind (the orientation of the isotherms) does not change with height, the projection of every wind vector on the normal to V_T is the same, i.e. $|V_{gi}||\nabla_p h|\cos\alpha_i$ is the same for all vectors in the layer.

The implication of this argument and of Figure 3.6 is that *any* geostrophic wind in a layer may be used as the advecting wind for the thickness provided that the isotherms in the layer do not change orientation with height. That this assumption is quite reasonable for typical weather situations is illustrated in Figure 3.7 for the case of Figures 3.1 and 3.2. Here, the low-level isotherms (represented by those at 850 mb) and the 500 mb isotherms are almost parallel over most of North America, although the gradients are rather different at the two levels. Not surprisingly, the isotherms are nearly parallel to the thickness contours (Fig. 3.2a). The resemblance of surface isotherms to thickness contours provides a useful principle in the analysis of surface fronts, which are placed along discontinuities in the gradient of thickness normal to the front (see Ch. 13).

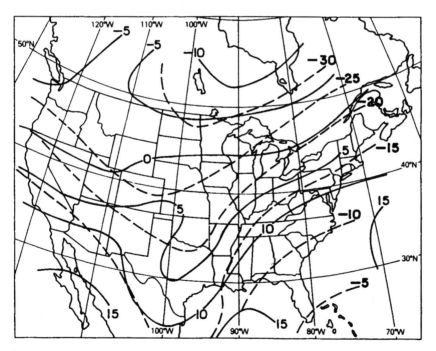

Figure 3.7 Isotherms at 850 mb (full curves labeled in °C) and at 500 mb (broken curves labeled in °C) for 1200 GMT 25 November 1979.

It can be seen from inspection of Figure 3.2 that the surface low tends to lie east of the 500 mb trough but that troughs and ridges exhibit very little westward slope with height above 700 mb. Westward slope of weather systems at middle latitudes is a consequence of the baroclinic structure of the atmosphere in which the temperature trough lags the surface low and temperature ridge lags the surface high. Indeed, this configuration is an inevitable consequence of the horizontal temperature advection in which warm air moves poleward east of the surface low and cold air is moved equatorward west of the surface low. If we are to accept the axiom that the temperature gradients are maximized near the surface and are weakest at mid levels, it is not be surprising that the surface temperature and 1000–500 mb thickness patterns are very similar. Moreover, because horizontal temperature gradients tend to maximize near the surface, the 500 mb geopotential height contours tend to resemble the shape of the surface isotherms and lie parallel to the surface fronts.

What is less obvious is why the surface temperature pattern should exhibit such a resemblance to the mid- and upper-level geopotential height pattern when the geopotential height is a function of the weight (and therefore the average density) of the air column above that level. The answer, of course, is that baroclinic systems do not consist of disassociated components but are linked from top to bottom by an internally consistent dynamic structure. We need only look at one aspect of this structure in order to deduce what is happening elsewhere in the system.

In subsequent discussion it will be convenient to use the 1000 mb geostrophic wind in place of $\vec{V}_g$ for advecting the 1000 to 500 mb thickness pattern, in which case $-\vec{V}_{g0} \cdot \nabla_p h$ is equal to the average thickness advection over the lowest 500 mb. The motivation for using the 1000 mb geostrophic winds arises from the current practice of the U.S. Weather Service to present the 500–1000 mb thickness pattern on the sea-level pressure field, as shown in Figure 3.2a. It would be equally justifiable to use V_{g5} in place of V_{g0} to obtain the 1000–500 mb geostrophic thickness and advection.

Applying the solenoid method discussed in Chapter 2 (see equations (2.19) and (2.20)), the geostrophic thickness advection is

$$- V_{g0} \cdot \nabla_p h = \frac{g \Delta Z_0 \Delta h}{f_0 \Delta n \Delta s} = \frac{g}{f_0} \frac{\Delta Z_0 \Delta h}{A_s} \tag{3.18}$$

where A_s is the area of the solenoid formed by intersecting thickness and 1000 mb height contours. Given that a spacing of 4 mb between sea-level pressure isobars is approximately equal to 30 m (3 dam) spacing between 1000 mb height contours, one can mentally relabel the sea-level isobars in Figure 3.2a to be 30 m interval contours on the 1000 mb surface. Therefore, for maps such as Figure 3.2a, $\Delta Z_0 = 30$ m and $\Delta h = 60$ m in equation (3.18), and the advection is approximately 1.4×10^{-2} m s^{-1}, or close to 1200 m day^{-1} for a (unit) $1° \times 1°$ latitude square advection solenoid. In Figure 3.2a, warm advection solenoids cover a broad area of the eastern United States. Strongest advections (smallest solenoids) occur over western Tennessee (TN) and Illinois just north of the low pressure center. Warm advection solenoids are also visible from the Texas panhandle to Kansas (KS). With some exceptions the warm advection in Figure 3.2a corresponds approximately to the region experiencing rising motion as shown in Figure 3.2b. At the center of the low, the geostrophic advection is zero (because $V_{g0} = 0$), whereas strong warm air advection is taking place over Indiana and Illinois (IN and IL). Note that the distribution of thickness solenoids tends to be broadly distributed with respect to the trough/ridge pattern, in contrast to the 500 mb vorticity advection solenoids, which tend to be more concentrated near the trough and ridge axes.

Although there are likely to be exceptions in any situation, ascent takes place generally in regions of warm advection and descent in regions of cold advection (see Ch. 4). The terms on the right-hand side of the thermodynamic equation ((3.9) or (3.14)) tend to be of roughly equal magnitude, although the magnitude of the advection term is usually larger in synoptic-scale weather patterns. The diabatic term can become intermittently large under some circumstances in which convection or surface heating are important. A general rule of thumb is that the advection term is about twice as large as the sum of the other two terms on the right-hand side of (3.16). An empirical expression based on experience is that

$$\partial h / \partial t = - C_*(V_{g0} \cdot \nabla_p h) \tag{3.19}$$

where C_* is an empirical constant whose value, as determined by observation, depends on the particular type of situation. A general rule is that $C_* \sim 0.5$, but it tends to be somewhat smaller during winter, especially over the ocean. C_* may be as small as 0.3–0.4, and larger for summer ($C_* \sim 0.6$). In cloudy or precipitating systems C_* may be larger than 0.5 because of latent-heat release. Given a value of $C_* = 0.5$, the effect of cold (warm) advection on the local temperature tendency is reduced 50% due to adiabatic warming (cooling). Consequently, thickness isopleths *appear* to be advected with about half the speed of the wind. Since C_* implicitly contains the effect of Q_{db}, not all this cancellation is due to vertical motion. The advection of cold air over warm water, for example, may produce a very large diabatic warming compared to the cooling caused by the advection. In some situations C_* can exceed 1.0, as would occur where there is warm (cold) advection in the presence of sinking (rising) vertical motion. This is also a relatively unusual circumstance and the thickness lines would move at a rate *faster* than the advecting wind speed. This case is referred to in discussions of 500 mb development, such as in Chapters 7 and 10.

Problems

3.1 Evaluate the sense and approximate magnitude of the tilting terms in the vorticity equation at 500 mb over Louisiana (LA) in Figure 3.2. Assume that the vertical motion at 500 mb is exactly the same as at 700 mb and that the vertical wind shear at this level can be expressed in terms of a vertical difference in wind speed between 700 and 300 mb. (For computational simplicity, note that the wind is essentially southerly at this location and that the horizontal gradient in ω is almost entirely in the east–west direction.)

3.2 (a) Locate the region of most rapid pressure falls in Figure 3.2e. Observe the 700 mb vertical motion chart (Fig. 3.2b) and compute for that location the near-surface (1000 mb) divergence. Assume that divergence at 1000 mb is approximated by the difference in ω between 1000 and 700 mb and that ω at 1000 mb is zero. What is the local rate of geostrophic vorticity tendency at that location (in s^{-1} per day) due to the development term of the vorticity equation? Note that there are about 1×10^5 seconds in a day and therefore $1 \times 10^{-10}\,s^{-2}$ is approximately equal to 1 vorticity unit per day. (You will need to know the absolute vorticity, which can be taken from the appropriate map.)

(b) For a sinusoidal wave of $L = 2400$ km, compute, using (3.2) and (3.3), the geopotential height tendency for the vorticity tendency that you computed in part (a). How does this estimate compare with the observed sea-level pressure tendency in part (a)?

3.3 Assuming sinusoidal patterns of geopotential height and identical amplitudes, what is the ratio of the relative vorticities and of the relative geostrophic vorticity advections between a wave of 2400 km and that of 4800 km? If the value of the 500 mb geostrophic vorticity advection for the shorter wave were 40 units per day, would you say that the magnitude of the geostrophic vorticity advection for the longer wave is significant?

Further reading

Astling, E. G. 1976. Some aspects of cloud and precipitation features associated with mid-latitude cyclones. *Mon. Wea. Rev.* **104**, 1466–73.

Charney, J. G. 1948. *On the scale of atmospheric motions.* Geophys. Publik. no. 17.

Ninomiya, K. 1971. Mesoscale modifications of synoptic situations from thunderstorm development as revealed by ATS III and aerological data. *J. Appl. Meteor.* **10**, 1103–21.

Petterssen, S. 1956. *Weather analysis and forecasting II.* New York: McGraw-Hill.

Saucier, W. J. 1955. *Principles and practice of synoptic analysis.* Chicago, IL: University of Chicago Press.

4

Quasi-geostrophic forcing of vertical motions and surface pressure tendency

In this chapter we will expand on the relationships developed in previous chapters by showing how simplified equations based on quasi-geostrophic theory allow one to diagnose the movement and development of weather systems and the distribution of precipitation. *Prognostic* equations, such as those for vorticity and temperature, describe the local rate of change with time of a variable and therefore the future state of that variable. Alternatively, one can determine the spatial distribution of meteorological variables at a given time using *diagnostic* equations. In this chapter we discuss two very important equations: one diagnostic (for vertical motion) and the other prognostic (for pressure tendency). We then proceed to simplify the equations in order to facilitate map analysis of lower-tropospheric vertical motion and surface pressure tendency. Our emphasis is not on exact solutions, but rather on the interpretation of conventional weather charts using quasi-geostrophic theory.

The key to cyclone development is the divergence pattern. At the surface, cyclogenesis is associated with convergence, which accounts for a spin up of cyclonic vorticity. The development term in the vorticity equation (3.6) contains the product of divergence and absolute vorticity. We note that the sign of this term is determined on the synoptic scale by the sign of the divergence, since the absolute vorticity must remain positive. The magnitude of the development term is strongly modulated by the magnitude of the absolute vorticity, which will be larger in cyclonic than anticyclonic areas.

Surface features migrate, develop or weaken, primarily as the result of the changing distribution of surface divergence, which we show in Chapter 3 to be the dominant term in the vorticity equation at the surface. The surface divergence pattern evolves in response to geostrophic advections at upper levels. Since the geostrophic wind is almost non-divergent, divergence is mainly due to the ageostrophic component of the wind, which is closely linked to the vertical motion.

The imposition of geostrophic balance on the winds allows vorticity and temperature to be related via the geopotential height field. In the quasi-geostrophic system, pressure (or geopotential height) and vorticity tendencies are closely related. (We can see this by allowing the fields of geopotential height to be sinusoidal, as in (2.16).) Changes in geopotential height therefore imply changes in vorticity and thickness. Accordingly, the local tendencies of vorticity and temperature are mutually constrained in the quasi-geostrophic system and this constraint imposes parallel changes in the wind and mass fields. Likewise, the vorticity tendency and vertical velocity fields are related via the vorticity equation.

4.1   Derivation of the quasi-geostrophic omega equation

In order to diagnose development (or vertical motion or divergence), we now derive a diagnostic equation for ω solely in terms of conventional fields of geostrophic advections. We can accomplish this by eliminating the local derivatives between (3.6a) and (3.14), which are the quasi-geostrophic vorticity and temperature equations, and solving for ω. Let us first write (3.16a), which is derived from (3.14) and (1.31), as

$$\frac{\partial}{\partial p}\left(\frac{\partial Z}{\partial t}\right) = -\bar{V}_g \cdot \nabla_p \frac{\partial Z}{\partial p} - \frac{R_d \bar{s}}{gp}\omega - \frac{R_d}{gpc_p}\dot{Q}_{db}. \tag{4.1}$$

With the aid of (2.7) the quasi-geostrophic vorticity equation (3.6a) can be written (neglecting friction) as

$$\nabla_p^2 \frac{\partial Z}{\partial t} = -\frac{f_0}{g} V_g \cdot \nabla_p(\zeta_g + f) + \frac{f_0^2}{g}\frac{\partial \omega}{\partial p} \tag{4.2}$$

after first multiplying by f_0/g. Now, a comparison of (4.1) and (4.2) indicates that the left-hand sides of the equations (the tendency terms) are both equal to

$$\frac{\partial}{\partial p}\nabla_p^2 \frac{\partial Z}{\partial t}$$

after taking the vertical derivative of (4.2) with respect to pressure and applying the Laplacian operator (∇_p) to (4.1). In taking the Laplacian of (4.1) the assumption is made that the static stability parameter ($\bar{s}$) is a spatially averaged value, giving it the property of being a local constant, as is the Coriolis parameter outside of the advection terms. In reality, horizontal variations in static stability are very important for cyclone development. We shall account for such variations by imagining that the atmosphere consists of various subregions, each with a spatially constant static stability. Thus, the tilde overbar above the static stability parameter signifies that a horizontal average has been made over a particular subregion but not over the whole map.

Having made these assumptions, the local (time) derivatives vanish after subtracting the vertically differentiated form of (4.2) from the Laplacian of (4.1). In the absence of friction and diabatic heating

$$\frac{R_d}{gp} \tilde{s} \, \nabla_p^2 \omega + \frac{f_0^2}{g} \frac{\partial^2 \omega}{\partial p^2} = -\frac{f_0}{g} \frac{\partial}{\partial p} [-V_g \cdot \nabla_p(\zeta_g + f)]$$

$$+ \nabla_p^2 \left(-V_g \cdot \nabla_p \frac{\partial Z}{\partial p} \right). \qquad (4.3)$$

The solution to the quasi-geostrophic *omega equation* (4.3) is obtained by specifying lower and upper boundary conditions. The solution is stable provided that the value of the static stability is positive. Customarily, the solution is obtained by choosing ω at the upper boundary ($p = 0$) to be zero. At the lower boundary, $\omega = \omega_s$; as shown in Chapter 2, ω_s is approximately equal to $(\partial p / \partial t)_s$, which is small compared to that at 700 mb, at least over flat terrain. Since (4.3) is a form of Poisson's equation, it yields no information on the boundaries, where $\omega \, (= \omega_s)$ is imposed on the solution. We can, however, diagnose the vertical profile of ω and its vertical derivative (the divergence) near the surface. As we have seen in Chapter 2, the vertical motion at 850 or 700 mb provides a means for assessing the near-surface divergence pattern in the development term of the vorticity equation.

Heating and friction terms were omitted in (4.3), since they constitute effects external to the quasi-geostrophic system. Later these terms will be treated as modifications to the omega equation. For the present, we ignore the frictionally forced vertical motion, $\omega_f \, (= \omega_f(p_f))$; it could be regarded as a lower boundary condition, although ω_f pertains to a level somewhat above the surface. In the absence of surface friction and diabatic heating, as well as orographic effects on the lower boundary condition for ω_s, the solution to the quasi-geostrophic omega equation (4.3) is called in this text the *dynamic* vertical motion (ω_d). The word "dynamic" signifies that ω_d is due solely to internal forcing by the two geostrophic advection terms in (4.3). On the large scale, the vertical profile of ω_d closely resembles the idealized bowstring model described in Chapter 2.

Before proceeding to solve (4.3), let us examine some of its properties. The vorticity advection term in (4.3) originates in the vorticity equation, as does the vertical Laplacian of ω on the left-hand side. The horizontal Laplacian of the temperature advection originates in the temperature equation, as does the horizontal Laplacian of ω. The right-hand side of (4.3) is referred to as the *forcing* by geostrophic advections and the vertical motion the *response* to geostrophic advections of mass (temperature) and momentum.

Equation (4.3) helps determine the vertical motions and therefore the divergence pattern using conventional charts of geopotential height and temperature; (4.3) also separates the geostrophic from the ageostrophic (divergent) part of the wind fields on the right- and left-hand sides. Other than vertical motion, static stability is the only important variable in the response. Forcing of ω_d consists entirely of geostrophic advections, which can be evaluated by the solenoid method. Written schematically,

$$ L(\omega_d) = F_V + F_T $$

where F_V and F_T are the two geostrophic forcing terms on the right-hand side of (4.3) and $L(\omega_d)$ represents the three-dimensional Laplacian on the left-hand side, which is the response. Since the Laplacian of a periodic function is proportional to the negative of that function (Ch. 2), positive values of F_V or F_T correspond to negative (ascending) values of ω_d.

Forcing and response are not separate but take place simultaneously without cause and effect. The two terms on the right-hand side of the quasi-geostrophic omega equation are presented separately because they pertain to apparently separate quantities, temperature and vorticity. Chapter 14 will show that the forcing terms can be combined mathematically into one term in the quasi-geostrophic system. It is customary to view forcing as being accomplished by geostrophic advections, and response by the vertical and ageostrophic wind components. In a sense, the geostrophic advections are responsible for moving the atmosphere away from balance (or at least away from its present state), while the ageostrophic/vertical component of the wind constantly tries to adjust the atmosphere to a new geostrophic balance.

Straightforward interpretation of (4.3) using patterns observed on conventional weather charts is impeded by the complexity of this equation. Note, however, that F_V is positive (and ω is negative) if the advection of vorticity increases with height (decreases with respect to pressure). Similarly, if the temperature advection is positive ($-V_g \cdot \nabla_p \partial Z/\partial p < 0$), the Laplacian of the temperature advection is negative, and therefore F_T is positive, $L(\omega_d)$ is positive and ω_d is negative.

Mathematically, each of the two quasi-geostrophic forcing terms can be treated independently as solutions to the left-hand side of the omega equations, with the sum of the two solutions for vertical motions being equal to

$$ \omega_d = \omega_V + \omega_T $$

82

where ω_V and ω_T refer, respectively, to components of dynamic vertical motion obtained separately from forcing due to the vertical derivative of the vorticity advection and from the Laplacian of the temperature advection.

It is convenient to interpret the forcing terms physically. (Forecasters sometimes attribute the vertical motion component forced by the Laplacian of the thickness advection to "overrunning".) Consider the situation along an upper-level thermal ridge, where the 1000–500 mb thickness is locally a maximum (e.g. east of the center of the surface cyclone in Figure 4.1b). If the advection of vorticity is positive at 500 mb and it is weaker or negative at lower levels (the former justified by the arguments presented in Chapter 3), the vorticity advection increases with height. Furthermore, if the 500 mb winds are geostrophic, the differential advection of positive vorticity in the absence of temperature advection implies that to maintain geostrophic

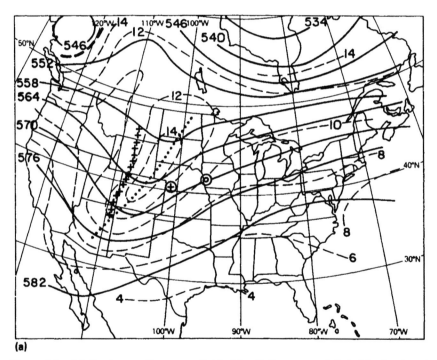

(a)

Figure 4.1 (a) Contours of geopotential height (full contours labeled in dam) and absolute vorticity (broken contours in intervals of $2 \times 10^{-5}\,s^{-1}$) at 500 mb for 0000 GMT 11 September 1986. The circled cross denotes the region of strong lower-tropospheric cold air advection and strong positive vorticity advection, and the double circle is the location of the surface cyclone center. The dotted curve marks the axis of the maximum absolute vorticity at 500 mb and the crosses show the axis of the thermal trough in the thickness field.

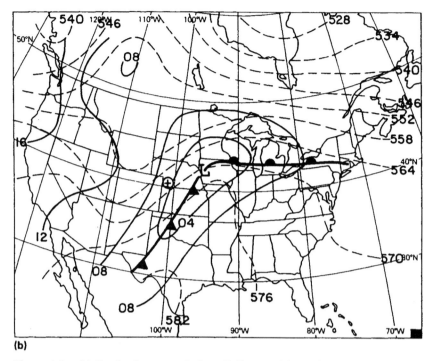

(b)

Figure 4.1 (b) Sea-level pressure isobars (full curves labeled in mb above 1000 mb) with fronts and 1000–500 mb thickness isopleths (broken contours labeled in dam) at 0000 GMT 11 September 1986. The circled cross indicates where there is both strong cold air advection and strong positive absolute vorticity advection. The blackened square at the lower right-hand corner is a unit advection solenoid.

balance the geopotential heights must be decreasing more rapidly at that level than below. Consequently, there must be a decrease in thickness with time due to the differential advection of lower heights aloft. Since this tendency cannot be accomplished by temperature advection at the point where the thickness is maximum, adiabatic cooling due to ascending motion is required to maintain geostrophic balance and account for the temperature tendency. This means that the advection of a deeper trough aloft rather than at the surface necessitates a decrease in thickness (temperature), which must be accomplished by rising motion (adiabatic cooling) in the absence of thickness advection.

Likewise, if one imagines a region where vorticity advection is zero but where there is warm advection in the 1000–500 mb layer, the local increase in temperature due to advection causes the geopotential heights to rise (and the relative vorticity to decrease) increasingly with height. According to the vorticity equation this is consistent with divergence at upper levels and

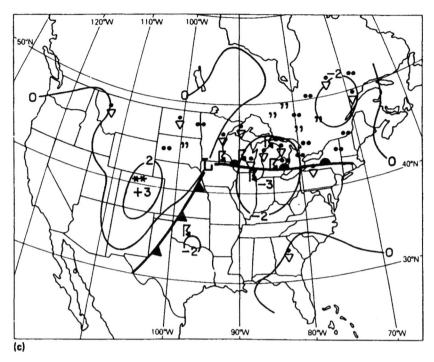

(c)

Figure 4.1 (c) Sea-level isallobars (full curves) at intervals of 2 mb (3 h)⁻¹, surface fronts and surface weather reports in conventional notation for 0000 GMT 11 September 1986.

therefore with ascending motion in accordance with Dines' compensation. The reverse is true for cold advection, but with height falls and convergence implied at upper levels. Of course, both temperature and vorticity advection are generally occurring at the same time and there can be augmentation or cancellation between the two forcing terms in the quasi-geostrophic omega equation; for example, cancellation occurs over the region just west of the surface low in Figure 4.1.

Consider the interpretation of (4.3) in the case of a typical mid-latitude wave/cyclone shown in Figure 4.1a. Although there are no vorticity advection solenoids east of the cyclone, significant positive and negative vorticity advections exist east and west of the upper trough, respectively. Strong positive vorticity advection covers the region of the surface cyclone, which is shown in the sea-level isobar chart in Figure 4.1b. Maximum positive thickness advection occurs northeast of the center of the cyclone. In the region between the center of the cyclone and the east coast of the United States, the temperature forcing (evidenced by the warm thickness advection) is large and positive north of the warm front; and the vorticity advection

85

varies from positive near the cyclone center to very small negative values further east. East of the cyclone there are significant pressure falls and precipitation (mostly north of the warm front) in the form of scattered showers or light rain (Fig. 4.1c).

Several aspects of this weather pattern are worth noting. First, the surface low is situated some distance east of the upper trough axis (center of maximum absolute vorticity), but not far from the region of maximum 500 mb vorticity advection. Positive vorticity advection is large near the center of the surface cyclone, which lies east of the 500 mb trough axis. Thus, the lag between surface and 500 mb troughs is essential if 500 mb vorticity advection is to occur in the vicinity of surface cyclones. Secondly, since the surface cyclone center is located between the upper trough and the down-stream ridge, warm air advection is allowed to take place near the down-stream 500 mb ridge and cold air advection near the 500 mb trough.

The configuration in geopotential height and thickness contours results in positive forcing by both terms in the quasi-geostrophic omega equation northeast of the surface cyclone, where there is significant precipitation (Fig. 4.1c), and negative southwest of the surface cyclone, where there is relatively fair weather. Since the upper winds tend to be strongest over the region of strongest horizontal thickness gradient, it is not surprising that the combined forcing by temperature and vorticity advections results in a dipole pattern of ω; with centers of ascent and descent located, respectively, northeast and southwest of the surface cyclone on the cold sides of the surface warm and cold frontal boundaries.

4.2   A simple model for ω_d

We now derive a simple expression relating the quasi-geostrophic forcing to vertical motion at 700 mb (which is a measure of the lower-tropospheric divergence and therefore of the development effect) To do this, we integrate the omega equation over the lower troposphere from 1000 to 500 mb. This integration allows us to make the approximations that vertical motion at 1000 mb is nearly zero and that the vorticity advection at 500 mb is very much stronger than at the surface (see Table 3.1). In keeping with the premise that patterns of height, vorticity, temperature and the advections of these quantities tend to vary sinusoidally in the horizontal, the assumption is now made that the dynamic component of the vertical motion also varies sinusoidally in x, y and p (as in the bowstring model). Thus, $\hat{\omega}(x, y)$ in equation (2.28) is equal to

$$\hat{\omega} \sin \left(\frac{2\pi x}{L_x} \right) \sin \left(\frac{2\pi y}{L_y} \right)$$

so that ω_d is expressed as

$$\omega_d = \hat{\omega}(x, y)\sin\left(\pi\frac{p - p_t}{p_*}\right)$$

$$= \hat{\omega}\sin\left(\frac{2\pi x}{L_x}\right)\sin\left(\frac{2\pi y}{L_y}\right)\sin\left(\pi\frac{p - p_t}{p_*}\right) \qquad (4.4)$$

where $\hat{\omega}$ is the amplitude of the vertical motion pattern ($\hat{\omega}_{max}$), which is maximized at $(p - p_t) = p_*/2$. This expression yields a sinusoidal variation in ω_d, in both the vertical and horizontal, and is consistent with the quasi-geostrophic omega equation and the bowstring model of Figure 2.8. A simple interpretation of the quasi-geostrophic forcing in (4.3) is now obtained by first differentiating ω_d with respect to x, y and p in (4.4) after multiplying across by p and setting $L_x = L_y = L$. This operation yields

$$-\left[\frac{R_d}{g}\tilde{s}\left(\frac{8\pi^2}{L^2}\right) + \frac{f_0^2}{g}\left(\frac{\pi^2 p}{p_*^2}\right)\right]\omega_d$$

$$= -\frac{f_0}{g}\frac{\partial}{\partial\ln p}\left[-V_g\cdot\nabla_p(\zeta_g + f)\right]$$

$$+ \nabla_p^2\left[-V_g\cdot\nabla_p\left(\frac{\partial Z}{\partial\ln p}\right)\right]. \qquad (4.5a)$$

As in deriving (3.16b) from (3.16a), the above equation is averaged using a vertical logarithmic averaging, as in (3.15), with the assumption (see Ch. 3) that the direction of the isotherms does not change with height between 1000 and 500 mb. This assumption is justified in the previous chapter and serves as a basis for other simple models discussed in this text. Taking the overbar to represent a logarithmic vertical average between p_b and p_t, we get

$$-\left[\frac{R_d}{g}\left(\frac{8\pi^2}{L^2}\right)\overline{\omega_d\tilde{s}} - \frac{f_0^2\pi^2}{gp_*^2}\overline{\omega_d p}\right]$$

$$= -\frac{f_0}{g}\frac{1}{\ln(p_b/p_t)}\{[-V_s\cdot\nabla_p(\zeta_g + f)]_{p_t} - [-V_g\cdot\nabla_p(\zeta_g + f)]_{p_b}\}$$

$$+ \frac{1}{\ln(p_b/p_t)}\nabla_p^2(-V_g\cdot\nabla_p h) \qquad (4.5b)$$

where we choose $p_t = 500$ mb and $p_b = 1000$ mb. Realizing that

$$\bar{p} \simeq \frac{\Delta p}{\ln(p_b/p_t)} \qquad \Delta p = 500\ mb \equiv \Delta p_s$$

87

and that the thickness advection $(-V_{g0} \cdot \nabla_p h)$ varies horizontally in a sinusoidal fashion, the above equation becomes

$$
-\left[\frac{R_d}{\bar{p}g} \bar{\tilde{s}} \left(\frac{8\pi^2}{L^2} \right) + \frac{f_0^2 \pi^2}{gp_*^2} \right] \overline{\omega_d(x, y)}
$$

$$
= \frac{f_0}{g\Delta p_s} [- V_{gs} \cdot \nabla(\zeta_{gs} + f) + V_{g0} \cdot \nabla_p(\zeta_{g0} + f)]
$$

$$
- \frac{8\pi^2}{\Delta p_s L^2} (- V_{g0} \cdot \nabla_p h). \tag{4.6}
$$

Here, we assume that $\overline{\omega_d p} = \bar{\omega}_d \bar{p}$ and $\overline{\tilde{s}\omega_d} = \bar{\tilde{s}}\,\bar{\omega}_d$, which is consistent with the fact that all parameters are assumed to vary linearly with the logarithm of pressure between 1000 and 500 mb. Further, logarithmic averages with respect to pressure of $\tilde{s}$, p and ω_d between 1000 and 500 mb (denoted by the overbar) are assumed to apply at the 700 mb level.

For simplicity, (4.6) is written

$$
- \bar{\omega}_d = - \omega_{d7}
$$

$$
= a\{[- V_{gs} \cdot \nabla_p(\zeta_{gs} + f)] - [- V_{g0} \cdot \nabla_p(\zeta_{g0} + f)]\} + b(- V_{g0} \cdot \nabla_p h)
$$

$$
= - (\bar{\omega}_V + \bar{\omega}_T) \tag{4.7a}
$$

where

$$
a \equiv \frac{f_0}{g\Delta p_s D} \qquad b \equiv \frac{8\pi^2}{L^2 \Delta p_s D} \tag{4.7b}
$$

$$
D \equiv \frac{R_d \bar{\tilde{s}}}{\bar{p}g} \left(\frac{8\pi^2}{L^2} \right) + \frac{f_0^2 \pi^2}{gp_*^2} = \frac{f_0^2 \pi^2}{gp_*^2} \left[1 + \frac{8R_d}{\bar{p}} \left(\frac{\bar{\tilde{s}}p_*^2}{f_0^2} \right) \frac{1}{L^2} \right] \tag{4.7c}
$$

$$
= \frac{f_0^2 \pi^2}{gp_*^2} \left(1 + \frac{L_R^2}{L^2} \right). \tag{4.7d}
$$

Here, we introduce a very important scale parameter, which emerges from the omega equation. That parameter is the *Rossby radius of deformation* (L_R), which is defined in this text as

$$
L_R \equiv \left[\left(\frac{8R_d}{\bar{p}} \right) \left(\frac{\bar{\tilde{s}}p_*^2}{f_0^2} \right) \right]^{1/2}. \tag{4.7e}
$$

In Chapter 11 we will discuss the significance of this parameter in cyclogenesis. For the present, it is sufficient to understand that L_R corresponds to a favored scale of development for the wave/cyclone. Originally defined by C. Rossby for shallow-water waves, L_R represented the scale at which rotation and buoyancy forces assume equal importance.

In (4.7), ω_d is expressed as the sum of two components: ω_V, due to forcing by vertical advection of vorticity, and ω_T, due to forcing by the Laplacian of thickness advection. Note that a minus sign is carried in front of ω_{d7}, so that positive values of the forcing terms correspond to ascending vertical motion.

It is understood in regard to (4.7) that the vertical motion ω_{d7} varies sinusoidally in x and y as do the advections on the right-hand side of the equation. It can be shown that the geostrophic vorticity for a sinusoidal height field specified by (2.14) imposes the same periodic variation on the geostrophic vorticity advection. On the other hand, a sinusoidal variation of the thickness contours, whose form is analogous to that of (2.14), is not mathematically equivalent to stating that the thickness *advection* has the same sinusoidal form. A proportionality between sinusoidal variation in the thickness *advection* and sinusoidal variation of 1000–500 mb thickness (h) can be made provided that one assumes either that the amplitude of the temperature perturbation is small compared with that of the geopotential height, or that the thickness perturbation possesses no y variation in Figure 2.3b, being everywhere equal to that at $y = L/4$. For the present, we will assume that the thickness advection pattern is sinusoidal in x and y. The properties of this equation will be further discussed in Chapter 11.

Equation (4.7) is not an exact solution of the quasi-geostrophic omega equation, but it does show a scaling of appropriate dimensions. Actually, (4.4) overspecifies the vertical motion because the solution yields an approximately sinusoidal profile of ω_d without having to impose one in the vertical. Nevertheless, (4.7) is a useful result because it illustrates the dependence of the solution of (4.3) on advection, static stability and wavelength. It also serves as an aid in evaluating the sign and approximate magnitude of ω_d in a simple, straightforward manner with the aid of conventional map products and without recourse to detailed calculations.

In summary, the model embodied in (4.7) is based on the following assumptions:

(a)   there is a sinusoidal dependence of the advection and vertical motion fields (e.g. (2.14) and (4.4)) with $L_x = L_y = L$;

(b)   the isotherms are parallel at all levels in the layer 1000–500 mb;

(c)   the geostrophic wind varies linearly with respect to the logarithm of pressure as in Figure 3.6;

(d)   static stability is independent of horizontal distance and height; and

(e)   the logarithmic mean level is the same for ω, p and $\bar{s}$ in the 1000–500 mb layer (it is easy to show, using equation (2.28), that $\overline{\omega_d} = 0.64\,\hat{\omega}_d$, $\omega_{d7} = 0.81\,\hat{\omega}_d$ and $\overline{\omega_d} = 0.8\,\omega_{d7}$).

89

Equation (4.7) pertains to the 700 mb vertical motion due to forcing by the difference of the 500 and 1000 mb absolute geostrophic vorticity advections and by the 1000–500 mb thickness advection by the 1000 mb geostrophic wind. A further restriction to be made in subsequent chapters, consistent with previous discussions, is that the 1000 mb geostrophic vorticity advection is negligible compared to that at 500 mb.

The parameters a and b represent amplification factors for the advections, allowing ω_d to depend on vertical and horizontal scale, static stability and latitude. Note that decreasing $\bar{s}$ increases D and decreases a and b, thus increasing the magnitude of ω_{d7} for given advections. For $L = 3000$ km, $f_0 = 10^{-4}\,\text{s}^{-1}$, $g = 9.8\,\text{m s}^{-2}$, $p_0 = 1000$ mb, $p = 700$ mb, $R_d = 287\,\text{J kg}^{-1}\,\text{K}^{-1}$ and $\bar{s} = 20$ K $(500\,\text{mb})^{-1}$, a and b are found to be equal to 8.2×10^5 mb s and 0.71 mb m^{-1}, respectively.

Let us now apply the concept of advection solenoids, as defined in Chapter 2, to interpret (4.7). Recall that a unit advection solenoid is defined as a $1° \times 1°$ latitude square formed by intersection of two differing but parallel contours of one scalar variable with two differing but parallel contours of a geopotential height field. Note also that the numerical values of the advection solenoids depend on the contour spacing which we take as that customarily used by the United States National Weather Service. The normalized values of a and b (multiplied by the values of their respective advections for unit advection solenoids) are 7.8 μb s^{-1} and 10.1 μb s^{-1}. Estimates of the normalized values of the amplifying factors a and b (referred to as C_a and C_b and scaled to units of μb s^{-1} per unit advection solenoid) are presented in Table 4.1 for three different values of the length scale parameter L_R, which is defined as (4.7e).

Values of the vertical motion derived from (4.7) can be obtained from conventional weather charts. Table 4.1 can be used to obtain the vertical motion for each forcing term by first determining the values of C_a and C_b in

Table 4.1 Magnitude of advection terms in linearized quasi-geostrophic omega and sea-level pressure tendency equations ((4.7) and (4.11)) in terms of their respective contributions to the dynamic vertical motion (μb s^{-1}) and the sea-level pressure tendency (mb (3 h)$^{-1}$) for unit advection solenoids at 40° latitude for a wavelength equal to $L_R = [(8R_d/\bar{p})(\bar{s}p_*^2/f_0^2)]^{1/2}$. The components of ω_{d7} or $(\partial p/\partial t)$, can be obtained by dividing the values given in the table by the number of unit advection solenoids inside an observed advection solenoid. Signs depend on sense of advection terms.

L_R (km)	C_a (μb s^{-1})	C_b (μb s^{-1})	$C_{a'}$ (mb (3 h)$^{-1}$)	$C_{b'}$ (mb (3 h)$^{-1}$)
2000	8	35	2	6
4000	8	7	10	5
6000	8	4	20	5

the table and dividing these coefficients by the number of unit solenoids contained within the respective observed advection solenoids. Thus,

$$|\omega_{d7}| = C_a/n_a + C_b/n_b.$$

The number of unit advection solenoids found within a given solenoid of 500 mb absolute vorticity advection by the 500 mb geostrophic winds (n_a) is divided into the coefficient C_a to obtain the contribution (ω_{V7}) to the dynamic vertical velocity. Similarly, the number of unit advection solenoids within a given solenoid of 1000–500 mb thickness advection by the 1000 mb geostrophic wind (which we approximate using the 4 mb intervals of sea-level isobars) is divided into C_b to obtain the contribution ω_{T7} to the dynamic vertical velocity. Thus, given the number of unit advection solenoids (n_a, n_b) and the appropriate wavelength L (which we can choose as L_R), ω_d is obtained using the coefficients in Table 4.1. A fortuitous property of the coefficients, given current U.S. Weather Service contouring conventions, is that equal-sized advection solenoids for 500 mb absolute geostrophic vorticity advection by the 500 mb geostrophic winds, and the 1000–500 mb thickness advection by the 1000 mb geostrophic winds with 30 m contour spacing, are roughly of equal importance in forcing the vertical motions.

The derivation of this equation is based on the premise of sinusoidal advection patterns. We pointed out in the previous chapter that the vorticity advections tend to be sharper and more focused than temperature advection with regard to the solenoids; whereas the thickness advections vary in size relatively slowly across the chart but enclose a larger region of solenoids than for absolute vorticity advection. This can be seen in Figure 4.1. One reason for the greater smoothness in the temperature field is the dependence of the vertical motion on the Laplacian of the temperature advection. A narrow concentration of thickness advection solenoids implies large Laplacian and, therefore, excessively large values of the vertical motion concentrated in relatively small regions.

As discussed in Chapter 3, the vorticity advection is much larger at 500 mb (Fig. 3.1c) than at 1000 mb (Fig. 3.1a) over most of the chart; therefore the vertical derivative of vorticity advection depends mainly upon the vorticity advection at 500 mb. In Figure 3.1c, the smallest positive vorticity advection solenoids at 500 mb are found over Arkansas (AR) and Mississippi (MS), while the strongest negative vorticity advections occur over Texas (TX) and southern Lake Michigan. Weak negative vorticity advection occurs over the Great Lakes and also over Colorado (CO). Figure 3.2a shows that warm advection prevails over most of the domain with greatest strength (smallest advection solenoids) over Indiana (IN) and Ohio (OH); weak cold advection solenoids are found over Louisiana (LA) and western Nebraska (NE).

Neither the computer analysis nor the subjectively inferred vertical motions are directly verifiable, but insight into the validity of the patterns may be gained by examining the weather depiction chart (Fig. 4.1c). Sustained precipitation is occurring poleward of the warm front with heaviest precipitation confined to a region of strong warm air advection. Rapid pressure falls (> 3.0 mb (3 h^{-1})) are centered over the warm frontal region, although there are large pressure falls south of the warm front in an area of weak forcing. Obviously, the atmosphere is more complex than expressed by (4.3) because the region of largest pressure falls extends into a region of weak forcing by the two advections represented in this equation.

On the vertical motion chart in Figure 3.2b, there is a wide region of ascent over the eastern part of the country, except for a small area south of the Great Lakes. Strong sinking motion is occurring over extreme western Texas (TX) and New Mexico (NM). Ascent is generally occurring where either or both advections, vorticity (Figs 4.1a & 3.1c) and thermal (Figs 4.1b & 3.2a), are significantly positive. The situation for descent and in the warm sector of the cyclone is more ambiguous. Strongest descent is shown in a region where there is very strong negative vorticity advection (western Texas and New Mexico).

Forcing terms may possess opposite signs and, as such, will tend to "cancel" each other on a map. Sometimes, the dominant forcing can be diagnosed by inspection of the size of the advection solenoids. An example is the region west of the position of the surface low in Figure 4.1 (denoted by the circled cross). There, temperature and absolute vorticity advections are of opposite sign and the forcing by the vorticity advection exceeds that of the temperature advection. Another example is over western Louisiana in Figures 3.1 and 3.2, where positive vorticity advection cancels negative vorticity advection. Alternatively stated, $|\omega_V| > |\omega_T|$. In other cases, the dominant effect may not be obvious. Chapter 8 examines alternative forms of the omega equations and suggests ways of determining the sign of the forcing in regions where the two forcing terms in (4.3) differ in sign.

Overall, the two quasi-geostrophic forcing terms in (4.7) usually possess the same sign. However, an implication of the above arguments is that the constant C_* in (3.19) may differ from 0.5 when one advection dominates or differs in sign from theory. Thus, when $C_* = 1.0$, the local thickness tendency is exactly equal to the geostrophic thickness advection because the quasi-geostrophic forcing terms cancel in the omega equation, leaving the vertical motion equal to zero. Ascent in the presence of cold temperature advection, for example, implies that the constant C_* in (3.19) may approach or exceed 1.0. On the other hand, if the temperature and absolute vorticity advections are of the same sign but the latter greatly exceeds the former, the magnitude of the vertical motion term in the temperature tendency equation may be larger than the magnitude of the temperature advection, and C_* may be negative. These situations are relatively infrequent, however.

4.3 Pressure tendency equation

One important manifestation of the vertical motion is that it strongly influences the weather; cloudiness and precipitation are associated with ascent and fair weather with descent. In addition, the vertical motion is a transformation of the vertical distribution of divergence, which is related to the vorticity and pressure tendencies associated with the development and decay of highs and lows. The movement and development of lows and highs can be regarded as a consequence of the evolution of the pressure (or geopotential height) pattern as manifested at the surface on conventional weather charts as a 3 h sea-level pressure change ending at the time of observation (e.g. Fig. 3.2e). Because the surface pressure tendency is closely related to surface vorticity tendency (uniquely for an idealized geostrophic and sinusoidal wave pattern as shown in Chapter 2), changes in one imply changes in the other. Moreover, since the local 1000 mb vorticity tendency is closely related to the pattern of divergence at the surface, the surface pressure tendency should closely resemble the vertical motion field at 700 mb, according to equation (3.6b).

An expression for the quasi-geostrophic pressure tendency can be obtained (as in the case of the omega equation) by starting with equations (4.1) and (4.2), which are the quasi-geostrophic temperature and vorticity equations. Realizing that the geopotential height tendency (defined as $\chi \equiv \partial Z/\partial t$) can be expressed explicitly and the vertical motion eliminated between the two equations, (4.1) is differentiated with respect to pressure (again keeping static stability constant in time and space) to yield

$$\frac{\partial^2}{\partial p^2}\left(\frac{\partial Z}{\partial t}\right) = \frac{\partial}{\partial p}\left(-V_g \cdot \nabla_p \frac{\partial Z}{\partial p}\right) - \frac{R_d \bar{s}}{gp}\frac{\partial \omega}{\partial p} - \frac{R_d}{gpc_p}\frac{\partial \dot{Q}_{db}}{\partial p}. \qquad (4.8)$$

After multiplying (4.8) by $f_0^2 p/R_d \bar{s}$, it is added to (4.2) to obtain the quasi-geostrophic geopotential tendency equation, i.e.

$$\left(\nabla_p^2 \chi + \frac{f_0^2 p}{R_d \bar{s}}\frac{\partial^2}{\partial p^2}\chi\right) = \frac{f_0}{g}\left[-V_g \cdot \nabla_p(\zeta_g + f)\right]$$

$$+ \left(\frac{f_0^2 p}{\bar{s}R_d}\frac{\partial}{\partial p}\right)\left[-V_g \cdot \nabla_p\left(\frac{\partial Z}{\partial p}\right)\right] - \frac{f_0^2}{\bar{s}gc_p}\frac{\partial}{\partial p}\dot{Q}_{db}$$

$$(4.9)$$

(where the last term on the right-hand side is the diabatic forcing term). For the present, the pressure (geopotential height) tendency equation retains diabatic heating, though its effect is not within the quasi-geostrophic framework. For $\dot{Q}_{db} = 0$, (4.9) closely resembles (4.3) in form, except that χ_d replaces ω_d

93

as the response variable in the Laplacian on the left-hand side. Vorticity and thermal advection terms constitute the quasi-geostrophic forcing, but the thermal advection is differentiated with respect to pressure, whereas the vorticity advection remains undifferentiated. In the absence of diabatic heating, χ is equal to χ_d, the quasi-geostrophic geopotential height tendency, which is analogous to ω_d in (4.7).

Equation (4.9) indicates that positive vorticity advection and increasing warm air advection (or decreasing cold air advection) with height correspond to geopotential height falls. Conversely, increasing cold air advection (or decreasing warm air advection with height) leads to height rises.

The problem in trying to evaluate the pressure (or geopotential height) tendency equation at the surface is that (4.9) is a Laplacian; this requires that the geopotential height or pressure tendencies be imposed at the lower and upper boundaries. While it is reasonable to imagine a surface at very high levels (e.g. 10 mb), where the geopotential height tendencies associated with a tropospheric disturbance would be negligible, specification of the tendency at the lower boundary poses a problem because the lower boundary tendency is generally non-zero.

Let us ignore, for the present, the diagnosis of surface pressure tendency and analyze the geopotential height tendency at an internal surface, such as 500 mb. Since we are interested only in evaluating the forcing terms in (4.9), we are free to choose the lower and upper boundary conditions arbitrarily. It is not difficult to conceptualize the effect of vorticity advection on the geopotential height tendency. If positive (negative) vorticity represents a relative valley (hill) in the geopotential height pattern, as we demonstrated with regard to Figure 2.5, it is clear that the advection of positive (negative) vorticity involves a decrease (increase) in geopotential height Z with time.

The effect of the vertical derivative of the thickness advection on the geopotential height tendency is more difficult to visualize. To illustrate, let us imagine a two-layer atmosphere, such as the one pictured in Figure 4.2, which consists of three pressure surfaces (1000, 500 and 10 mb). Since geopotential height tendencies at the top and bottom surfaces must be fixed independently, we are free to set them equal to zero. Let us also set the absolute vorticity advection at 500 mb equal to zero by considering the tendency at the location of a vorticity maximum. Geopotential height at 500 mb could nevertheless change in response to a vertical variation of temperature or thickness advection in the column. Now, advection of higher thickness values below this point from the left requires that the 500 mb surface must rise (increasing its geopotential height). Note that this arrangement also requires a decreasing thickness with time above 500 mb, a consequence of the fact that we have constrained the 1000 and 10 mb geopotential height surfaces to remain constant. *Thus, decreasing warm air advection or increasing cold air advection with height require geopotential height rises,*

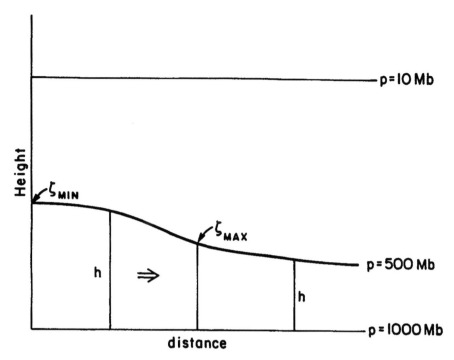

Figure 4.2 Schematic height–distance cross section showing the geopotential height of the 1000, 500 and 10 mb pressure surfaces (full curve lines), or, the 1000–500 mb thickness (h; vertical full lines) and the location of the 500 mb maximum and minimum absolute vorticity. The bold-faced arrow denotes the direction of the geostrophic wind in the 1000–500 mb layer. The figure illustrates that higher thickness values are being advected below 500 mb into the region of the 500 mb vorticity maximum.

and the reverse for decreasing cold air advection or increasing warm air advection with height for geopotential height falls.

If the lower geopotential height surface is allowed to vary, the situation becomes a bit ambiguous. The effect of warm air advection at low levels lowers the 1000 mb surface while also raising the 500 mb surface. In general, however, maximum height changes at 500 mb will be greater than those at 1000 mb and, consequently, the changes we have just stated for fixed lower and upper boundaries still apply to the case of ordinary surface pressure changes. Warm (cold) air advection in the lower troposphere tends to correspond to surface convergence (divergence) and therefore to surface pressure falls (rises), according to (4.3) and (3.6). Maximum temperature advections in the troposphere, however, tend to be found near the surface, particularly on the cold sides of surface fronts. *In classic cyclone situations, in which there is strong warm or cold air advections on the cold sides of the surface warm or cold frontal boundaries, the magnitudes of the temperature gradients and the advections tend to diminish with height,* with resulting consequences for the geopotential height tendencies at middle levels.

Implied in Figure 4.2 is a reversal in the sign of the temperature advection at upper levels. Hydrostatics demand that *decreases* in geopotential height at 1000 mb, corresponding to decreases in surface pressure, must be accounted for by a net decrease in density (*warming*) of the total atmospheric column. As discussed in Chapter 3, temperature advection tends to be highly correlated with advection of the same sign throughout the lower troposphere, but decreasing in magnitude with height. If a large temperature advection were to maintain the same sign over the entire atmosphere, the implied hydrostatic surface pressure changes due to warming of the column would be unrealistically large. Partial cancelling of the effects of temperature advection by temperature advection of the opposite sign at a higher level (and by vertical motion) is thus called for; this idea is similar to that of Dines' compensation. Reversal in the sign of temperature advection with height is possible because of the westward tilt of weather systems with height. More precisely, however, the compensation in advection of opposite sign tends to occur near the tropopause, where there are much larger horizontal temperature gradients than at middle levels. We will discuss this point in more detail in regard to cyclogenesis in Chapter 10. A more substantive argument involving 500 mb development will be presented in Chapter 7.

Consider now the effect of diabatic heating in (4.9). Let us imagine a mid-level diabatic heat source such as might be produced by latent-heat release in convective clouds. Equation (4.9) indicates that, above the heat source (where $\partial \dot{Q}_{db}/\partial p > 0$), the diabatic term ($F_3'$) is negative, and therefore χ is positive. Similarly, below the heat source there are negative geopotential tendencies where ($\partial \dot{Q}_{db}/\partial p < 0$). Falling geopotential heights below the heat source and rising heights above lead to the formation of a low below and a high above, in accordance with observations.

However, consider the case of a *surface* heating maximum, where $\dot{Q}_{db}$ is a maximum at the lower boundary. In this case $\partial \dot{Q}_{db}/\partial p$ is positive above the surface and $\partial Z/\partial t$ is positive. It is widely observed, however, that differential surface heating produces a surface "heat low" and an upper anticyclone. An example of this on a large scale is the Saharan anticyclone, which lies above the desert heat low. The negative tendency at the lower boundary is not expressed by (4.9), but is dictated by the constraint that surface heating decreases the total density, and therefore mass, of the vertical column. Further discussion of the effects of diabatic heating are presented in Chapter 9.

In order to resolve the lower boundary ambiguity, we will make use of the quasi-geostrophic vorticity tendency equation (3.6a), expressed as a surface pressure tendency equation, and diagnose the near-surface divergence (e.g.

at 1000 mb) with the omega equation in the form of (4.7). Thus, we return to (3.6a) expressed in the following form:

$$\frac{\partial \zeta_{g0}}{\partial t} = -V_g \cdot \nabla_p(\zeta_{g0} + f) + f_0 \left(\frac{\omega_0 - \omega_7}{\Delta p_3}\right) \qquad (4.10)$$

where Δp_3 is equal to 300 mb and ω_0 is taken to be zero for the case of horizontal terrain slope. A final step in the derivation is to equate the 1000 mb geostrophic vorticity tendency to the local sea-level pressure tendency through use of equations (3.2) and (3.3). The result is an equation that relates (through $-\omega_{d7}$) the forcing terms in (4.7a) to the 1000 mb geostrophic absolute vorticity tendency and therefore to the sea-level pressure tendency $(\partial p/\partial t)_s$.

Ignoring diabatic heating, the surface pressure tendency equation is written as

$$-\left(\frac{\partial p}{\partial t}\right)_s = \frac{f_0 L^2}{8\pi^2 K_0 g}[-V_{g0} \cdot \nabla_p(\zeta_{g0} + f)] - \frac{f_0^2 L^2}{8\pi^2 K_0 g \Delta p_3}\omega_7 \quad (4.11a)$$

where $\Delta p_3 \equiv 300$ mb. Substituting for ω_7, this becomes

$$-\left(\frac{\partial p}{\partial t}\right)_s = \frac{f_0 L^2}{8\pi^2 K_0 g}[-V_{g0} \cdot \nabla_p(\zeta_{g0} + f)]$$

$$+ \frac{f_0^2 L^2 a}{8\pi^2 K_0 g \Delta p_3}\{[-V_{gs} \cdot \nabla_p(\zeta_{gs} + f)] - [V_{g0} \cdot \nabla_p(\zeta_{g0} + f)]\}$$

$$+ \frac{f_0^2 L^2 b}{8\pi^2 K_0 g \Delta p_3}(-V_{g0} \cdot \nabla_p h) \qquad (4.11b)$$

that is

$$-\left(\frac{\partial p}{\partial t}\right)_s = \frac{f_0 L^2}{8\pi^2 K_0 g}\left(1 - \frac{f_0 a}{\Delta p_3}\right)[-V_{g0} \cdot \nabla_p(\zeta_{g0} + f)]$$

$$+ \frac{f_0^2 L^2 a}{8\pi^2 K_0 g \Delta p_3}[-V_{gs} \cdot \nabla_p(\zeta_{gs} + f)]$$

$$+ \frac{f_0^2 L^2 b}{8\pi^2 K_0 g \Delta p_3}(-V_{g0} \cdot \nabla_p h) \qquad (4.11b)$$

which is given approximately by

$$-\left(\frac{\partial p}{\partial t}\right)_s \simeq a'[-V_{gs} \cdot \nabla_p(\zeta_{gs} + f)] + b'(-V_{g0} \cdot \nabla_p h) \qquad (4.11c)$$

where

$$a' \equiv \frac{f_0^3 L^2}{8\pi^2 K_0 g^2 \Delta p_3 \Delta p_5} \frac{1}{D}$$

$$b' \equiv \frac{f_0^2}{K_0 g \Delta p_3 \Delta p_5} \frac{1}{D} \qquad\qquad (4.11d)$$

and where D has the same meaning as in (4.7).

Equation (4.11) is a linearized version of the pressure tendency equation applied at the surface or sea level (which is close to 1000 mb in this model). Surface pressure falls occur in response to positive vorticity advections at 500 mb (and to a lesser extent at 1000 mb) and to warm thickness advections, modulated respectively by the factors a', b' and a''. All of these parameters are positive, although the negative sign in a'' (4.11d) is a consequence of the partial cancellation between the 1000 mb vorticity advection appearing in both the 1000 mb vorticity equation (4.10) and the linearized omega equation (4.7). In general, the constants in these linearized equations (a, b, a', b' and a'') are about the same magnitude. (Henceforth the surface advection term and its coefficient (a'') will be ignored.)

For a wavelength of 3000 km the parameters a' and b' are about 7 mb $(3\ h)^{-1}$ for unit advection solenoids. Values of a and b, multiplied by their respective advections for unit solenoids to obtain ω_{d7}, are referred to as C_a and C_b in Table 4.1; and the parameters a' and b' multiplied by their unit advections to obtain $(\partial p/\partial t)_s$ are called $C_{a'}$ and $C_{b'}$ in the table. Evaluation of (4.11) using Table 4.1 is exactly analogous to that for (4.7), except that the coefficients differ from those for ω_{d7}. Thus,

$$\left|\left(\frac{\partial p}{\partial t}\right)_s\right| = \frac{C_{a'}}{n_a} + \frac{C_{b'}}{n_b}.$$

where the signs of the coefficients are determined by inspection of the advections.

In Table 4.1 $C_{a'}$ and $C_{b'}$ are, respectively, the coefficients for the vorticity and thickness advections and n_a and n_b are the number of unit advection solenoids for these advections. (Experience shows that choosing values of $7\ \mu b\ s^{-1}$ for C_a and C_b, respectively, and 7 mb $(3\ h)^{-1}$ for $C_{a'}$ and $C_{b'}$, respectively, for the constants in Table 4.1 results in credible values of vertical motion and surface pressure tendency for most weather patterns.) Constituted in this linear form, the surface pressure tendency equation contains the same 500 mb absolute vorticity advection and 1000–500 mb thickness advection as used to diagnose ω_{d7}.

Let us now see how this equation can be evaluated for a typical weather situation, using the example of Figure 4.1. In this example, rapid sea-level pressure falls and disturbed weather are generally associated with small solenoids of 500 mb positive vorticity and thickness advections. (The same observation can be made in regard to Figure 3.2e.) The equation allows us to diagnose the sign and magnitude of the surface pressure tendency using the solenoid method and conventional maps of thickness advection by the 1000 mb geostrophic wind, and the absolute vorticity advection by the 500 mb geostrophic wind. Smallest temperature advection solenoids are found just north of the warm front and west of the cold front. In the former region, ascent is being forced almost exclusively by the temperature field and there is considerable cloudiness and precipitation.

West of the cold front, positive vorticity advection is maximized (circled cross) where there is cold advection. Recall, however, our recent discussion concerning opposing signs in the quasi-geostrophic forcing. At this location the two terms in (4.11) possess different signs. Maximum pressure rises occur further west where the positive vorticity advection is absent but temperature advection remains strongly negative. The patterns of pressure falls indicate that the surface low moves along the thickness gradient toward the region of pressure falls, which are found north of the warm front. Thus, the pattern of solenoids not only indicates regions of ascent and descent and pressure rises and falls, but also suggests the current motion of lows and highs.

In view of (4.11) it is obvious why cyclones move in a direction a little on the cold side of their warm front and why highs move toward the cold front, the latter being generally toward the southeast. These are regions of maximum forcing of surface pressure falls and rises. According to the magnitude analysis of the vorticity equation presented in Chapter 3, changes in the surface pressure pattern occur largely as the result of forcing by the flow aloft acting through the forcing terms in (4.11). *Rather than migrate or be steered by the upper winds, surface features move largely because they reconstitute themselves continuously according to the distribution of pressure falls and rises, rather more like a wrinkle on a bedsheet than a bubble moving in a stream of water.* (The latter is expressed more closely by 1000 mb vorticity advection.) These patterns can become considerably modified during the evolution of the pressure change pattern.

Figure 4.1 also makes it clear why it is necessary to have a phase lag between the 1000 mb cyclone and the 500 mb trough to intensify the surface cyclone. Surface pressure falls will occur ahead of 500 mb troughs and rises ahead of 500 mb ridges in the regions of maximum vorticity advections. The phase lag between upper and lower patterns allows the surface pressures at the center of surface cyclones to decrease with time, since the thickness advection term cannot produce pressure falls at the cyclone center (where V_0 is zero). Another way of looking at the role of the phase lag is to picture a typical open wave cyclone, such as that in Figure 4.1, and realize that the frontal structure requires cold air (lower thicknesses and 500 mb heights) to the west and southwest of the surface cyclone and warm air (higher thicknesses and 500 mb heights) to the east. This arrangement of thickness contours consistent with the fronts necessitates the existence of a 500 mb trough to the west and a ridge to the east of the surface low, resulting in positive vorticity advection over the cyclone; cold air advection to the west and slightly equatorward of the surface cyclone; and warm air advection to the east and slightly poleward of the surface cyclone. The result is a dipole pattern of ascent and surface pressure falls, and descent and surface pressure rises around the surface cyclone. The presence of a dipole pattern of surface pressure changes and vertical motions east of the upper trough implies some cancelling near the cyclone center of the two quasi-geostrophic forcing terms in (4.7) and (4.11).

It follows that the phase lag between surface and 500 mb troughs or ridges is not fortuitous but is a necessary condition in a baroclinic atmosphere. As we will see in Chapter 7, this arrangement also leads to development of the upper trough and to a distortion in the geopotential height and temperature fields such that in time the phase lag disappears and development ceases. In terms of (4.11), surface pressure falls can occur at the center of the surface cyclone as the result of forcing of the surface divergence pattern by vorticity advection at upper levels, although both temperature and vorticity are intimately related in the quasi-geostrophic system. It is worth restating that the development term in the vorticity equation contains the product of both the absolute vorticity and divergence. Although the quasi-geostrophic forcing dictates the evolution of the divergence patterns, surface development will be more rapid where there is an existing vorticity maximum, for reasons that were discussed in regard to the development (divergence) term in the vorticity equation in Chapter 3. Thus, spin up of cyclonic vorticity and surface pressure falls is more efficient in the presence of an existing surface disturbance.

Use of (4.7) and (4.11) can be made in qualitative interpretation of vertical motion and surface pressure tendency. Since current models do not permit the computation of reliable vertical motions from initial conditions, qualitative inspection of this equation can provide insight into the forcing of vertical motions using map analyses. Because of friction, orography and diabatic heating, the actual vertical motion profile may differ greatly from the idealized bowstring model for ω_d, since the total vertical motion (ω) is composed of components for dynamic, frictional, orographic and diabatic heating, respectively subscripted d, F, M and db in this text. Thus, we may write

$$\omega(p) = \omega_d(p) + \omega_F(p) + \omega_M(p) + \omega_{db}(p).$$

Finally, it must be emphasized that the 1000 mb geopotential height tendency in Figure 4.2 is assumed to remain constant for purposes of discussing the role of the two forcing terms in equation (4.9). This is equivalent to imposing a lower boundary condition of zero height tendency in the equation. In fact, the 1000 mb geopotential height (or surface pressure tendency) is usually non-zero. To solve this equation exactly, the geopotential height tendency at 1000 mb and at the top boundary must be supplied from other considerations. For example, suppose the surface pressure change is negative (and is zero at the top of the atmosphere). These boundary conditions would yield an additional component of negative geopotential height tendency whose magnitude would decrease with height from its value at the bottom to zero at the top boundary. If this lower boundary condition is sufficiently large with respect to the two forcing terms in equation (4.9), its effect on the solution could be very important even at mid levels. The question of how and where the total column warming would be produced such as to satisfy the condition of surface pressure falls is reserved for a later chapter. In any case, it is clear that equation (4.9) can not tell the whole story.

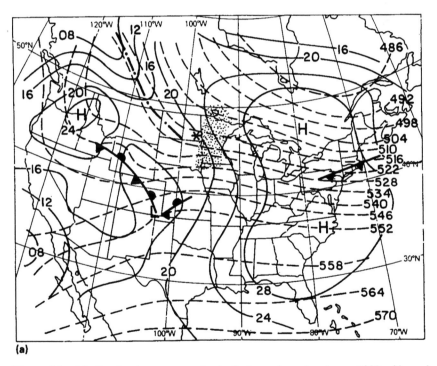

(a)

Figure 4.3 (a) Sea-level pressure chart (isobars in mb above 1000 mb) and 1000–500 mb thickness contours (labeled in dam) for 0000 GMT 17 January 1979. The star denotes the site of initial surface cyclone development along the southern end of a trough (chain-curve) extending from Canada. Shading denotes continuous precipitation.

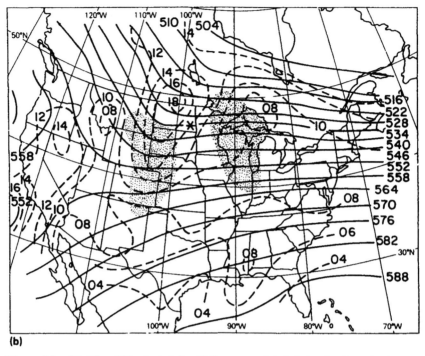

(b)

Figure 4.3 (b) The 500 mb height (full curves labeled in dam) and absolute vorticity (broken curves labeled at intervals of $2 \times 10^{-5}\,\mathrm{s}^{-1}$) for 0000 GMT 17 January 1979. The initial site of surface development is denoted by a star and shading denotes the region where surface pressure falls or rises exceed 2.0 mb (3 h^{-1}).

Rainfall analysis

One can assess the precipitation potential of a weather situation in terms of the diagnosed vertical motions and the precipitable water (W), defined as

$$W = \frac{1}{\rho_w g} \int_0^{p_s} q\,dp$$

where q is the specific humidity and ρ_w the density of liquid water ($= 1.0 \times 10^3\,\mathrm{kg\,m}^{-3}$). Precipitable water, expressed as the equivalent liquid water depth of water vapor over an entire column of air above a unit surface, is typically 1–5 cm at mid-latitudes. A useful formula for assessing the instantaneous rainfall rate (R), expressed in cm day^{-1}, is

$$R = -0.3\,\omega_7 W$$

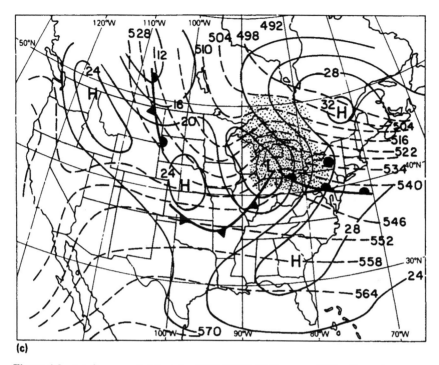

(c)

Figure 4.3 (c) Same as (a) but for 1200 GMT 17 January 1979. Central pressure of the low has fallen to 1012 mb. Shading denotes a region of continuous precipitation and the full circle is the location of the surface low center 12 h later (0000 GMT 18 January).

where W is in cm of precipitable water and ω_7 is in μb s^{-1}. This empirical relationship holds only where the vertical motion is upward and where the mean 1000–500 mb relative humidity exceeds about 70%. Current U.S. Weather Service practice is to furnish maps of precipitable water over the United States every 12 h. Accordingly, with the aid of these charts, the geostrophic advections and the constants provided in Table 4.1, one can make a rough estimate of the distribution of precipitation rate given (4.7).

4.4 An example of cyclogenesis as forced by the upper flow pattern

We now illustrate the idea of surface pressure changes occurring in response to 1000–500 mb thickness and 500 mb absolute vorticity advections. The case presented in Figure 4.3 is one in which cyclogenesis occurred at the site of a very weak surface perturbation (the star) in response to forcing by absolute vorticity advection at 500 mb. Figure 4.3a shows a very weak

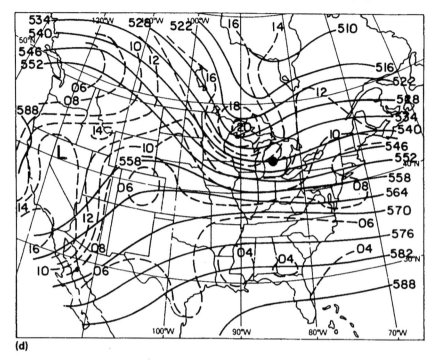

(d)

Figure 4.3 (d) Same as (b) but for for 1200 GMT 17 January 1979 but without indication of surface pressure changes. The full circle denotes the location of the surface cyclone center shown in (c).

lee-side surface trough situated along the Rocky Mountains and extending southward from Canada. Despite the presence of a well-defined low pressure center over Colorado at this time, important development began farther north along a trough, which is denoted by the chain curve in Figure 4.3a. Surface reports show that the sea-level pressures were falling at a rate of about 2 mb $(3 \text{ h})^{-1}$ just east of the surface trough (the area marked by shading) and there was a small area of continuous precipitation. In Figures 4.3a and b, localized but rather large values of 500 mb vorticity advection and 1000–500 thickness advection are indicated by the small advection solenoids over the region of the star. Twelve hours later a small but well-developed cyclone has formed in the region of maximum vorticity advection, exhibiting a classic open wave from a broad area of light and moderate snow and rain along and just north of the warm front, as denoted by shading in Fig. 4.3c. Over 12 h the central pressure of the cyclone decreased by 6 mb to 1012 mb, while the pressure at the location of the cyclone in Figure 4.3c decreased by 17 mb from 1029 mb.

Smallest vorticity and thermal advection solenoids are found just east of the storm center and, consequently, the storm moved eastward rapidly with strong ascent and significant precipitation confined to an area just east of the low. Strong negative vorticity and thickness advections behind the low correspond to rising surface pressures. Movement of the surface low is approximately parallel to the thickness contours.

The example presented in Figure 4.3 illustrates that *surface cyclogenesis occurs in response to favorable baroclinic forcing aloft, represented by advections of temperature and vorticity, and at the site of an existing (though perhaps very weak) maximum of absolute cyclonic vorticity at the surface.* Cyclone *movement*, however, tends toward regions of greatest surface pressure falls, which are strongly modulated by temperature advection. Regions of strong surface pressure falls can be quickly evaluated by examining the pattern of advection solenoids of 500 mb absolute vorticity advection and of 1000–500 mb thickness advection by the 1000 mb geostrophic wind.

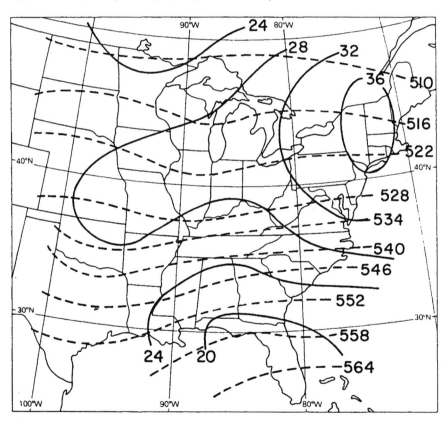

Figure 4.4 (a) Sea-level pressure isobars (mb above 1000 mb) and 1000–500 mb thickness contours (broken curves in dam). (b) The 500 mb geopotential height contours (full curves labeled in dam) and absolute vorticity (intervals of $2 \times 10^{-5} \, s^{-1}$) for the case shown in (a).

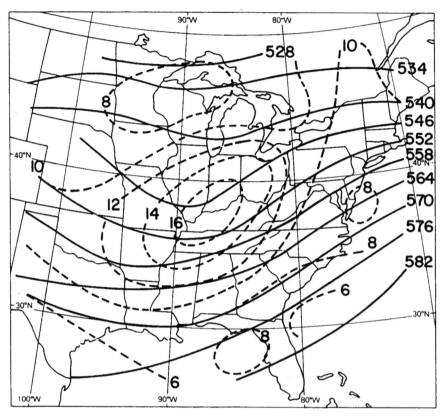

Figure 4.4 (b) The 500 mb geopotential height contours (full curves labeled in dam) and absolute vorticity (intervals of $2 \times 10^{-5} \, s^{-1}$) for the case shown in (a).

Problems

4.1 Calculate, using the values for unit advection solenoids, the vertical motion (ω_{d7}) and the surface pressure tendency from the linearized omega and pressure tendency equations in the region just north of the warm front in Figure 4.1. Assume that the wavelength of the system (L) is 4000 km and equal to L_R. (Take the blackened square in Figure 4.1 as the area of a unit advection solenoid.) Why do you suppose that there are large pressure falls in the warm sector despite the absence of strong quasi-geostrophic forcing?

4.2 Show that for the same geostrophic forcing the response of the vertical motion (in the linearized ω equation (4.7)) is much less when the static stability is very large than when it is very small. Show that the surface pressure tendency goes to zero at the Equator in the linearized pressure tendency equation (4.11). What does this suggest as a constraint on baroclinic development at low latitudes?

4.3 Explain why strong temperature gradients on the cold sides of surface frontal boundaries (such as those shown in Figure 4.1) imply a phase lag between the surface low and the 500 mb trough. (Use a sketch to show that, given an open wave cyclone and the thickness pattern consistent with the arrangement of fronts, it is necessary to have both the thickness and 500 mb troughs to the west of the surface low.) Why does such an arrangement necessitate forcing of surface pressure tendencies over the surface cyclone?

4.4 Figures 4.4a and b show a rather ill-defined disturbance but strong quasi-geostrophic forcing. Compose a pattern of dynamic vertical motion and surface pressure tendency from the appropriate quasi-geostrophic advection patterns which you diagnose using the solenoid method. Use Table 4.1 but for convenience (since accuracy is not attainable in any case) assume that the coefficients C_a and C_b are identical and have the magnitude of 10 μb s^{-1} for a unit advection solenoid, and that the coefficients $C_{a'}$ and $C_{b'}$ are also identical and have the magnitude of 10 mb (3 h)$^{-1}$ for a unit advection solenoid. Use a piece of tracing paper and construct the fields on a separate pair of blank maps. Write a short paragraph discussing what effect the quasi-geostrophic forcing is having on the surface pressure field.

Further reading

Holton, J. R. 1979. *An introduction to dynamic meteorology*, 2nd edn (Int. Geophys. Ser., Vol. 23). New York: Academic Press.

Iskenderian, H. 1988. Three-dimensional air flow and precipitation structure in a non-deepening cyclone. *Wea. Forecast.* **3**, 18–32.

Ninomiya, K. 1971. Mesoscale modifications of synoptic situations from thunderstorm development as revealed by ATS III and aerological data. *J. Appl. Meteor.* **10**, 1103–21.

Reed, R. J. 1963. *Experiments in 1000 mb prognosis*. Nat. Meteor. Center, Wea. Bur., ESSA (NOAA). Tech. Memo no. 26. U.S. Dept of Commerce.

Sanders, F. 1971. Analytic solutions of the nonlinear omega and vorticity equations for a structurally simple model of disturbances in the baroclinic westerlies. *Mon. Wea. Rev.* **99**, 393–407.

Younkin, R. J., R. A. LaRue and F. Sanders 1965. The objective prediction of clouds and precipitation using vertically integrated moisture and adiabatic vertical motions. *J. Appl. Meteor.* **4**, 3–17.

Zwack, P. and B. Okossi 1986. A new method for solving the quasi-geostrophic omega equation by incorporating surface pressure tendency data. *Mon. Wea. Rev* **114**, 655–66.

5

Quasi-geostrophic energetics

Available potential and kinetic energy

The atmosphere is basically a stable heat engine within which local instabilities manifest themselves as transient eddies, which grow and decay in response to the transfer of heat and momentum across horizontal and vertical temperature gradients. Motion occurs in response to a continuing process of restoring imbalances created by the differential heating of the atmosphere, i.e. by the conversion of potential to kinetic energy. For example, one can imagine a tank of liquid, as in Figure 5.1, in which a vertical partition separates dense fluid (shaded) on the left from lighter fluid on the right. After removal of the partition, the dense fluid begins to sink and move toward the right at lower levels in the tank, creating the horizontal and vertical circulation shown in the middle panel of Figure 5.1. Finally, the dense fluid lies completely beneath the lighter one, as in the right-hand figure.

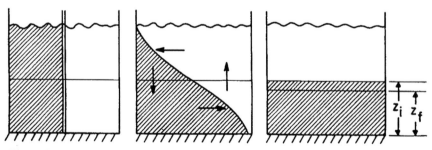

Figure 5.1 Schematic illustration of conversion of potential to kinetic energy and subsequent dissipation by friction in a tank of fluid. On the left, fluids of slightly differing densities (higher density shaded) are at rest and separated by a vertical partition (double line). The partition is removed and the denser fluid slides downward and toward the right, creating a circulation (middle figure). The rearrangement of the fluid involves a lowering of its center of mass from a height Z_i to Z_f above the base of the tank. In the final state (right-hand side), the fluid is once again at rest after having lost all its kinetic energy (an amount per unit mass of $g(Z_i - Z_f)$) to friction.

In the absence of friction the motion would continue indefinitely, surging back and forth like a pendulum. Subject to the influence of friction, however, the dense fluid would eventually come to rest in the state shown by the third panel of the figure. Since the center of mass of the system sinks from a level Z_i initially to Z_f in the final state, the total kinetic energy created – and subsequently destroyed by friction – is just the difference between the two potential-energy states, $mg(Z_i - Z_f)$, where m is the total mass of the liquid. Because the density difference between the two fluids is small, the total change in potential energy is relatively small compared to the initial or final potential-energy states.

The motion created by removing the partition could be maintained against the dissipation by friction in the tank if one were to add dense fluid continuously at the upper left part of the tank and withdraw lighter fluid from the lower right side. Rather than add or subtract fluid, which is cumbersome, a more appropriate experiment for the atmosphere would be to heat one side and cool the other side of the tank, thereby maintaining a density variation due to difference in temperature of the fluid. Furthermore, if the tank were a rotating annulus with heat added to the rim (the left wall) and removed by cooling at the center (the right wall), motion would take the form of wave-like undulations, resembling those on upper air wind charts, within a certain range of rotation rates and horizontal temperature gradients. Whether or not wave motion is established or a simple ("Hadley") circulation – rising in the heated region and sinking on the cooler side – prevails is a function of a non-dimensional parameter called the "thermal Rossby number" (R_T), which is proportional to the ratio of the horizontal difference in temperature between the Pole and the Equator and the rotation rate of the tank. If waves form in the rotating fluid, the energy of these perturbations would be derived from the latitudinal temperature gradient across which non-uniform lateral fluxes of heat and momentum would be taking place. These fluxes yield longitudinal temperature gradients, which represent the potential-energy reservoir available for conversion to the kinetic energy of wave motions.

The motion of the fluid can be expressed as kinetic energy and the potential for driving the fluid in terms of an available potential energy. Such quantities as available potential and kinetic energy, and their budgets, are extremely useful concepts in meteorology. They provide a means for assessing development of waves and cyclones on various scales up to global. Components in the budgets of potential and kinetic energy relate directly to patterns seen on conventional weather charts. Moreover, disturbances in general can be regarded as manifestations of energy generation, transport, conversion and dissipation. In this regard, local rates of change of kinetic and potential energy provide an alternative view of spin up or spin down of the vorticity and of the local change in the thickness or temperature of the perturbations.

Perturbation quantities

Consider first the total kinetic energy per unit of a parcel:

$$k = \tfrac{1}{2}(u^2 + v^2 + w^2).$$

In accordance with hydrostatic balance, which implies that $\tfrac{1}{2}w^2$ (and its changes with time) are insignificant compared to those for the horizontal wind components, it is sufficiently accurate to consider only horizontal kinetic energy $\tfrac{1}{2}(u^2 + v^2)$. We may also wish to know the kinetic energy of the perturbations, as opposed to that of an atmospheric base state. If that base state is the mean wind speed averaged over a domain of length, time or area, the deviation from the mean can be considered as a measure of the perturbation. For example, any scalar, such as wind speed, can be separated into the mean and "eddy" (perturbed) components using the definition

$$u = \bar{u} + u'$$

where the overbar denotes the average and the prime (perturbation) denotes a deviation from the average; note that the mean of the perturbation quantity is identically zero ($\overline{u'} = 0$). For the product of two quantities (e.g. uv), the following identity holds:

$$\overline{uv} = \bar{u}\,\bar{v} + \overline{u'v'}.$$

Analogous identities apply to any form of averaging. Several of these averaging operators are used in this text. For averaging over a length of time T,

$$(\quad)' = (\quad) - (\overline{\quad})$$

where the overbar operator is defined as

$$(\overline{\quad}) = \frac{1}{T}\int_0^T (\quad)\,dt.$$

For a horizontal average along a latitude circle of length L,

$$(\quad)^* = (\quad) - (\overset{\cdots}{\quad})$$

where the dotted overbar is defined as

$$(\overset{\cdots}{\quad}) = \frac{1}{L}\int_0^L (\quad)\,dx.$$

For an area on a horizontal surface,

$$(\quad)'' = (\overset{\frown}{\quad}) - (\quad)$$

where the circumflex operator is defined as

$$(\overset{\frown}{\quad}) = \frac{1}{\sigma} \int_\sigma (\quad) \, dx \, dy$$

and the symbol σ refers to a surface area.

5.1 Available potential energy

In 1903 M. Margules formulated a thought experiment based on a model of the atmosphere similar to that shown in Figure 5.1. He imagined that the atmosphere, with its various perturbations involving variations of temperature on isobaric surfaces, was rearranged adiabatically and without friction such that the final state is one in which there was no variation of pressure on an isentropic surface (and consequently no variation of temperature on an isobaric surface). In requiring the atmosphere to move adiabatically from its perturbed to its base state, there is a lowering of the center of mass of the atmosphere, a decrease in potential energy and a gain in kinetic energy.

The total potential energy of a column is very large, however, about 2000 times that of its kinetic energy. Indeed, the total potential energy (P) will not be zero unless the temperature of the column reaches absolute zero. In Figure 5.1 it is clear that, after rearrangement of mass, the fluid still possesses a finite potential energy, although no further release of kinetic energy is possible. The difference between the potential energy of an atmospheric column and that of its base state represents the realizable form of potential energy that can be converted to kinetic energy; this energy is called the *available potential energy* [A].

Available potential energy was first defined rigorously for an adiabatic base state by Lorenz in 1955, who later applied the concept to the general circulation. Here, we will circumvent the various assumptions made by Lorenz, Van Miegham, P. Smith and others, and proceed directly to our assumption that the base state is that of the mean state of the atmosphere (or a very large domain). Since available potential energy is as much symbolic as a real quantity, we are free to choose this base state as being a near-equivalent to that originally proposed by Lorenz.

Let us first look at the meaning of available potential energy with a concrete image. Imagine that this book of mass m is suspended above the desk top. The book has a potential energy with respect to the desk top or the floor or the lower level of the street outside. Released, the book will fall at an increasing speed V (due to gravitational acceleration) to the desk top a distance Δz below and gain kinetic energy by an amount $(\frac{1}{2}mV^2)$, equal to the decrease in gravitational potential energy $m(g\Delta z)$ at any point up to impact. Had the book fallen a greater distance to the floor or to the street outside, a still greater kinetic energy would be achieved by the book at impact. The desk top, floor or street represent possible base states for the book, depending on the likely minimum state for the potential energy. Realizable (available) potential energy therefore constitutes a hypothetical difference of potential energy between the present state of the atmosphere and a reference base state. Although that base state may be fictitious, it serves to scale the potential energy to magnitudes comparable to realistic changes in the kinetic energy. Resting on the floor the book still possesses an *unavailable* amount of potential energy.

It is convenient to define the available potential energy on the basis of the perturbed state of the temperature field in relation to its globally averaged state. This is *one* way of defining $[A]$, but not necessarily the only or the best way. As stated above, Lorenz defined atmospheric available potential energy as the difference in potential energy between the existing and adiabatic states, as in the Margules formulation. The adiabatic state, however, can no more be achieved in the real atmosphere than that of constant temperature. Thus, calculations of the available potential energy must rely on specifying the base state from an actual perturbed state. The base state constitutes an idealized mean atmosphere in which no temperature gradients can exist on pressure surfaces such as a globally averaged atmosphere.

Unlike the falling of an object under the force of gravity, available potential energy may be created by either increasing or decreasing the total potential energy of a column. For example, if we cool the polar regions in a thought experiment, the average temperature of the atmosphere decreases along with the total potential energy but the horizontal temperature gradient increases, providing additional potential energy of the perturbed state available for conversion to kinetic energy. In terms of the kinetic energy equation, the strength of the the temperature perturbations increases. We may think of the temperature perturbations on isobaric surfaces as a measure of available potential energy.

The available potential energy is the difference between existing potential energy and that which would result if the temperature field were adiabatically rearranged to become that of the mean state. Let us define the available potential energy per unit mass $[A]$ as the potential energy of the perturbed state per unit mass, $(R_d/p\hat{s})\frac{1}{2}T''^2$, which depends upon the variance of temperature on isobaric surfaces. Note that the greater the stability $(\hat{s})$, the more work is required to achieve this rearrangement. In a manner similar to that of Lorenz, we define the available potential energy $[A]$ per unit surface area for the total volume as

$$\frac{R_d}{g} \int_\sigma \int_0^{p_*} \frac{1}{2} \frac{\widehat{\tilde{T}''^2}}{\tilde{s}} \, \mathrm{d}\ln p \, \mathrm{d}\sigma = [A].$$

Thus, it is not the temperature itself but the horizontal temperature *gradients* (expressed as a variance of temperature on quasi-horizontal surfaces) that create the motions manifested as perturbations. This is also analogous to saying that vertical motions and pressure tendencies are forced by advections of temperature and vorticity rather than by the temperature and vorticity themselves.

We now derive an equation for the rate of change of available potential energy. Starting with the temperature equation (3.10) and multiplying each term by $TR_d/\tilde{s}p$, we obtain

$$\frac{R_d}{p\tilde{s}} \frac{\partial}{\partial t} \frac{1}{2} T^2 \equiv \frac{\partial e}{\partial t} = -V_g \cdot \nabla_p e + \omega T \frac{R_d}{p} + \dot{Q}_{db} \left(\frac{T}{\tilde{s}}\right)\left(\frac{R_d}{c_p p}\right) \qquad (5.1)$$

which is an equation for the potential energy per unit mass (e) ($= (R_d/p\tilde{s})\frac{1}{2}T^2$). The second term on the right-hand side of (5.1) indicates that rising warm air and sinking cool air result in a decrease in potential energy (which can reappear as an increase in kinetic energy). If (5.1) is integrated over the entire mass of the atmosphere (requiring the advection terms to vanish), one obtains an expression for the rate of change of *potential energy per unit surface area* [P], which is

$$\frac{\partial [P]}{\partial t} = \frac{R_d}{g} \int_0^{p_*} \left(\widehat{\omega T} + \frac{\widehat{\dot{Q}_{db} T}}{c_p \tilde{s}}\right) \mathrm{d}\ln p \qquad (5.2)$$

or

$$\frac{\partial [P]}{\partial t} = \frac{\partial [P_{\text{mean}}]}{\partial t} + \frac{\partial [P_{\text{eddy}}]}{\partial t}$$

$$= \frac{R_d}{g} \int_0^{p_*} (\widehat{\omega'' T''} + \hat{\omega}\hat{T}) \mathrm{d}\ln p$$

$$+ \frac{R_d}{g c_p} \int_0^{p_*} \left(\widehat{\dot{Q}''_{db} \frac{T''}{\tilde{s}}} + \widehat{\dot{Q}_{db}} \frac{\hat{T}}{\tilde{s}}\right) \mathrm{d}\ln p \qquad (5.3)$$

where the integration over area σ is understood. In the domain average, the flow is uniform and the vertical motion (ω) and heating rate $\dot{Q}_{db}$ are set equal to zero. By definition, therefore, $\partial [A_{\text{mean}}]/\partial t = 0$ since $[A_{\text{mean}}]$ is defined as zero. Therefore, for an atmosphere in which the area average is taken over the entire Earth, and the change in the mean state is zero,

113

$$\frac{\partial[A]}{\partial t} = \frac{\partial[P_{\text{eddy}}]}{\partial t} = \frac{R_d}{g} \int_0^{p_s} \widehat{\omega'' T''} \, d\ln p + \frac{R_d}{gc_p} \int_0^{p_s} \widehat{\dot{Q}''_{\text{db}} \frac{T''}{\tilde{s}}} \, d\ln p \quad (5.4a)$$

This is to be compared with the equivalent tendency equation for the change in eddy kinetic energy (K_E), which is

$$\frac{\partial[K_E]}{\partial t} = -\frac{R_d}{g} \int_0^{p_s} \widehat{\omega'' T''} \, d\ln p + \frac{1}{g} \int_0^{p_s} \widehat{V''_g \cdot F''} \, dp. \quad (5.4b)$$

Summing (5.4a) and (5.4b) gives an expression for the total rate of change of eddy kinetics plus available potential energy, which is

$$\frac{\partial[K]}{\partial t} + \frac{\partial[A]}{\partial t} = \frac{1}{g} \int_0^{p_s} \widehat{V''_g \cdot F''} \, dp + \frac{R_d}{gc_p} \int_0^{p_s} \widehat{\dot{Q}''_{\text{db}} \frac{T''}{\tilde{s}}} \, d\ln p. \quad (5.5)$$

Equation (5.5) (predicated on the assumption that changes in the base state with time are negligible) reflects the fact that the total energy balance is maintained against frictional losses of kinetic energy (represented by the first term on the right-hand side of (5.5)) and by differential heating due to diabatic processes (the second term on the right-hand side of the equation). For the entire atmosphere, energy generation occurs largely through heating at lower latitudes and cooling at higher latitudes.

We can ascribe a physical interpretation to (5.1)–(5.4). Terms involving the products of vertical motion and temperature represent warm air (cold air) rising (sinking), corresponding to the creation of kinetic energy with time at the expense of a corresponding decrease in potential energy. The ascent of warm or descent of cold air is, when integrated over a volume of the atmosphere, equal to a net (domain-averaged) cross-isobaric flow toward lower pressure, a process that lowers the center of mass in the sense illustrated by the simple example of Figure 5.1. Friction, of course, usually decreases the kinetic energy with time.

It is important to note that these equations do not refer to point values in the atmosphere but are meaningful only when averaged over a large area, although maps of the components, either as vertically integrated quantities or on isobaric surfaces as deviations from area means, can provide insight into the location and strength of energy sources and sinks. For example, inspection of Figures 3.2a and b suggests that ascent is occurring in regions of higher thickness (warmer air), at least over most of the area. Such an arrangement signifies that kinetic energy is being generated with time and that baroclinic development is occurring. Thus, it is reasonable to conjecture without extensive calculations that the disturbance shown in Figures 3.1 and 3.2 is producing kinetic energy at the expense of a decrease in potential energy, which is equivalent to saying that baroclinic processes within the

wave domain are contributing to an intensification with time of the disturb-
ance. Ascent of warm air or descent of cold air ($\widehat{\omega''T_v''} < 0$) forms a circulation
that is called a *direct circulation*. Conversely, ascent of cold air or descent of
warm air corresponds to an *indirect circulation*. Note that direct circulations
tend to reduce horizontal temperature gradients.

5.2 Energy transformation; eddy and zonal components

It is useful to treat separately eddy and zonal kinetic- and potential-energy
tendencies in order to isolate the effects of wave perturbations in the
energetics. Eddy potential and kinetic energies (which pertain to the waves)
are defined in relation to a zonally uniform flow, but they possess a longitudinal gradient of temperature and wind speed.

Let us now consider the available potential and kinetic energies as consisting of two types: a zonal component due to the latitudinal variation of the
zonally averaged temperature or wind speed, and an eddy (or wave) component due to variations along a latitude circle. We will now derive equations for the eddy available potential and eddy kinetic energies with respect
to averages about a latitude circle and their deviations from the global base
state. Since the latitudinal means can vary in the meridional direction, the
potential energy of a zonal mean contains the available potential energy of
that mean with respect to a global (area-averaged) mean. Thus, the eddy
available potential energy with respect to the zonal mean is a smaller
quantity than that defined for a perturbation from a global base state
because the zonal average itself has an available potential energy with
respect to the global mean. Let us consider the following definitions:

$$[A] = \frac{R_d}{g\sigma} \int_0^{p_s} \int_\sigma \tfrac{1}{2} \frac{T''^2}{\tilde{s}} \, d\sigma \, d\ln p.$$

and

$$[K] = \frac{1}{g\sigma} \int_0^{p_s} \int_\sigma \tfrac{1}{2} (u^2 + v^2) d\sigma \, dp.$$

Now, the area-averaged eddy available potential energy ($[A_E]$), defined with
respect to a mean around a latitude circle, becomes

$$[A_E] \equiv \frac{R_d}{g\sigma} \int_0^{p_s} \int_\sigma \tfrac{1}{2} \frac{\overline{T^{*2}}}{\tilde{s}} \, d\sigma \, d\ln p$$

and the eddy kinetic energy ($[K_E]$) in a similar fashion is

$$[K_E] \equiv \frac{1}{g\sigma} \int_0^{P_s} \int_\sigma \tfrac{1}{2} (\overset{\dots}{u^{*2}} + \overset{\dots}{v^{*2}}) \, d\sigma \, dp.$$

The zonal available potential energy is then

$$[A_Z] \equiv \frac{R_d}{g\sigma} \int_0^{P_s} \int_\sigma \tfrac{1}{2} \left(\overset{\prime\prime}{\frac{\overset{\dots}{T}}{\tilde{s}}} \right)^2 d\sigma \, d\ln p$$

and the zonal kinetic energy is

$$[K_Z] \equiv \frac{1}{g\sigma} \int_0^{P_s} \int_\sigma \tfrac{1}{2} (\overset{\dots}{u}^2 + \overset{\dots}{v}^2) \, d\sigma \, dp.$$

In order to simplify the derivation we continue to use quasi-geostrophic theory, in which the basic equations of motion (1.28a, b) with friction terms are written

$$\frac{\partial u_g}{\partial t} + \frac{\partial (u_g u_g)}{\partial x} + \frac{\partial (u_g v_g)}{\partial y} = -g \frac{\partial Z}{\partial x} + fv + F_x \tag{5.6a}$$

and

$$\frac{\partial v_g}{\partial t} + \frac{\partial (u_g v_g)}{\partial x} + \frac{\partial (v_g v_g)}{\partial y} = -g \frac{\partial Z}{\partial y} + fu + F_y \tag{5.6b}$$

Further, it is assumed that there is a wave train such as depicted in Figure 2.3b. No net meridional flow is permitted across latitude circles, a reasonable assumption at mid-latitudes, and the meridional component of the wind at any level vanishes along the northern and southern borders. The wave train continues around the latitude circle, so that the mean of zonal derivatives around a latitude circle vanishes. In the interior the wave can possess any shape with no restriction on asymmetries in the wind or temperature fields. In this system the x derivatives of mean zonal quantities vanish as does the zonally averaged meridional wind; the y derivatives of the mean zonal quantities are retained, however. These constraints imply that: (1) $X(0, y) = X(L, y)$ where X represents the variables u, v, T, u_g, v_g and $\dot{Q}_{db}$, and L the wavelength of the disturbance; (2) $v = 0$ at $y = (0, L/2)$; (3) $\partial(X)/\partial x = 0$ along the top and bottom borders; and (4) the zonal average meridional geostrophic wind is zero ($\overset{\dots}{v_g} = 0$). Recall that the dotted overbar refers to a zonal average; we will use asterisks to denote a departure from the zonal average. The averaging operator for distance defines the averaging and the decomposition betweeen the mean *zonal* and *eddy* components for the variables u, v, ω, T and $\dot{Q}_{db}$.

Without further derivation, we now present the tendency equations governing the changes in zonal and eddy kinetic and available potential energies integrated over the entire domain and expressed in terms of unit surface area. The tendency equation for the zonal kinetic energy in the volume (per unit surface area) is

$$\frac{\partial[K_Z]}{\partial t} = \frac{1}{g\sigma} \int_0^{p_*} \int_\sigma (\overset{\cdots}{u_g^* v_g^*}) \frac{\partial \overset{\cdots}{u_g}}{\partial y} \, d\sigma \, dp$$

$$- \frac{R_d}{g\sigma} \int_0^{p_*} \int_\sigma (\overset{\cdots}{\omega} \, \overset{\cdots}{T}) d\sigma \, d\ln p$$

$$+ \frac{1}{g\sigma} \int_0^{p_*} \int_\sigma \overset{\cdots}{u_g} \, \overset{\cdots}{F_x} \, d\sigma \, dp \tag{5.7}$$

where a time average, is implied for all of the terms in the tendency equations. The total kinetic-energy tendency equation for the domain is

$$\frac{\partial[K]}{\partial t} = -\frac{R_d}{g\sigma} \int_0^{p_*} \int_\sigma (\omega T) d\sigma \, d\ln p + \frac{1}{g\sigma} \int_0^{p_*} \int_\sigma (V_g \cdot F) d\sigma \, dp. \tag{5.8}$$

The equation for the eddy kinetic energy is

$$\frac{\partial[K_E]}{\partial t} = -\frac{1}{g\sigma} \int_0^{p_*} \int_\sigma \overset{\cdots}{u_g^* v_g^*} \frac{\partial \overset{\cdots}{u_g}}{\partial y} \, d\sigma \, dp$$

$$- \frac{R_d}{g\sigma} \int_0^{p_*} \int_\sigma \overset{\cdots}{\omega^* T^*} d\sigma \, d\ln p$$

$$+ \frac{1}{g\sigma} \int_0^{p_*} \int_\sigma (\overset{\cdots}{V_g^* \cdot F^*}) \, d\sigma \, dp. \tag{5.9}$$

Note that the eddy and zonal forms of the kinetic-energy equation, respectively (5.9) and (5.7), contain an additional term not present in the total kinetic-energy equation (5.8). This term represents the interaction between the eddies and the zonal flow.

In a similar manner, the area-averaged available potential-energy equations for the eddy and zonal components are derived from the basic definitions of these quantities. These equations are written for the zonal component. The equation for the rate of change of zonal available potential energy is

$$\frac{\partial [A_Z]}{\partial t} = \frac{R_d}{g\sigma} \int_0^{P_s} \int_\sigma \frac{1}{\tilde{s}} \overset{\cdots\cdots}{v_g^* T^*} \frac{\overset{\cdots}{\partial T}}{\partial y} \, d\sigma \, dp$$

$$+ \frac{R_d}{g\sigma} \int_0^{P_s} \int_\sigma (\overset{\cdots}{\omega} \overset{\cdots}{T}) \, d\sigma \, d\ln p$$

$$+ \frac{R_d}{gc_p\sigma} \int_0^{P_s} \int_\sigma \frac{\overset{\cdots}{T}}{\tilde{s}} \overset{\cdots}{\dot{Q}_{db}} \, d\sigma \, d\ln p \qquad (5.10)$$

for the total flow

$$\frac{\partial [A]}{\partial t} = \frac{R_d}{g\sigma} \int_0^{P_s} \int_\sigma (\omega T) \, d\sigma \, d\ln p + \frac{R_d}{gc_p\sigma} \int_0^{P_s} \int_\sigma \left(\frac{T\dot{Q}_{db}}{\tilde{s}}\right) d\sigma \, d\ln p. \quad (5.11)$$

and for the eddy available potential energy

$$\frac{\partial [A_E]}{\partial t} = -\frac{1}{g\sigma} \int_0^{P_s} \int_\sigma \frac{R_d}{\tilde{s}} \overset{\cdots\cdots}{(v_g^* T^*)} \frac{\overset{\cdots}{\partial T}}{\partial y} \, d\sigma \, d\ln p$$

$$+ \frac{R_d}{g\sigma} \int_0^{P_s} \int_\sigma \overset{\cdots\cdots}{(\omega^* T^*)} d\sigma \, d\ln p$$

$$+ \frac{R_d}{gc_p\sigma} \int_0^{P_s} \int_\sigma \left(\frac{\overset{\cdots\cdots}{T^* \dot{Q}_{db}^*}}{\tilde{s}}\right) d\sigma \, d\ln p \qquad (5.12)$$

Note again that the eddy and zonal equations, unlike the mean flow equation, contain an extra term that pertains to interactions between the eddies and the zonal flow.

5.3 Energetics of disturbances and the general circulation: a brief overview

The previous equations (5.7)–(5.12) can be written schematically as sums of terms involving energy conversion rates between different forms of kinetic and available potential energy. Thus,

$$\partial [K_Z]/\partial t = [K_E, K_Z] + [A_Z, K_Z] + D_Z \qquad (5.13)$$

and

$$\partial [K_E]/\partial t = [K_Z, K_E] + [A_E, K_E] + D_E \qquad (5.14)$$

The bracketed terms in (5.13) and (5.14) are defined in (5.7) and (5.9) as

$$[K_E, K_Z] = \int_m \overline{u_g^* v_g^*} \, \frac{\partial \overline{u_g}}{\partial y} \, dm \qquad (5.15a)$$

$$[A_Z, K_Z] = - R_d \int_m \frac{\overline{\omega} \; \overline{T}}{p} \, dm \qquad (5.15b)$$

$$[D_Z] = \int_m \overline{u_g F_x} \, dm \qquad (5.15c)$$

$$[A_E, K_E] = - R_d \int_m \frac{\overline{\omega^* T^*}}{p} \, dm \qquad (5.15d)$$

$$[D_E] = \int_m \overline{(V_g^* \cdot F^*)} \, dm \qquad (5.15e)$$

where m is the mass of the atmosphere ($dm = dp/g$) per unit surface area. (The area-average (circumflex) operator is understood in these equations.) These terms are the integrated conversion rates (per unit surface area) of eddy kinetic energy to zonal kinetic energy (5.15a); of zonal available potential to zonal kinetic energy (5.15b); of frictional dissipation of zonal kinetic energy (5.15c); of eddy potential to eddy kinetic energy (5.15d); and of frictional dissipation of eddy kinetic energy (5.15e).

Equations (5.10) and (5.12) can be written schematically as

$$\partial[A_Z]/\partial t = [A_E, A_Z] + [K_Z, A_Z] + [G_Z] \qquad (5.16)$$

and

$$\partial[A_E]/\partial t = [A_Z, A_E] + [K_E, A_E] + [G_E] \qquad (5.17)$$

where the terms in (5.16) and (5.17) are

$$[A_E, A_Z] = R_d \int_m \frac{1}{\tilde{s}} \, \overline{(v_g^* T^*)} \, \frac{\partial \overline{T}}{\partial y} \, dm \qquad (5.18a)$$

$$[G_Z] = \frac{R_d}{c_p} \int_m \frac{\overline{\dot{Q}_{db} \; \overline{T}}}{\tilde{s}p} \, dm \qquad (5.18b)$$

$$[G_E] = \frac{R_d}{c_p} \int_m \left(\frac{\overline{\dot{Q}_{db}^* T^*}}{\tilde{s}p} \right) dm \qquad (5.18c)$$

119

Here, (5.18a) is the conversion rate (per unit surface area) of eddy available potential energy to zonal available potential energy, and (5.18b) and (5.18c) are, respectively, the generation of zonal available potential energy by zonal diabatic heating and the generation of eddy available potential energy due to eddy diabatic heating. Note that it is redundant to define all the terms in the energy equations since some terms are equal to the negative of their inverse, e.g.

$$[A_Z, A_E] = -[A_E, A_Z].$$

5.4 The energy cycle; eddy and zonal energy exchanges

Let us now examine the individual terms in (5.15)–(5.18) with reference to the rates of exchanges of energy between the zonal and eddy components and between available potential energy and kinetic energy. Figure 5.2 expresses the time-averaged energy transfer flow within the atmosphere system and between the atmosphere and its sources and sinks of energy. The boxes refer to the energy components defined in the previous section, and the arrows the overall direction and magnitude of the conversion rates as expressed in (5.15)–(5.18).

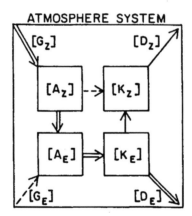

ATMOSPHERE SYSTEM

Figure 5.2 Schematic energy and energy conversion diagram of the atmospheric system. Bracketed symbols within the boxes refer to mean zonal (subscript Z) and eddy (subscript E) components for kinetic energy $[K]$ and available potential energy $[A]$. Energy inputs in the form of diabatic heating $[G]$ and frictional dissipation $[D]$ are shown. Arrows denote the most likely direction of the conversion between energy components for a large-scale mid-latitude region averaged over the passage of many disturbances. Bold-faced arrows denote a generally large rate of conversion, thin arrows a small rate of conversion and broken arrows a very small rate of conversion or one that is not consistently in one direction.

Zonal available potential energy ($[A_Z]$), which is embodied in the meridional gradient of temperature, is added at the top left corner of the diagram. This source of energy is provided by differential solar heating of the Earth's surface. That energy is converted, for the most part, to eddy available kinetic energy ($[A_E]$) through the conversion $[A_Z, A_E]$. As we can see from inspection of (5.18a), this type of energy conversion is accomplished by the correlation between meridional wind velocity and zonal temperature perturbations acting in a mean meridional temperature gradient. This occurs when a wave is situated within a meridional temperature gradient and warm air is carried poleward and cold air equatorward. There is also a small energy conversion between $[A_Z]$ and $[K_Z]$ through the term $[A_Z, K_Z]$, which occurs as the result of warm air rising at low latitudes and sinking at higher latitudes (5.15b).

The principal internal energy conversion is that between zonal and eddy available potential energy ($[A_Z, A_E]$) and between eddy available potential energy and eddy kinetic energy ($[A_E, K_E]$). The latter, expressed by (5.15d), is accomplished by warm air rising and cold air sinking in the disturbances. Some eddy available kinetic energy is generated as the result of diabatic heating in the eddies ($[G_E]$), such as might occur due to the release of latent heat by convection within the warm sector of cyclones. Eddy kinetic energy ($[K_E]$) is destroyed by frictional dissipation (as expressed by the term $[D_E]$) (5.15e), but there is a small conversion rate to zonal kinetic energy. The zonal kinetic-energy generation from eddy kinetic energy, as expressed by (5.15a), occurs as the result of a correlation between the u and v components of the wind within a meridional gradient of zonal wind speed. Such a conversion is accomplished by waves that are asymmetric with respect to the v component of the wind, as in the case of troughs and ridges, which typically tilt from northeast to southwest in the Northern Hemisphere. Finally, some kinetic energy of the zonal flow is dissipated by friction, as expressed by (5.15c).

The conversions $[A_E, K_E]$ and $[A_Z, A_E]$ together represent the baroclinic processes in the atmosphere. These conversions, of which the former is by far the larger, occur as a direct result of cyclones and anticyclones. Conversion of energy between zonal and eddy kinetic energy ($[K_Z, K_E]$ is referred to as barotropic conversion because a temperature gradient is not explicitly stated in equation (5.15a). By contrast, the baroclinic terms imply both a horizontal gradient of temperature and vertical motion. A recurring theme throughout this text is that mid-latitude cyclogenesis is intimately dependent on the presence of temperature gradients (e.g. Ch. 10), which represent the available potential energy.

Conversions between potential and kinetic energy, whether zonal or eddy, appear in both the kinetic- and potential-energy equations, but they possess the opposite signs. Over the averaging domain, the mean zonal vertical

121

motion ($\dot{\omega}$) is relatively small, so that $[A_Z, K_Z]$ is small. Summing (5.13), (5.14), (5.16) and (5.17) results in the cancellation of all internal energy transfer terms and produces the following equation

$$\frac{\partial}{\partial t} ([A_Z] + [A_E] + [K_Z] + [K_E]) = [G] + [D] \qquad (5.19)$$

which states that the total rate of change of kinetic plus available potential energy over the domain occurs as the result of the difference between diabatic energy generation ($[G]$) and frictional dissipation ($[D]$) rates. Over a long period and a large domain, such as the globe during a season or a year, the right-hand side of (5.19) balances, so that $[G] + [D] = 0$. In this instance there is no net gain in total energy and a *steady state* can be said to exist with all forms of $[G]$ balancing all forms of $[D]$. Of course, some of the dissipation is converted back to heat and therefore to potential energy. An interesting consideration is whether the frictional energy sink is significant as a potential-energy source. If frictional dissipation were completely converted to potential energy, the energy cycle would effectively lack an energy sink and the atmosphere would warm up. Heat created by friction, however, is relatively small and entirely dissipated at the surface by radiational cooling to space.

The greatest energy resides in the mean zonal potential energy ($[A_Z]$), with about one-tenth that amount of energy in the eddy kinetic energy ($[K_E]$). Conversion rates between the two are a few watts per square meter or less when averaged over the globe but can be substantially larger (tens of watts per square meter) for very intense disturbances. These values are nevertheless a very small fraction of the solar constant (1367 W m^{-2}). Generation of eddy kinetic energy ($[K_E]$) (and destruction of available potential energy) occurs through the rising of warm air and the sinking of cold air. The idea that direct circulations act to reduce horizontal temperature gradients is an important one, which emerges in regard to frontal circulations discussed in Chapters 13 and 14.

Terms involving correlation between ω and T are equivalent in an integrated sense to cross-isobaric flow in which kinetic energy is created by ageostrophic motion toward lower geopotential heights on constant-pressure surfaces. The primary agents for effecting this energy transfer are the waves, cyclones and anticyclones of middle latitudes. The magnitude of $[A_E, K_E]$, the conversion rate between the eddy available potential energy and eddy kinetic energy, represents a measure of cyclogenesis. In fact, this term tends to be the largest of the conversion terms in Figure 5.2, and very intense cyclones may temporarily provide a significant fraction of the global energy conversion, as much as 20% in extreme cases. The conversion rate is a maximum at mid-troposphere (where ω is a maximum).

The sense of the flow of energy from creation of zonal available potential energy by diabatic heating to dissipation by both the eddies and the zonal flow is depicted in Figure 5.2. Maintenance of the zonal available potential-energy field ($[A_Z]$) is accomplished by the zonal diabatic heating ($[G_Z]$). The zonal temperature field furnishes available potential energy to the eddies, thereby contributing to the eddy available potential energy ($[A_E]$) and, to a much lesser extent, to the kinetic energy of the zonal flow ($[K_Z]$). In turn, the eddy available potential energy ($[A_E]$) produces eddy kinetic energy ($[K_E]$) – manifested by cyclones and anticyclones – which is lost to friction ($[D_E]$). Frictional losses by the zonal flow ($[D_Z]$) are relatively small. The conversion from eddy kinetic to zonal kinetic energy (the barotropic energy conversion term) is also very small. Generation of zonal available potential energy due to differential heating is important, although the generation of eddy available potential energy by eddy diabatic heating ($[G_E]$) is relatively small when averaged over the whole Earth. The latter can nevertheless play an extremely important role in individual disturbances and over limited parts of the globe.

Mean zonal diabatic heating occurs principally at lower latitudes by cumulus convection and in the middle and upper troposphere by differential radiational cooling. Diabatic warming by cumulus convection is intermittently important in middle latitudes. Generation of eddy available potential energy ($[A_E]$) by eddy diabatic heating ($[G_E]$) due to cumulus convection can be quite large in some cases of explosive cyclogenesis. Eddies in the temperature pattern are augmented by cumulus convection, which feeds back to the baroclinic conversion terms via increased baroclinicity and enhanced vertical motions. Optimum distribution of warming in a disturbance occurs when the diabatic heating takes place in relatively warm regions, leading to an increase in the correlation between T^* and $\dot{Q}^*_{db}$. Thus, creation of eddy available potential energy due to the release of latent heat in convective clouds will be more efficient in intensifying a disturbance if latent-heat release occurs in warmer regions of the disturbance.

The rate of diabatic available potential-energy generation is also greater when the static stability is low, such as might occur when there is large-scale ascent and condensation in weakly stable stratified air. The ratio of eddy diabatic heating to stability ($\dot{Q}^*_{db}/\bar{s}$) can be considered as a measure of the *efficiency* of eddy diabatic heating in effecting an energy conversion.

Creation of zonal from eddy kinetic energy is accomplished by the perturbations, primarily in the high troposphere near the level of the jet stream. Zonal kinetic energy is transferred to or from the eddies when the eddy wind components u^* and v^* are positively correlated and situated in a gradient of zonal wind speed. The correlation term ($\overline{u^* v^*}$) is a measure of the poleward flux of eddy angular momentum, whose convergence at latitudes 40–50° provides for the maintenance of the zonal westerlies and the jet stream against internal losses due to turbulent dissipation. A positive value signifies that the waves are asymmetric in the sense of having stronger westerlies on the eastern side of the wave where the air is moving poleward. This arrangement is consistent with the observed latitudinal tilt of waves in the westerlies such that the poleward part of the trough leads (i.e. is located further east of) the equatorward part. A pronounced latitudinal tilt (northeast–southwest in the Northern Hemisphere) is often observed with regard to non-developing waves (e.g. Fig. 2.5) but this orientation frequently reverses itself in developing systems (e.g. Fig. 2.4).

A normal latitudinal wave tilt, in the presence of a zonal westerly current increasing poleward, implies a conversion from eddy to zonal kinetic energy. This barotropic energy conversion is generally very small when averaged over the hemisphere, at least in comparison with conversions denoted by the double-faced arrows in Figure 5.2. Non-linear growth of mid-latitude waves in the westerlies is very much dependent upon transfer up and down the scales of motion, although the ultimate sink of the synoptic-scale kinetic energy is in turbulence and surface friction.

An important transfer of available potential energy from the zonal flow to the eddies occurs through the term $[A_Z, A_E]$ by means of a positive correlation of meridional wind speed (v^*) and T^* acting in a zonal meridional temperature gradient, values decreasing poleward. Since the zonal mean temperature gradient typically decreases poleward, troughs in the westerlies, whether tilted or symmetric, transport warmer air poleward and cold air equatorward, thus converting zonal to eddy available potential energy. Indeed, troughs and ridges are manifestations of this meridional transfer. Largest values of $(\overline{v^* T^*})$ tend to be found in the lower and middle troposphere, although there is some indication that the conversions are large near the tropopause.

Friction usually constitutes a sink of energy. If evaluation of the energy equations could take account of all scales of motion including turbulence, all kinetic-energy dissipation would be attributed to viscosity at the lower boundary, the ground. However, given data and grid size limitations, dissipation usually includes kinetic-energy dissipation scales too small to measure with conventional observations. This is referred to as subgrid-scale energy dissipation. The friction terms have not been shown explicitly in this chapter except in schematic form. As represented, they govern only the surface dissipation, not internal dissipation in the free atmosphere. From equation (3.4), the frictional drag force is written

$$- g[V_g \cdot (\partial \tau / \partial p)] \tag{5.20a}$$

where

$$\tau_0 = r^2 \rho_s C_d | V_{g0} | V_{g0}. \tag{5.20b}$$

Since the frictional dissipation rate (e.g. (5.15c)) comprises the product of wind speed and frictional drag force, kinetic energy dissipation due to friction is effectively proportional to the cube of the surface wind speed, and therefore the frictional dissipation rate is highly sensitive to the surface wind speed. The greater loss of eddy kinetic energy than the mean flow kinetic energy to friction reflects the fact that the zonal flow at the surface is very weak, while the surface wind speed in cyclones can be very large.

Evaluations of the energy budget terms, using real data, may lead to ambiguous results. Values of the terms obtained may differ significantly from one study to another. Some of the weaker energy conversions can differ in sign, as in the case of the thin or broken arrows in Figure 5.2. Calculated energy conversion rates are somewhat sensitive to the mode of time and space averaging, to assumptions regarding the neglect or manner of calculating the boundary terms, to the use of geostrophic winds, to definitions of available potential energy, to errors arising from insufficient data and to the size of the domain. In regard to domain size, the lateral boundary terms may be especially important when the averaging is restricted to the area of a small disturbance. Large lateral boundary terms are significant because the energy of the system is receiving or transferring energy to or from outside. If the domain encloses a single disturbance, large boundary terms may signify the transfer of energy to or from larger-scale disturbances.

The index cycle

An obvious feature of the atmosphere is its variability. Periods of storminess and large wave amplitudes alternate with weather patterns that are relatively quiescent and zonal. Although these oscillations are far from regular, variations in the strength of the zonal current occur with periods of about 3–8 weeks with an amplitude that is most pronounced in the winter months. One way of expressing such variations in the wind field is to compute the "zonal index", which is the mean strength of the zonal geostrophic wind at middle latitudes. Customarily the zonal index has been computed for the latitude belt 35–55° from the mean geopotential heights around each of these latitude circles. Low-index situations occur when the pattern is perturbed by large eddies; high-index ones occur when the flow is mostly zonal.

There is evidence that the eddy available potential energy reflects the zonal cycle. Although the arrows indicating the direction of the energy conversions in Figure 5.2 do not imply a causal or temporal progression, it is nevertheless true that there is low eddy available potential energy during high-index periods. Throughout the year the atmosphere undergoes heating at low latitudes and cooling at high latitudes, creating zonal potential energy expressed by the term $[G_z]$. Periodically, the atmosphere represented by this build-up of zonal potential energy becomes unstable and breaks down into

125

wave disturbances, which are manifested by an increase of eddy available potential energy ($[A_E]$). During this stage, upper air patterns are very wave-like, and heat and angular momentum transfer are large in the north–south direction. Cyclogenesis is also frequent around the globe, although there may be a particular geographic region over which the breakdown of the zonal flow and subsequent cyclogenesis seem to originate. (The concept of downstream development is treated in Chapter 12.) With the increase in cyclogenesis there is also an increase in the eddy kinetic energy largely associated with ascent in warm and descent in cold regions, as expressed by the conversion $[A_E, K_E]$. The time lag between breakdown of the zonal flow and the onset of enhanced cyclogenesis may be only a few days.

This eddy activity increases the meridional flux of heat manifested by an increase in the lateral baroclinic term $[A_Z, A_E]$, decreases the meridional temperature gradient as the eddies achieve maturity, and increases static stability as poleward-moving warm air flows over cold. Accordingly, there follows eventually a decrease in the lateral baroclinic and vertical baroclinic energy conversions, respectively, of zonal to eddy available potential energy and of eddy potential to eddy kinetic energy $[A_Z, K_E]$. At this stage the atmosphere exhibits large vigorous eddies, which begin to decay because of increased eddy frictional losses, specifically an increase in $[D_E]$ due to an increase in the barotropic decay processes expressed by $[K_E, K_Z]$ and to an increase in overall static stability. This stage may involve large circular eddies that do not move very rapidly or remain stationary, a situation known as "blocking". Blocking is discussed elsewhere in this text in relation to various topics, such as equivalent barotropic motion and downstream development. The existence of large eddies and weakened meridional temperature gradients result in a decrease in the conversion rates of zonal to eddy forms, and from eddy available potential to eddy kinetic energy. As the eddies begin to decay, the pattern moves from blocking toward zonal. There is an accompanying increase in the meridional temperature gradient, aided by differential heating on a global scale as reflected in the term $[G_Z]$. The stage is set for the inception of a new energy cycle.

5.5 Barotropic growth and decay of waves

The breakdown of relatively zonal flow with meridional shear into a train of waves with high amplitude may take place suddenly and be accompanied by a rapid transfer of zonal to eddy kinetic energy. This type of rapid wave growth is associated with *barotropic instability*, and with a significant contribution by the barotropic conversion term in the energy balance equations discussed earlier.

Barotropic instability derives its energy from the zonal shear ($\partial U/\partial y$). A

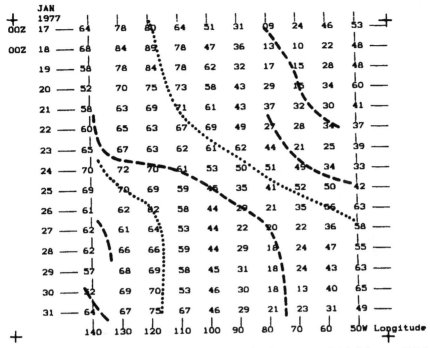

Figure 5.3 Hovmoller diagram showing the 500 mb geopotential heights at 40 °N (in dam above 500 dam) versus day of the month during January 1977 and versus longitude (°W). Dotted lines denote the ridge axes and broken lines the trough axes. (Figure courtesy of Gregory Forbes.)

necessary mathematical condition for the exchange between zonal and eddy kinetic energy is that the eddy components of the wind be correlated $((\overline{u^* v^*}) \neq 0)$, resulting in a meridional flux of relative angular momentum. Theoretical accounts suggest that waves amplify when they move with phase speeds less than a maximum zonal wind speed and greater than the minimum zonal wind speed, and kinetic energy is fed from the basic current into those waves. Growth rates can vary with the speed of the zonal current. Although large zonal wind shear and zonal absolute vorticity is a normal situation in the middle and upper troposphere at 30–55° latitude, especially during winter, the growth of waves to significant amplitude within a time frame of days or less will occur only when the zonal shear achieves sufficiently large values.

It seems quite likely that, at middle latitudes, barotropic instability becomes important in the initial growth of waves during situations of high zonal index when the zonal flow is relatively strong and therefore the

127

latitudinal shear is greatest. Barotropic decay occurs after waves have reached maximum intensity, and it is manifested by a poleward movement of the wave and a transfer of kinetic energy to longer waves.

We can regard the formation, movement and decay of troughs and ridges with a *Hovmoller diagram* (Saucier, 1955), an example of which is shown in Figure 5.3. Here, the slopes of the ridge and trough axes (the change in longitude with time) represent the phase speed of the features. Trough or ridge lines oriented vertically in the figure are quasi-stationary, while those that slant forward in time at the greatest angle represent waves moving eastward the most rapidly. Even within the relatively short two-week period covered by the Hovmoller diagram, there is a variation in index. Initially, there is only one trough and one ridge in the longitude band 50–140°W. Beginning on 20 January both a trough and a ridge form near 90°W and move eastward. While the trough amplitude increases, another trough and ridge pair move into the latitude band and a third weak trough forms on 24 January near 130°W. Subsequently, the pattern returns to that of one trough and one ridge on 29 January. The presence of a single trough–ridge pair on 17 January and again on 29 January does not constitute a complete zonal (high-index) cycle because the amplitude of the wave is rather large during these two periods of lower frequency. Nevertheless, the oscillatory nature of wave generation from a zonal current and the wavelength that is formed are evident in this brief record.

Formation and decay of troughs in the westerlies are associated with barotropic energy conversions and, in the former, with barotropic instability. That the conversion between eddy and zonal kinetic energy is normally relatively weak compared to that between eddy and zonal potential energy is an indication that barotropic instability is not a very important factor in cyclogenesis at middle latitudes. (This is not true for the Tropics. There, the growth of easterly waves over the tropical Atlantic Ocean may depend, in their early stages of development, on the barotropic transfer of kinetic energy from the zonal flow to the waves.) At middle latitudes the creation of eddy kinetic energy from available potential energy is accomplished largely by the baroclinic processes, whereas barotropic conversion acts weakly to transfer energy from the waves to the zonal flow. During cyclogenesis, baroclinic (rather than barotropic) processes dominate in creating eddy kinetic energy.

Problems

5.1 (a) Give a qualitative argument that indicates that the pattern of temperature and diagnosed vertical motion in Figure 4.1 implies that there is conversion of eddy potential to eddy kinetic energy.

(b) If cumulus convection were to occur in this pattern, where would its diabatic heating be most efficient at increasing the eddy potential energy over the domain?

5.2 In regard to friction, why would intense cyclones at the surface decay more rapidly than weak ones once the conversion of eddy potential to eddy kinetic energy has ceased to occur.

5.3 Suggest a parameter based on definitions of eddy and zonal kinetic energy that could describe the index cycle and the extent of perturbed flow.

Further reading

Colucci, S. J. 1987. Comparative diagnosis of blocking versus nonblocking planetary-scale circulation changes during synoptic-scale cyclogenesis. *J. Atmos. Sci* **44**, 124–39.

Danard, M. B. 1964. On the influence of released latent heat on cyclone development. *J. Appl. Meteor.* **3**, 27–37.

Holopainen, E. and C. Fortelius 1987. High-frequency transient eddies and blocking. *Mon. Wea. Rev.* **44**, 1632–45.

Illari, L. and J. C. Marshall 1983. On the interpretation of eddy-fluxes during a blocking episode. *J. Atmos. Sci.* **40**, 2232–42.

Kenny, S. E. and P. J. Smith 1983. On the release of eddy available potential energy in an extratropical cyclone system. *Mon. Wea. Rev.* **111**, 745–55.

Kidson, J. W. 1985. Index cycles in the Northern Hemisphere during the Global Weather Experiment. *Mon. Wea. Rev.* **113**, 607–23.

Kidson, J. W. 1986. Index cycles in the Southern Hemisphere during the Global Weather Experiment. *Mon. Wea. Rev.* **114**, 1654–63.

Kung, E. C. 1977. Energy sources in mid-latitude synoptic-scale disturbances. *J. Atmos. Sci* **34**, 1352–65.

Lin, S. C. and P. J. Smith 1979. Diabatic heating and generation of available potential energy in a tornado-producing extratropical cyclone. *Mon. Wea. Rev.* **107**, 1169–83.

Lorenz, E. N. 1955. Available potential energy and maintenance of the general circulation. *Tellus* **7**, 157–67.

Lorenz, E. N. 1967. *The nature and theory of the general circulation of the atmosphere.* WMO Publ. no. 218.TP.115.

Oort, A. H. 1964. On the estimates of the atmospheric energy cycle. *Mon. Wea. Rev.* **82**, 483–93.

Palmen, E. and E. O. Holopainen 1962. Divergence, vertical velocity and conversion between potential and kinetic energy in an extratropical disturbance. *Geophysica* (Helsinki) **8**, 89–113.

Saltzman, B. 1957. Equations governing the energetics of the larger scales of atmospheric turbulence in the domain of wave number. *J. Meteor* **14**, 513–23.

Saucier, W. 1955. *Principles of meteorological analysis.* Chicago: University of Chicago Press.

Smith, P. J. 1980. The energetics of extratropical cyclones. *Rev. Geophys. Space Phys.* **18**, 378–86.

Winston, J. S. and A. F. Krueger 1961. Some aspects of a cycle of available potential energy. *Mon. Wea. Rev.* **89**, 307–18.

6

Evolution and motion of mid-tropospheric waves: barotropic viewpoint

The large-scale flow pattern at upper levels differs considerably from that at the surface. Upper levels are characterized by waves (troughs and ridges) while low levels exhibit cellular patterns (highs and lows) and weaker height and vorticity gradients. The increase in zonal wind speed with height up to approximately the level of the tropopause is a consequence of the baroclinic nature of the atmosphere, in particular the poleward decrease in temperature. This poleward temperature gradient necessitates a westerly shear according to thermal wind considerations (1.34), with a resulting increase in speed with height up to the tropopause of the intensity of the vorticity pattern (Fig. 1.1).

As is illustrated in Chapter 2, waves are a manifestation of a vortex superposed on the basic current. At the surface the basic zonal current is relatively weak and therefore the vorticity advection is weak. The presence of a basic non-divergent zonal current superposed on a vortex allows the vorticity to be advected with roughly the speed of the basic current. Vorticity advection increases with height up to the tropopause.

At the tropopause, the meridional temperature gradient reverses and the zonal (westerly) wind speed diminishes with height. The level and latitude at which the westerly wind speed is maximized is loosely associated with the polar "jet stream". A more detailed examination of tropopause structure is reserved for Chapter 10.

In the rotating tank apparatus, discussed at the start of Chapter 5, both the fluid in the tank and the waves move at a faster speed than that of the rotation (relative westerly winds for a counterclockwise rotation). The fluid exhibits narrow regions of relatively fast motion, resembling jet streaks in the real atmosphere. Waves translate in the direction of the rotation, with the belt of most rapidly moving fluid coinciding with a very sharp horizontal temperature gradient (front), colder toward the pole and toward the trough axes.

Forecasters customarily focus on the wind and geopotential height patterns at the surface and at 500 mb to diagnose the movement and intensification of disturbances. Historically, the emphasis on weather maps at these
two levels is based on the development during the 1950s of simple prediction
models that treated the 500 mb level as an equivalent barotropic surface (to
be defined shortly) and the 1000 mb level (or sea level) as responding to
vorticity advections primarily at 500 mb. Emphasis in this chapter is on the
500 mb geopotential height pattern and the movement of troughs and ridges
at upper levels.

6.1 Conservation of absolute vorticity; constant absolute vorticity

In the idealized vertical motion profile depicted in Figure 2.8, horizontal
divergence vanishes at some level in the middle troposphere. Although the
real atmosphere is more complicated than this model, there is at least a
statistical minimum of divergence at middle levels not far from 500 mb;
consequently, the flow behaves at that level as though the absolute vorticity
were conserved. As expressed by equation (3.6c), parcels conserving geostrophic absolute vorticity are constrained to move on an isobaric surface which
is at the level of non-divergence. The conservation of absolute vorticity of
the total wind

$$d(\zeta + f)/dt = 0 \qquad (6.1a)$$

is approximated at the level of non-divergence because the tilting, vertical
vorticity advection and friction terms in the vorticity equation are normally
small in the middle troposphere. Absolute vorticity isopleths move approximately with the advecting wind because at the level of non-divergence
absolute vorticity is neither created nor destroyed.

For example, an air parcel moving equatorward with conservation of
absolute geostrophic vorticity and with an initial *relative* vorticity of zero
units acquires positive relative geostrophic vorticity in exchange for a
decrease in f. If $(\zeta + f)$ is conserved following the movement of an air parcel,
then

$$(\zeta + f) = \text{constant}$$

or

$$(\zeta_1 + f_1) = (\zeta_2 + f_2) \qquad (6.1b)$$

where the subscripts 1 and 2 refer, respectively, to successive times t_1 and t_2

along the trajectory. In gaining positive relative vorticity, the parcel acquires either cyclonic curvature or cyclonic shear or both (2.3). The parcel may either continue equatorward, gaining additional cyclonic shear, or acquire cyclonic curvature and eventually turn poleward due to the cyclonic rotation. After recurvature toward the pole the parcel experiences a decrease with time of ζ and reaches its initial latitude, where it once more possesses zero relative vorticity. Subsequently, the parcel moves poleward, acquiring negative relative vorticity and eventually executes an anticyclonic rotation and moves equatorward. The motion of this *constant absolute vorticity* trajectory would look something like a sinusoidal geopotential height contour depicted in Figure 2.3b.

An interesting point emerges in the explanation of apparent deepening or weakening. As the Coriolis parameter in the expression for geostrophic relative vorticity decreases in response to an equatorward movement of a perturbation, the denominator in the Laplacian of geopotential height in (2.8) also decreases leaving the effect of latitude change on the geopotential height perturbation due to a

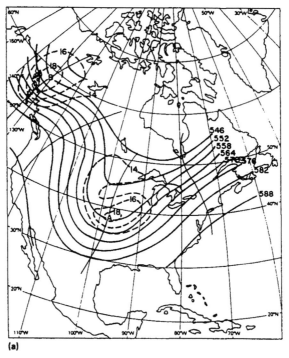

(a)

Figure 6.1 (a) 500 mb geopotential height (full contours labeled in dam) and absolute vorticity isopleths (broken curves labeled in units of $10^{-5}\,\mathrm{s}^{-1}$) for 0000 GMT 26 September 1985. Vorticity centers are labelled A and B; thin lines denote trough and ridge axes.

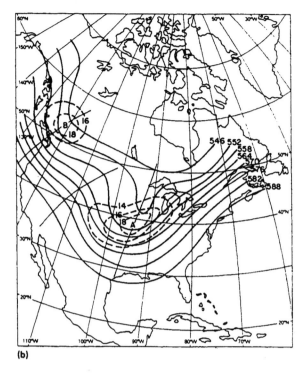

(b)

Figure 6.1 (b) 500 mb geopotential height (full contours labeled in dam) and absolute vorticity isopleths (broken curves labeled in units of $10^{-5} s^{-1}$) for 1200 GMT 26 September 1985.

conservation of absolute vorticity somewhat ambiguous. However, because the wavelength of perturbations tends to increase with decreasing latitude, (2.13) suggests that the effects of changing wavelength and changing value of the Coriolis parameter in front of the Laplacian would tend to cancel, leaving the apparent deepening concept intact.

The role of Earth's vorticity

Consider a typical wave pattern at 500 mb as in Figure 6.1a. Because the vorticity isopleths are superposed on a basic current and are somewhat circular, advection solenoids are formed immediately upstream and downstream from the centers of maximum vorticity, which are found near the axes of the troughs. According to arguments posed in Chapter 2, relative vorticity advection tends to be larger (with respect to the Earth's vorticity advection) with shorter wavelengths for a given amplitude of the perturbation. Con-

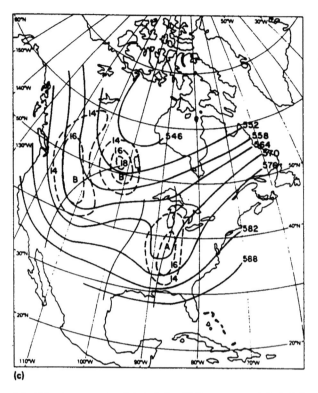

(c)

Figure 6.1 (c) 500 mb geopotential height (full contours labeled in dam) and absolute vorticity isopleths (broken curves labeled in units of $10^{-5} \, s^{-1}$) for 1200 GMT 27 September 1985.

sequently, short-wave advection patterns exhibiting numerous advection solenoids of geostrophic absolute vorticity are more often found where the wavelengths are not very long.

In a typical wave pattern, air to the east of the trough axis flows eastward and poleward and air to the west of the trough axis flows eastward and equatorward. Air originating west of the trough may subsequently cross the trough axis provided that its zonal speed is greater than that of the propagation speed of the wave. Similarly, air parcels move westward with respect to the wave axis if the propagation speed of the trough axis is faster than the eastward component of the air.

At the level of non-divergence, parcels crossing the wave axis carry the absolute (and therefore the relative) vorticity pattern eastward, whereas air moving with a poleward component east of the wave axis (or an equatorward component west of the wave axis) loses (or gains) the relative cyclonic vorticity in moving toward the ridge (trough). Consequently, the

134

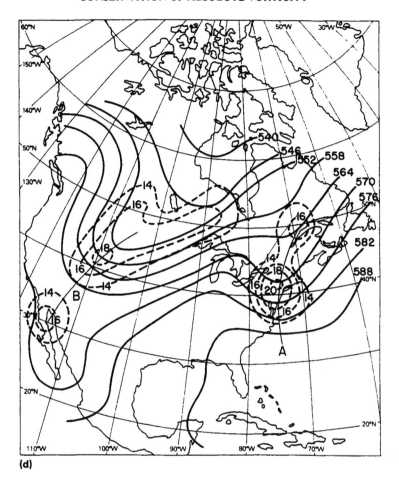

(d)

Figure 6.1 (d) 500 mb geopotential height (full contours labeled in dam) and absolute vorticity isopleths (broken curves labeled in units of 10^{-5} s^{-1}) for 0000 GMT 28 September 1985.

wave moves eastward due to the advection of relative vorticity, which (in the quasi-geostrophic framework) implies that parcels crossing a trough axis carry relative cyclonic vorticity (lower geopotential heights) downstream. This causes the trough to move eastward but not as fast as the advecting current because the poleward advection of relative vorticity ahead of the trough results in the Earth vorticity (f) increasing in exchange for a decrease in relative cyclonic vorticity (ζ). The same argument can be posed with respect to a ridge. (Recall that relative vorticity is directly related to the geopotential height perturbation via geostrophic balance.)

135

Stated in terms of (2.8) and (3.6c),

$$\frac{q}{f_0} \nabla_p^2 \frac{\partial Z}{\partial t} = -\frac{g}{f_0} V_g \cdot \nabla_p(\nabla_p^2 Z) - v_g \beta. \tag{6.1c}$$

The first term on the right-hand side of this expression is the advection of relative geostrophic vorticity by the geostrophic wind, which tends to be positive (negative) east (west) of trough axes. Accordingly, the eastward movement of the wave is accomplished by the advection ahead of the wave of positive spin (relative vorticity); the wave continuously reconstitutes itself downstream by advection of its relative vorticity, which is related to the geopotential heights via the geostrophic relationships. The second term on the right-hand side, the beta effect, represents the meridional advection of Earth vorticity, which is negative in the poleward flow east (and positive in the equatorward flow west) of trough axes. A similar argument can be posed with respect to ridges. Consequently, the beta effect, by counteracting the advection of relative vorticity, slows down the eastward motion of troughs and ridges. Since β is approximately constant with latitude at middle latitudes (about 1.5×10^{-11} s^{-1} m^{-1}), the beta effect depends mainly on v_g. For a meridional wind component of 10 m s^{-1}, the Earth vorticity advection is 1.5×10^{-10} s^{-2}.

Longer waves generally possess a greater meridional displacement between trough and ridge (but a smaller v component for a given amplitude), whereas with shorter, smaller-amplitude waves, the relative vorticity is larger for the same amplitude of the wave. In terms of the scale arguments posed in Chapter 2, the relative vorticity advection is inversely proportional to the cube of the wavelength, where advection of the Earth's vorticity is simply proportional to the v component. Alternatively stated, the beta effect is less efficient in slowing down the eastward movement of the short than the long waves. For an idealized sine wave traveling in a non-divergent current of mean $\hat{U}$ with phase speed c, where

$$v_g = \hat{v} \sin \left[\left(\frac{2\pi}{L_x} \right) (x - ct) \right] \cos \left(\frac{2\pi y}{L_y} \right); \qquad u = u_g = \hat{U}$$

the solution to the linearized quasi-geostrophic vorticity equation at the level of non-divergence is the well-known Rossby wave equation (for a two-dimensional wave), which is

$$c = \hat{U} - \beta L^2 / (8\pi^2) \tag{6.2}$$

for the square-wave case, where $L_x = L_y = L$. Equation (6.2) states that, for the case of a uniform westerly current ($\hat{U} > 0$), the speed of the waves at the level of non-divergence is slightly slower than that of the basic zonally averaged current ($\bar{U}$), the difference arising from the second term on the right-hand sides of (6.2) and (6.1c). This term quantifies the "beta wind", which expresses the modifying influence of the change in Coriolis parameter (f) with latitude. Given an appropriate value of β, the beta wind correction is about 3 m s^{-1} for a wavelength of $3 \times 10^6 \text{ m}$. Thus, the magnitude of the second term on the right-hand side of (6.2) is usually relatively small compared with $\bar{U}$, but in some instances, when the flow is highly perturbed, $\bar{U}$ may be relatively small while L is large. Under such circumstances the wave may move very slowly or even *retrogress*. It is commonly observed that short waves move eastward faster than long waves in mid-latitudes, while highly perturbed flows and very long-wave patterns tend to remain stationary, constituting blocking (see Ch. 12). In such blocking patterns, such as eddies possessing closed or "cut off" geopotential height contours, the vorticity isopleths also tend to be aligned nearly parallel to the height contours and have few advection solenoids. Understandably, such a pattern does not translate eastward very rapidly.

Vertical integrity of waves

It is generally observed that waves move at about the same eastward speed at all levels. If, as (6.2) indicates, the phase speed of a wave is only a little less than that of the mean zonal current at the level of non-divergence and if the wind speed increases with height up to the tropopause (as is illustrated in Figure 1.1), air parcels must generally overtake the wave at high levels and be overtaken by the waves in the lower troposphere. At the level of non-divergence, the total time rate of change in absolute vorticity following a parcel is zero and the wave moves with a speed slightly less than that of the zonally averaged wind speed at that level.

Above the level of non-divergence, the mean zonal wind ($\bar{U}$) is greater than the phase speed of the wave (c), i.e. $\bar{U} - c > 0$, and the air moves through the trough to the downstream ridge from a region of positive to a region of negative relative vorticity. Therefore, since the absolute vorticity evidently decreases following the movement of a parcel from trough to ridge, it must experience divergence, according to (3.6a). Moreover, Dines' compensation implies ascending motion. We have seen, however, that positive vorticity advection ahead of troughs is consistent with upward vertical motion and divergence (convergence) in the high (low) troposphere between trough and downstream ridge. We can turn the argument around and say that divergence ahead of the trough and convergence behind it retards the eastward movement of a trough by destroying positive vorticity advected ahead of the trough and creating positive vorticity behind the trough, thereby partially canceling the effects of differential advection of vorticity in the vertical. Similarly, in the lower troposphere, air is overtaken by the trough and thus the air parcels experience a gain in absolute vorticity with time east of the trough axis due to low-level convergence. *This arrangement exactly compensates for the slower (faster) zonal wind speeds (with respect to the wave's phase speed) in the lower (higher) troposphere.*

This argument underscores the fact that patterns of convergence and divergence in the atmosphere enable waves to translate at a constant phase speed in the vertical. A different situation exists at low latitudes where, except for regions of deep convection, weather systems tend to be decoupled in the vertical because of the absence of strong large-scale divergence patterns that are coherent over a deep atmospheric layer. Such decoupling in the vertical is characteristic of a barotropic atmosphere. The point here is that the motion of troughs and ridges is consistent with the observed vertical profile of divergence and convergence and with the existence in the mean of a level of non-divergence. Thus, the increase with height of the westerly wind at middle latitudes is related to the pattern of convergence and divergence found in the vicinity of waves in the westerlies.

Apparent deepening and weakening

It is not uncommon to find short waves superposed on longer ones. In such instances, as depicted in Figure 6.1, the short wave (which is represented by the vorticity center B) is embedded within the longer one, whose trough axis is centered over the center of the continent. Short waves also tend to possess relative vorticity maxima and minima whose magnitudes equal or exceed that of the Coriolis parameter. Such short waves appear to move through the longer ones, with the resultant geopotential perturbation reflecting the distribution of relative vorticity according to (2.8). While larger-scale geopotential height perturbation remains relatively fixed in place, smaller-scale vorticity centers move through the larger wave. Clustering of smaller vorticity maxima constitutes a region of relative geopotential height minimum (a long-wave trough axis). Therefore, the larger-scale geopotential height pattern may be composed of smaller perturbations, whose resultant vorticity pattern determines the configuration of the larger-scale system.

Waves tend to be steered by the basic current; smaller relative vorticity maxima and minima embedded within longer ones are steered by a smooth current embedded within the larger-scale flow. Thus, short waves at 500 mb tend to move rapidly through longer ones, which may remain quasi-stationary. In effect, therefore, an appropriate value of the current speed ($\bar{U}$) for steering shorter waves is that in the longer ones.

The absolute vorticity maximum (labeled B) in Figure 6.1 is associated with a small-amplitude wave over western Canada upstream from a major trough. Although the latter is prominent, the absolute vorticity values (A and B) are of approximately equal magnitude. Wave B moves southeastward (along with a perturbation in the height field), gradually becoming larger in amplitude. While the upstream wave *appears* to deepen, because its relative vorticity increases with time due to its equatorward displacement, the absolute vorticity remains approximately the same as it moved from 60°N in Figure 6.1a to about 42°N after 60 h. Intensification in the amplitude of wave B is due to a gain of 3 units of relative vorticity (from a value of about 6 units in Figure 6.1a to 9 units in Figure 6.1d). This gain in relative vorticity constitutes an increase of about 50% in the amplitude of the relative vorticity perturbation and therefore a 50% increase in the amplitude of the geopotential height perturbation.

Equatorward movement of the upstream vorticity maximum, combined with poleward movement of the downstream one, results in the upstream wave becoming the dominant one in the height pattern. Sixty hours after the time of Figure 6.1a, wave B has become the dominant perturbation in the 500 mb geopotential height field. Let us define a change in amplitude of the perturbation in the geopotential height field that is likely due to latitude displacement of the vorticity centers as "apparent weakening or deepening". Thus, perturbation B appears to intensify (while A appears to weaken) because of latitudinal displacement of the vorticity maxima. The resultant motion of the long-wave trough was effectively unchanged over 36 h.

6.2 Equivalent barotropic model

Since the vertical integral of the divergence from the bottom to the top of the atmosphere vanishes in the absence of surface vertical motion (2.25), it seems reasonable to suppose that the vertically averaged vorticity tendency $\overline{(\mathrm{d}(\zeta + f)/\mathrm{d}t)}$ possesses a form similar to (3.6c). If this is so, the vertically averaged geostrophic absolute vorticity would behave almost as if it were on a surface of non-divergence. We can prove this by taking the vertical average between $p = 0$ and p_s of the geostrophic vorticity tendency equation (3.6a) with respect to pressure as we did in deriving (3.16). Thus,

$$\frac{1}{p_s} \int_0^{p_s} \frac{\mathrm{d}(\zeta + f)}{\mathrm{d}t} \, \mathrm{d}p = \int_0^{p_s} f_0 \frac{\partial \omega}{\partial p} \, \mathrm{d}p = \frac{f_0 \omega_s}{p_s} = \frac{\overline{\mathrm{d}(\zeta + f)}}{\mathrm{d}t}. \qquad (6.3)$$

The vertically averaged pattern of winds and geopotential contours closely resembles that at 500 mb. To relate the vertical structure of the wind field to the vorticity tendency at the mean level, a simple parametric relationship between geostrophic wind and height is presented. This relationship leads to the so-called "equivalent barotropic model". In this formulation, the vertical distribution of the geostrophic wind velocity does not change its direct-

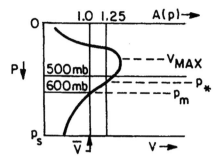

Figure 6.2 Schematic distribution of wind speed versus pressure in middle latitudes. The equivalent barotropic weighting function $A(p)$ possesses a profile similar to that of the wind speed (V) and is labeled accordingly along the upper axis. The mean pressure level (p_m) and the equivalent barotropic level (p_*) are also labeled. $\bar{V}$ represents the pressure-average wind speed, $\bar{V} = V_m$.

ion with height, and the speed of the wind at any level is taken as some fraction $A(p)$ of that at the mean level p_m, as shown in Figure 6.2. Mathematically, this can be expressed as follows:

$$u_g = A(p)u_m \qquad v_g = A(p)v_m \qquad V_g = A(p)V_m. \qquad (6.4a)$$

Generally,

$$(\quad)_m \equiv (\overline{\quad}) \equiv \frac{1}{p_s}\int_0^{p_s}(\quad)dp \qquad (6.4b)$$

which implies that $u_g/v_g = $ constant (no change in direction with height). Further,

$$\zeta_g = A(p)\,\zeta_{gm} \qquad (6.4c)$$

where the subscript g (for geostrophic) will now be omitted for convenience. The hodograph for this model wind profile obeying (6.4) is shown in Figure 6.3. Since the westerly winds increase with height at mid-latitudes, the average value of $A(p)$ typically increases with height, reaching the level of its vertically average value $(A(p_m) = 1.0)$ just above 600 mb and values greater than 1.0 in the upper troposphere. (At the reference level, $p = p_m$, $V = V_m$ and $A(p_m) = 1.0$.) The value of $A(p)$ can be obtained from large-scale averages, and some climatological values are presented in Table 6.1.

Substituting (6.4) into (3.6a), integrating from the surface to $p = 0$ using the pressure-averaging operator in (6.4b) and realizing that the vertical average of $A(p)$ from 0 to p_s (A_m) is 1.0, one obtains the following:

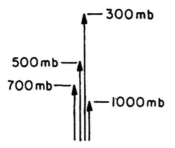

Figure 6.3 Schematic hodograph of wind vectors satisfying equivalent barotropic assumptions of equation (6.4).

$$\frac{\partial \zeta_m}{\partial t} + \overline{A^2(p)} \; V_m \cdot \nabla_p \zeta_m + V_m \cdot \nabla_p f = \frac{f_0 \omega_s}{p_s} \qquad (6.5a)$$

where terms subscripted by m are independent of pressure. Multiplying (6.5a) by $\overline{A^2(p)}$ yields

$$\overline{A^2(p)} \frac{\partial \zeta_m}{\partial t} + \overline{A^2(p)} \; V_m \cdot \nabla_p [\overline{A^2(p)} \, \zeta_m + f] = \frac{\omega_s f_0}{p_s} \overline{A^2(p)} \qquad (6.5b)$$

or

$$\frac{\partial \zeta_*}{\partial t} + V_* \cdot \nabla_p(\zeta_* + f) = \frac{\omega_s f_0}{p_s} \overline{A^2(p)}. \qquad (6.6)$$

The definition of the starred quantities in (6.6) is that $V_* = \overline{A^2(p)} \, V_m$ and $\zeta_* = \overline{A^2(p)} \, \zeta_m$. The starred level is the equivalent barotropic level where $A(p) = \overline{A^2(p)}$. (Note that the square of the mean value $A(p)$ is not generally equal to the mean of the square of $A(p)$, which can differ from the value of 1.0)

Table 6.1 Values of weighting factors in equivalent barotropic and two-parameter baroclinic models (Buch, 1954).

p(mb)	$A(p)$	$F(p)$ (mb)	$N(p)$ (mb)	$B(p)$
1000	0	0	1000	− 1.00
900	0.32	19	982	− 0.76
700	0.75	95	882	− 0.31
500	1.22	155	688	0.24
300	1.61	117	400	0.67
200	1.66	62	230	0.81
100	1.34	6	71	0.46
0	0	0	0	− 1.00

Equation (6.6) is identical in form to (3.6c) except for the lower boundary term containing ω_s, which accounts for the mean atmospheric divergence due to vertical motion induced at the surface, such as by orographic motion. It should be noted that the value of $(\zeta_* + f)$ is conserved following the advecting flow V_*, provided that the vertical motion at the surface is zero. The starred level differs slightly from the mean level at which $p = p_m$. According to climatological data p_m is near to or slightly above 600 mb. At the starred level where $p = p_*$, $A(p) = \overline{A^2(p)} = 1.25$; this value can be found very close to 500 mb where $A(p)$ in Table 6.1 has the value of 1.22. The term *equivalent barotropic model* refers to the fact that, were the atmosphere barotropic (no horizontal temperature gradients on pressure surfaces) $\omega(p)$ would be constant with height, $A(p)$ would be 1.0 at every level and equation (6.5a) would have the same form as (3.6c) (except for the lower boundary term) and apply to all levels. Stated differently, the value of $A(p)$ for a baroclinic atmosphere is equivalent to that for a barotropic atmosphere $(A(p) = 1.0)$ at the mean level p_m, which is close to the level where absolute vorticity $(\zeta_* + f)$ is conserved (for $\omega_s = 0$).

In summary, in an atmosphere with a unidirectional shear, the equivalent barotropic level is very close to 500 mb. The equivalent barotropic model given by (6.6) is identical to the non-divergent version of the geostrophic vorticity equation (3.6c) (in the absence of surface boundary effects) and is also consistent with the Rossby wave equation (6.2) in which $\tilde{U}$ would apply at the equivalent barotropic level, $p = p_*$.

6.3 Vertical motion and vorticity advection in the equivalent barotropic system

The equivalent barotropic model implies a vertical wind shear and a horizontal temperature gradient; isotherms are everywhere parallel to the wind vector (see the hodograph in Figure 6.3) and hence there are *no* temperature advections. This restriction on advection of temperature in the quasi-geostrophic omega equation (Ch. 4) can be illustrated by the following derivation. First, write the vorticity equation (3.6a) after substituting the relationships in (6.4) as

$$A(p) \frac{\partial \zeta_m}{\partial t} + A(p)\, V_m \cdot \nabla_p[A(p)\, \zeta_m + f] = f_0 \frac{\partial \omega}{\partial p}. \qquad (6.7a)$$

Multiplying (6.5a) by $A(p)$ gives

$$A(p) \frac{\partial \zeta_m}{\partial t} + A(p)\{V_m \cdot \nabla_p[\overline{A^2(p)}\, \zeta_m + f]\} = \frac{f_0\, \omega_s A(p)}{p_s} \qquad (6.7b)$$

which is substituted in (6.7a) to give

$$\frac{\partial \omega}{\partial p} = -D_p = -\frac{1}{f_0}[A(p)\,\overline{A^2(p)} - A^2(p)]\,V_m \cdot \nabla_p \zeta_m + \frac{\omega_s A(p)}{p_s} \quad (6.8a)$$

or

$$\omega = -\left(\frac{F(p)}{f_0}\right)(-V_m \cdot \nabla_p \zeta_m) + \frac{\omega_s N(p)}{p_s} \equiv \omega_d(p) + \omega_M(p) \quad (6.8b)$$

where

$$F(p) = \int_0^{p_s}[A(p)\,\overline{A^2(p)} - A^2(p)]\mathrm{d}p \quad \text{and} \quad N(p) = \int_0^{p_s} A(p)\,\mathrm{d}p.$$

Values of these quantities can be derived from climatological wind data; typical values are shown in Table 6.1.

Equation (6.8) bears some resemblance to the quasi-geostrophic omega equation (4.3) in the sense that ω_d is related to the advection of vorticity. Here it is explicitly expressed simply in terms of the *relative* vorticity advection, which constitutes the sole dynamic forcing term. Note that (6.8) *does not account for temperature advection* as can be seen in Figure 6.3. Neglect of the temperature advection term is tantamount to suppressing the feedback between development at the surface and at 500 mb, and therefore to an inhibition of cyclogenesis.

A further limitation of the equivalent barotropic model is that the various weighting factors ($A(p)$, $F(p)$ and $N(p)$) must be determined from time and space averages, which do not account for the effects of scale and static stability.

Despite its limitations (it is no longer used by the U.S. Weather Service) the equivalent barotropic formulation is highly instructive because it expresses the essential aspects of wave motion at 500 mb. Moreover, the vertical motion prescribed by (6.8) conforms approximately to observation, i.e. ascending motion in regions of positive vorticity advection. For a geostrophic vorticity advection of $10 \times 10^{-10}\,\mathrm{s}^{-2}$ and $f = 10^{-4}\,\mathrm{s}^{-1}$, the dynamic vertical motion at 700 mb ($F(p) = 95$ mb; Table 6.1) is about $1\,\mu\mathrm{b}\,\mathrm{s}^{-1}$. The concept of steering at the level of non-divergence arises from the assumption that the absolute vorticity is conserved. In the quasi-geostrophic system the absolute vorticity centers are moved along a constant-pressure surface by advection, while the relative vorticity maintains a geostrophic relationship with the geopotential height field.

Early numerical weather prediction models were based on the concept of an equivalent barotropic surface, which was loosely associated with the

500 mb level. The discovery that the weather could be predicted with some accuracy as the result of making the geostrophic approximation to the winds and assuming a layer of non-divergence in the mid-troposphere provided a primary theoretical boost for objective weather forecasting during the 1940s. In general, these early forecasts were made by computer, but relatively efficient graphical methods were also developed for semi-objective weather prediction where sophisticated computation facilities were lacking.

One of the most ingenious of these methods was introduced by R. Fjortoft. Although manual methods for executing these models are no longer operational in the United States, they are still useful for pedagogic reasons because they demonstrate, often simply and clearly, the essential processes that account for the movement and change of weather patterns.

Fjortoft showed that graphical solutions to the quasi-geostrophic barotropic vorticity equation (3.6c) are obtainable by a combination of addition, subtraction and smoothing of the height fields. Rewriting (3.6c) in finite difference notation using (2.11) one obtains

$$\frac{4g}{f_0 d^2} \frac{\partial}{\partial t} [\tilde{Z} - Z(0)] = -\frac{4g}{f_0 d^2} V_g \cdot \nabla_p [\tilde{Z} - Z(0)] - V_g \cdot \nabla_p f \qquad (6.9)$$

where $\tilde{Z}$ represents the average geopotential height a distance d around the point at which $Z = Z(0)$. Thus, $\tilde{Z}$ is a smoothed geopotential height field. Equation (6.9) simplifies, after multiplying by $f_0 d^2/4g$, to

$$\frac{\partial}{\partial t} [\tilde{Z} - Z(0)] = -V_g \cdot \nabla_p [\tilde{Z} - Z(0)] - V_g \cdot \nabla_p \frac{f_0 d^2}{4g} f. \qquad (6.10)$$

In the last term on the right-hand side of (6.9), f_0 is replaced by f, and a part of the term $(f_0 \nabla_p f)$ reduces simply to $f \, \partial f/\partial y = 2\Omega f \cos \theta \, d\theta/dy$, where θ is the latitude. Expressing this as $f^2 \cot \theta \, d\theta/dy$, and defining

$$G \equiv \int_0^\theta \frac{f d^2}{4g} \cot \theta \, d\theta \qquad (6.11a)$$

one can write

$$\partial(\tilde{Z} - Z)/\partial t = -V_g \cdot \nabla_p (\tilde{Z} - Z + G) \qquad (6.11b)$$

where $Z(0)$ is henceforth understood to be simply Z. (Note that (6.11b) represents a vorticity–geopotential height transformation: the terms in parentheses on the left and right of the equation respectively represent the relative and absolute vorticities. Since both f and $\partial f/\partial y$ determine the G field (the beta effect) in (6.11b), the development of the steering equations is slightly contradictory to the assumption that f is a local constant in (2.11). The result is, nevertheless, in basic accord with the quasi-geostrophic framework.

Realizing that the advection of a field by a wind that blows parallel to the contours of that field is zero, (6.11b) can be transformed, using the identities that $V_g \cdot \nabla_p Z = 0$, $\tilde{V}_g \cdot \nabla_p \tilde{Z} = 0$ and $V_G \cdot \nabla_p G = 0$, into the expression

$$\partial(\tilde{Z} - Z + G)/\partial t = -(\tilde{V}_g + V_G) \cdot \nabla_p(\tilde{Z} - Z + G). \qquad (6.11c)$$

Here, $V_G = (g/f)k \times \nabla_p G$ is the geostrophic wind of the "G field" (analogous to the beta wind term in (6.2)), expressed by (6.11a) and V_g is the geostrophic wind of the $\tilde{Z}$ field. The quantity $(\tilde{Z} - Z)$ is a measure of the relative vorticity or the amplitude of a perturbation on a smoothed current. Alternatively stated, it represents the difference between the average (smoothed) geopotential height a radius d from a point and the geopotential height at that point.

Fjortoft's central idea was that V_g is replaced by a steadier, more conservative, wind field $\tilde{V}_g$, which changes more slowly in time and in space. This smooth field then steers the perturbations. Smoothing is accomplished using (2.10) and the interval d is chosen according to the scale of the features one is trying to predict. Both the magnitude of the G field as well as the degree of smoothing depend on the value of d.

Fjortoft's reasoning is based on observations that relatively smooth and featureless flow patterns appear to change less rapidly than those with perturbations embedded within them. The local height change is obtained by the advection of $(\tilde{Z} - Z + G)$, which is analogous to an absolute vorticity, by the wind field $\tilde{V}_g + V_G$. The quantity $(\tilde{Z} - Z)$ represents the relative vorticity of the perturbation field with respect to the smoothed height pattern. For example, given a value of $(\tilde{Z} - Z)$ equal to 60 m, a Coriolis parameter of $1 \times 10^{-4}\,s^{-1}$ (approximately 40° latitude), a smoothing interval of 600 km and $g = 9.8\,m\,s^{-2}$, the relative perturbation vorticity from (2.11) is about $6.5 \times 10^{-5}\,s^{-1}$.

G has the units of geopotential height, increasing poleward. The wind field V_G, the beta wind, blows from east to west with a speed inversely proportional to the spacing of the G contours, increasing with increasing size of the smoothing interval d. Spacing between 60 m contours in the G field is somewhat greater than the latitudinal extent of the United States in the case where $d = 600$ km. For a value of d equal to 600 km the magnitude of the beta wind is about 2 m s^{-1}, which is roughly comparable to the speed of the beta wind in the Rossby wave equation (6.2) for a wavelength of 2400 m. Thus the beta wind as defined in (6.2) is close to that given by the geostrophic wind of the G field. In the case of the short waves, the beta wind is almost insignificant in retarding the eastward movement of troughs and ridges.

Since perturbations with wavelengths smaller than d are effectively elimi-nated, Fjortoft recommended the value of d to be one-quarter the wave-length of the perturbation, or 600 to 1000 km for typical wave disturbances. In general, a smoothing interval of this magnitude appears to be optimal for prediction of wave movement in middle latitudes.

The predicted height field is obtained by graphical subtraction of the advected height field $(\tilde{Z} - Z + G)$ from the unchanged $(\tilde{Z} + G)$ field. Fjortoft showed that successive smoothing must be applied to obtain an exact solution of (6.11) by graphical methods; in practice this refinement is felt to be unnecessary considering all the uncertainties involved in the theoretical approach. Therefore, assuming that $\tilde{Z}$ does not change appreciably over a 24 h period in comparison to the Z field, one can obtain the 24 h forecasted height field by advecting $(\tilde{Z} - Z + G)$ with the wind field $(\tilde{V}_g + V_G)$ and reconstituting graphically the predicted Z pattern (e.g. 24 h later). This operation is equivalent to the steering of the shorter waves by the zonal flow (analogous to $\tilde{U}$ in (6.2)) (corrected for the beta effect) or by the flow associated with longer waves. Stated alternatively, $(\tilde{Z} + G)$ defines a geostro-phic *steering* current in the geopotential height field with geostrophic wind velocity $(\tilde{V}_g + V_G)$. Note that the G field is simply the equivalent of the Coriolis parameter scaled to units of geopotential height.

Allowing that the time change $(\Delta_t/\Delta t)$ of $(\tilde{Z}_5 + G)$ is negligible over periods of about a day, equation (6.11) can be rewritten to show the rate of change of the 500 mb geopotential height (Z_5), which is

$$\partial(Z_5)/\partial t \approx -(V_g + V_G) \cdot \nabla_p(\tilde{Z}_5 - Z_5 + G) \qquad (6.12a)$$

or

$$\Delta_t Z_5 \approx -(V_g + V_G) \cdot \nabla_p(\tilde{Z}_5 - Z_5 + G) \, \Delta t \qquad (6.12b)$$

where $\Delta_t Z_5$ denotes a local change in Z_5 over a time interval Δt (e.g. 12 or 24 h).

The virtue in the Fjortoft method is that it can be used to advect the perturbed height pattern longer distances and over greater time periods than is possible using the unsmoothed flow. The method seems to yield reasonable forecasts of 500 mb patterns over periods of 24 h *provided that substantial development is not taking place*. Equally important is its use as a conceptual tool in understanding the movement of features at 500 mb. The following section illustrates how the steering flow can be qualitatively deduced by inspection of the 500 mb geopotential height contours.

6.4 Illustrations of 500 mb steering

Let us consider two schematic 500 mb geopotential height patterns in Figures 6.4 and 6.5. The former represents a pure symmetric wave and the latter an asymmetric one. Both figures contain a single wave with the axis of the trough in the center of the frame. A smoothing interval (d) is chosen to be exactly one-quarter of the wavelength. Smoothing is accomplished at each point in the domain by taking an average of the geopotential heights a distance d from the point, as shown in the figures. The smoothed geopotential height field (shown in Figs 6.4b and 6.5b) is called $\tilde{Z}$ and the geostrophic wind of the smoothed field is $\tilde{V}_g$. For a symmetric wave, the resulting smoothed geopotential field is simply a zonal pattern of contours that reflect a latitudinally varying wind speed but do not contain the original perturbation. The broken contours represent the pattern of $\tilde{Z} - Z$, which is the difference between the smoothed and unsmoothed fields. As discussed in Chapter 2, $\tilde{Z} - Z$ is equivalent to the relative geostrophic vorticity, except that the choice of the smoothing interval influences its numerical value. Perturbations with wavelengths much smaller than $4d$ are largely attenuated by the smoothing and therefore no wavelengths much less than about $4d$ appear in the $\tilde{Z} - Z$ field. According to (6.12), a forecast can be made by moving the $\tilde{Z} - Z + G$ contours with the speed of the steering wind, which is $\tilde{V}_g + V_G$. For the present, let us ignore the effect of the beta wind, which is given by the G field. If β were zero, we could predict the future Z field by moving the perturbation in the $\tilde{Z} - Z$ field with the geostrophic wind of the steering flow $\tilde{V}_g$. Clearly, in the case of Figure 6.4, the $\tilde{Z} - Z$ contours are displaced eastward by a distance that is inversely proportional to the spacing of the smoothed contours $\tilde{Z}$ (assuming that f is a constant over the domain). At the end of the forecast, the displaced $\tilde{Z} - Z$ contours are subtracted from the smoothed $(\tilde{Z})$ field to yield Z_{24}, which would exhibit a similar wave as the original, but at a location downstream in the smoothed steering current.

In the case of an asymmetric wave (Fig. 6.5), the smoothing operation produces a different pattern from that of Figure 6.4. Because of the tighter gradient of geopotential heights on the east side of the wave (and therefore a stronger poleward geostrophic wind component), the Z field contains an asymmetry reflecting the stronger geostrophic winds and geopotential height gradient on the east side of the wave. The resulting $\tilde{Z} - Z$ pattern exhibits a maximum on the eastern flank of the wave (as is the original vorticity maximum in the Z field), unlike the symmetric case in which the vorticity maximum is located along the trough axis and the minimum in the ridge axis. The steering current therefore carries the perturbation $(\tilde{Z} - Z)$ eastward *and* poleward. The trough itself follows the movement of the relative vorticity maximum.

Now let us consider the beta effect on smoothing. The G field is exactly

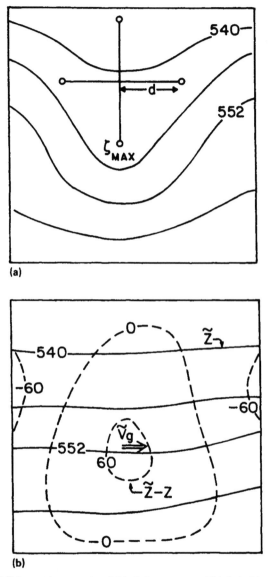

(a)

(b)

Figure 6.4 (a) Schematic geopotential height contours (Z labeled in dam) at 500 mb and the location of the absolute vorticity maximum (ζ_{max}) for a symmetric wave. The cross depicts the smoother of length d for obtaining the smoothed field ($\tilde{Z}$). (b) Smoothed height contours $\tilde{Z}$ (full curves in dam) and perturbation quantity ($\tilde{Z} - Z$, labeled in m), showing also the direction of the smoothed wind velocity $\tilde{V}_g$ (bold-faced arrow) for the steering field.

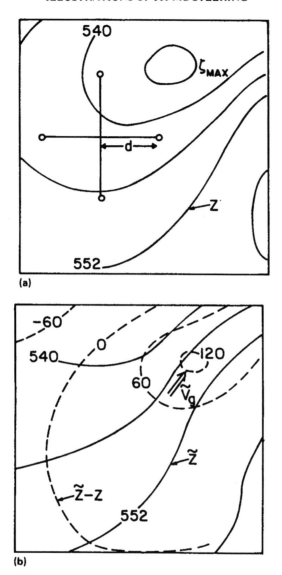

Figure 6.5 Same as Figure 6.4 but for asymmetric case.

analogous to the Coriolis parameter in the original vorticity equation used to derive (6.12). The G field defines a geostrophic wind V_G, which blows toward the west with increasing speed as the smoothing interval (d) is increased. This geostrophic wind (V_G) is analogous to the beta wind correction in the Rossby wave equation (6.2), which also increases with increasing scale length, L. In the symmetric wave case, the G wind slows the eastward movement of the perturbation by an amount roughly equal to the beta correction in (6.2) with $L = 4d$ and $\tilde{U}$ equal to the mean speed of the smoothed geostrophic wind speed $\tilde{V}_g$. Unlike the symmetric wave, the asymmetric wave in Figure 6.5 moves poleward as well as eastward, resulting in larger values of G and smaller values of the relative vorticity perturbation $(\tilde{Z} - Z)$, as is discussed earlier in this chapter.

An actual example of an asymmetric wave that moved poleward and underwent apparent weakening is presented in Figure 6.6. In this sequence a 500 mb trough, initially deep (1200 GMT 19 September 1978; Fig. 6.6a), is situated over western North America. Subsequently the trough moved eastward and northward with a progressive decrease in amplitude, although the absolute vorticity maximum retained a value of about 18–19 units. By the end of the period (Fig. 6.6d), vorticity advection solenoids are found almost entirely north of 40° latitude, representing a poleward displacement of the vorticity maximum of about 10° of latitude during three days (the tracks of the vorticity maxima in Figure 6.6d). The reason for the northward displacement of the vorticity can be seen in the steering current (Fig. 6.6e), which was computed using a smoothing interval of 660 km. This steering current possesses a northward component as the result of the bias in the large-scale current produced by the strong anticyclonic and southwesterly flow along the east coast of the United States. Neglecting the beta effect (the G field), the predicted movement of the trough axis between 20 and 21 September (the dotted trajectories in Figure 6.6e) is very accurate using the equivalent barotropic steering by the Fjortoft steering method. Trajectories representing the location of the maximum relative vorticity (the maximum the $\tilde{Z} - Z$ field) move to positions downstream in the $\tilde{Z}$ pattern. Subtraction of the 24-hour displaced $\tilde{Z} - Z$ field from that of the fixed pattern of $\tilde{Z}$ yields the predicted pattern of Z_{24}.

In reality, the $\tilde{Z}$ field does change with time during 12 or 24 hours. The temporal evolution of the larger-scale (wavelengths greater than $4d$) flow pattern can be predicted in a similar manner to that of the perturbed flow using a larger smoothing interval than that for the shorter wave. For qualitative visual evaluation of the evolution of a 500 mb pattern, it is sufficiently accurate to predict the movement of the feature by imagining a smoothed current; advecting the feature with the smoothed current velocity; and making a simple correction for the beta effect.

There are a number of simple variants on the patterns shown in Figures 6.4 and 6.5. Consider a wave in which the geopotential height gradient is stronger on the west than on the east side of the wave (Fig. 6.7a). This wave is asymmetric with a vorticity maximum within the northwesterly flow. The steering current $(\tilde{Z})$ has a bias such that the vorticity pattern is advected equatorward and the wave moves with an equatorward component, increas-

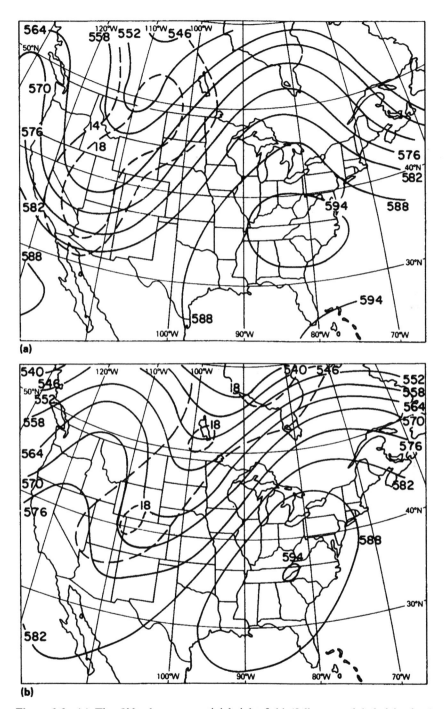

Figure 6.6 (a) The 500 mb geopotential height field (full curves labeled in dam) for 1200 GMT 19 September 1978. Absolute vorticity isopleths (units in $1 \times 10^{-5}\,\mathrm{s}^{-1}$) are drawn as broken curves for values greater than or equal to $14 \times 10^{-5}\,\mathrm{s}^{-1}$. (b) Same as (a) but for for 1200 GMT 20 September 1978.

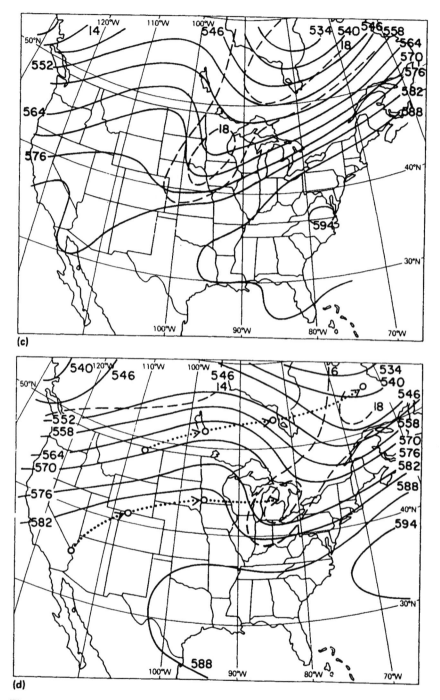

Figure 6.6 (c) Same as (a) but for for 1200 GMT 21 September 1978. (d) Same as (a) but for for 1200 GMT 22 September 1978. Dotted streamlines with circles denote trajectories of positive vorticity centers at 24 h intervals.

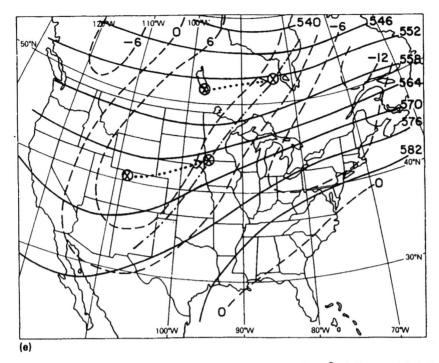

(e)

Figure 6.6 (e) The 500 mb smoothed geopotential height field $\bar{Z}_5$ (full curves labeled in dam) for smoothing interval (d) of 660 km at 1200 GMT 20 September 1978. The broken contours are the values of $\bar{Z}_5 - Z_5$ and the pairs of crossed circles represent the 24 h displacement of the trough axis based on the advection of $(\bar{Z}_5 - Z_5)$ with the geostrophic wind velocity of the $\bar{Z}_5$ field; the beta effect is neglected. The observed trough axis at 1200 GMT 21 September 1978 is shown by the chain curve.

ing its relative vorticity and intensifying its geopotential height perturbation (apparent deepening). The opposite bias, that of stronger winds on the eastern side of the vortex (Fig. 6.7b) corresponds to the case of Figure 6.6, in which the vorticity center moves eastward and poleward.

Another example is that of a cut-off vortex, in which closed contours appear in the geopotential height pattern. If the vortex is symmetric with respect to the east–west direction, but there are stronger westerly winds on the equatorward side than easterly winds poleward of the vortex, a resultant westerly steering current is over the perturbation (Fig. 6.7c). Thus, eastward movement tends to be very slow if the vortex were nearly symmetric in the north–south direction and the easterly and westerly flow components are nearly identical. Such vortices might even *retrogress*, either because the easterly flow is stronger than the westerly flow or because the beta wind is stronger than the resultant westerly steering component by $\bar{V}_g$.

153

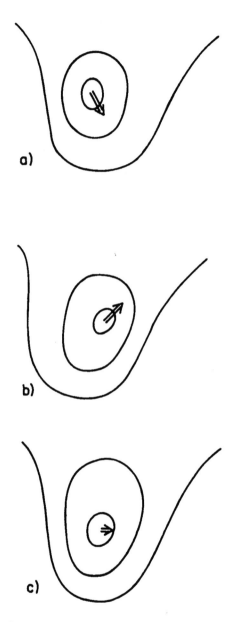

Figure 6.7 Schematic examples of 500 mb geopotential height patterns, showing the movement of vortices (bold-faced arrows) for various asymmetries: (a) stronger geostrophic wind speed on west side of vortex; (b) stronger geostrophic wind speeds on east side of vortex; and (c) east–west symmetry with strongest geostrophic winds equatorward of vortex.

Other asymmetries can be imagined: vortices with stronger winds on the western side and stronger easterlies poleward of the center than westerlies equatorward of the center would move equatorward and westward. The latter often occurs near 200 mb in the vicinity of deep troughs that extend toward low latitudes. These troughs tend to shed vortices, which then move westward in the Tropics.

It may be difficult to visualize a 500 mb steering current in complex situations, e.g. where there are several small vorticity centers and a highly distorted geopotential height pattern. A vortex may simply represent a perturbation in a larger-scale flow, which is steering that vortex. Although the steering calculation is scale-dependent (depending on d), increasing the smoothing interval to include a stronger, larger-scale steering current might not result in a faster eastward movement of the feature. Capturing more of the larger-scale westerlies by increasing d also involves an increase in westward retardation by the beta wind.

In the sequence shown in Figure 6.6, the latitudinal displacement of the 500 mb trough by the steering current is evident in the geopotential height pattern. Poleward movement occurred because of a tighter geopotential height gradient on the eastern side of the trough. This movement of the trough led to the absolute vorticity maximum moving north of the United States border and therefore to a poleward shift of the vertical motion and pressure tendency pattern. Apparent weakening, associated with the poleward displacement of the vorticity maximum, decreased the relative vorticity perturbation and therefore decreased the geopotential height perturbation. This weakening of the height perturbation without weakening of the absolute vorticity maximum illustrates an aspect of barotropic decay of waves referred to in Chapter 5. In Chapters 7 and 10 we will see that cyclogenesis on the eastern sides of mid-latitude troughs inevitably leads to an intensification of the 500 mb absolute vorticity maximum east of the trough axis and therefore to a poleward movement of the 500 mb trough once cyclogenesis has ceased.

Problems

6.1 The following exercise is designed to illustrate the basic idea behind barotropic steering (the Fjortoft method). The object is to make a 24 h forecast of the 500 mb geopotential height pattern. Necessary materials are several pieces of tracing paper, a 500 mb geopotential height map (such as those presented in Figure 6.7), a pencil and a light table; the latter can be simulated during the daytime by a window that faces outdoors. Perform the following steps:

(a) Overlay a piece of tracing paper on a 500 mb chart and trace the geopotential height contours, labeling them. Repeat the operation on a second, identical piece of tracing paper, being careful that the edges of the tracing paper on your own reference marks coincide with those on the original.

(b) Choose a displacement interval (d) equal to about one-quarter of the wavelength of the weather system to be predicted, e.g. 600 km (6° of latitude).

(c) Obtain a smoothed $\tilde{Z}$ field by the following method. Displace the two copies of the 500 mb chart in the x direction a distance of $2d$. Overlay a piece of blank tracing paper, centering it exactly half-way between the two 500 mb charts. Now, construct the average $\tilde{Z}_x$ field for the x direction on the tracing paper by adding the two fields together graphically. An average field is obtained by drawing through every other intersection of the two original fields and labeling the contours as half the sum of the intersecting contours of the original 500 mb charts.

(d) Repeat the previous step using the original two 500 mb maps but now, using a second piece of tracing paper, displace the charts a distance $2d$ in the y direction. Obtain the graphically averaged $\tilde{Z}$ field in y.

(e) Superpose exactly the two averaged $\tilde{Z}$ charts (one in x and the other in y) and graphically average these charts (the products of steps (c) and (d)), labeling every other intersection as before. (The new product is the total $\tilde{Z}$ field.)

(f) Graphically subtract the original Z from the $\tilde{Z}$ field to obtain $(\tilde{Z} - Z)$.

(g) Now, perform a 24 h time displacement of the $(\tilde{Z} - Z)$ field along the contours of the $\tilde{Z}$ pattern by advecting the former with the geostrophic wind of the latter. Do this in 6 h intervals for a few points on each contour line of the $\tilde{Z} - Z$ pattern. First, estimate the 6 h displacement downstream for each point and then with a pair of dividers measure the contour spacing at the mid-point. (It would be useful to construct a nomogram relating contour spacing to geostrophic wind speed and to distance traveled over a 6 h period for different latitudes.) A rough approximation, however, is $X_s = 29/(\Delta n' \sin \theta)$, where X_s is the 24 h distance displaced (in degrees of latitude) along the contour line of $\tilde{Z}$, θ is the latitude and n' is the spacing between $\tilde{Z}$ contours, measured in degrees of latitude. Next, correct the predicted $(\tilde{Z} - Z)$ pattern for the beta effect by using the Rossby wave equation (6.2) with $L = 4d$. Displace *westward* the final $(\tilde{Z} - Z)$ pattern a distance equal to the magnitude of the beta wind speed times 24 h. (Note, this operation does not account exactly for the beta effect, but it will be a close approximation for zonal steering.)

(h) Finally, subtract graphically the 24 h $(\tilde{Z} - Z)_{24}$ pattern from the constant $\tilde{Z}$ field to obtain the predicted Z_{24} contours. Compare your result with the actual 500 mb height field and discuss.

6.2 (a) Compute, using the equivalent barotropic model (6.8), the vertical motions at 500 mb equivalent to a unit (1° latitude square) advection box for relative vorticity, where $\omega = - \left(\dfrac{F(p)}{f_0} \right) (- V_m \cdot \nabla_p \xi_m)$ from (6.8b) and Table 6.1.

(b) Assuming that the relative vorticity advection at 500 mb in Figure 2.4 is approximately equal to the absolute vorticity advection, make a rough estimate of the 500 mb equivalent barotropic vertical velocity in the vicinity of the shaded solenoid in the figure and compare that result to the predicted values at that location in Figure 2.9. You can accomplish this by using the result for the unit advection solenoid, which was determined in part (a).

(c) By qualitative inspection of Figures 2.4 and 2.9, find some points in the vicinity of the storm system where the barotropic vertical motions are clearly in error as to sign. Suggest a possible reason for the failure of the barotropic model vertical motion calculations in those areas.

6.3 Draw a schematic geopotential height pattern in which a cutoff vortex at 500 mb moves with both a westward and equatorward component of motion.

Further reading

Buch, H. 1954. *Hemispheric wind conditions during the year 1950.* Final Report, Part 2, Contract AF 19(122)-153, Dept Meteor., M.I.T., Cambridge, MA, pp. 32–80.

Fjortoft, R. 1952. On a numerical method of integrating the barotropic vorticity equation. *Tellus* **4**, 179–94.

Haltiner, G. J. 1971. *Numerical weather prediction.* New York: Wiley.

Haltiner, G. J. and F. L. Martin 1957. *Dynamical and physical meteorology.* New York: McGraw-Hill.

Lorenz, E. N. 1972. Barotropic instability of Rossby wave motion. *J. Atmos. Sci.* **29**, 258–64.

Randall, W. J. and J. L. Stanford 1985. The observed life cycle of a baroclinic instability. *J. Atmos. Sci.* **42**, 1364–73.

7

Simple dynamic models of wave/cyclone motion development: baroclinic viewpoint

Troughs and ridges at 500 mb move great distances over periods of a week or more without a significant change in their intensity. This suggests evidence that, in the mean, equivalent barotropic conditions are satisfied at this level. This is especially true of short waves, which are sometimes referred to as "mobile waves". Except during cases of explosive cyclogenesis at the surface, the evolution of the 500 mb trough into a closed cyclone takes place more slowly than at the surface.

Nevertheless, intensification of upper-level troughs and ridges does occur. The relationship between development at the surface and at mid-levels is an essential aspect of cyclogenesis and of baroclinic instability. Let us define *developmental* as referring to geopotential height or vorticity changes at a point that are not associated with the advection of vorticity. We will now show how development occurs at 500 mb and how that development is related to the surface geopotential height patterns.

7.1 Baroclinic development at 500 mb: a two-parameter model

A major limitation of barotropic models is that they are unable to describe the development or decay of weather systems. The absence of development in a barotropic atmosphere is linked to the absence of temperature (thickness) advection. The one-parameter equivalent barotropic model, described in Chapter 6, accounts for the spatial redistribution of mid-level vorticity by advection of the absolute vorticity pattern. The model does not permit intensification or weakening, although it does implicitly account for development at the surface via the implied surface divergence pattern. As discussed in Chapter 3, a balance of vorticity near 500 mb exists between the local derivative and the advection of absolute vorticity. In the equivalent barotropic model, the vertical motion – and therefore divergence or convergence – is related to only one forcing function, the advection of relative vorticity.

Let us now examine the conditions under which non-advective geopotential height changes occur at 500 mb (we will assume that these conditions apply generally to the middle troposphere). As discussed in Chapter 4, the height tendency equation (4.9) tells us that the 500 mb geopotential height tendency is negative if warm advection increases with height or cold advection decreases with height over the middle troposphere. The temperature gradients (and therefore the temperature advections) below the upper troposphere tend to have their largest magnitude in the lower troposphere, particularly on the cold side of fronts. Therefore, 500 mb height falls associated with the vertical derivative of the temperature advection occur in regions of cold air advection near the surface. The reverse is true for 500 mb height rises with respect to warm air advection in the lower troposphere; these rises will occur where the warm air advection is strong near the surface. Accordingly, one expects to find development at 500 mb occurring most strongly on the cold sides of surface fronts, where advection of temperature is relatively large.

In order to express the effects of temperature advection on 500 mb wave development, a model atmosphere is now considered in which the geostrophic winds *can* change direction with height, unlike the model presented in Chapter 6. We will choose the simplest case of temperature advection in which the direction of the isotherms is constant with height. Since isotherms typically do not change their orientation rapidly with height, at least in the lower troposphere, we will constrain the direction of the geostrophic wind shear (but not necessarily the direction of the geostrophic wind) to be constant with height. The hodograph satisfying the assumption of constant shear with height (no change with height in the orientation of the isotherms) resembles that in Figure 7.1a. Consider a mean level (subscripted m) and let

$$V_g = V_{gm} + B(p) V_T \qquad (7.1a)$$

$$\zeta_g = \zeta_{gm} + B(p) \zeta_T \qquad (7.1b)$$

where the thermal wind velocity (V_T) is defined as ($V_{gm} - V_{g0}$), and ζ_T, defined as ($\zeta_m - \zeta_0$), is called the *thermal vorticity*. (Henceforth, the subscript g will be dropped when the subscript m is used.) The mean-level quantities subscripted by the letter m are defined exactly as in the previous chapter, with p_m corresponding to a mean level that is not far below 500 mb. The vertical distribution of B (see Table 6.1), based on large-scale averages of real

159

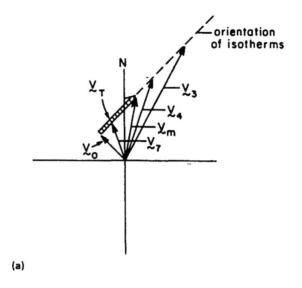

(a)

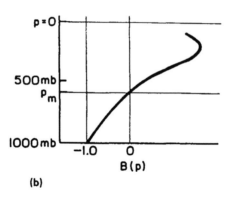

(b)

Figure 7.1 (a) Wind hodograph for the two-parameter model. The thermal wind between the mean level and 1000 mb is shown hatched. The orientation of the isotherms is given by the broken line. (b) Schematic vertical distribution of the parameter $B(p)$.

wind velocity, is shown schematically in Figure 7.1a. At the mean level, $B(p_m)$ is equal to zero, i.e.

$$B_m = \frac{1}{p_s} \int_0^{p_s} B(p)\,dp = 0$$

whereas at 1000 mb, $B_0 = -1.0$. It is clear from inspection of Figure 7.1b that temperature advection is permitted in this model.

Let us now derive an equation for this two-level baroclinic model. Proceeding as in section 6.2., the quasi-geostrophic vorticity equation (3.6a) is integrated with respect to pressure from the surface to the top of the atmosphere using relationships (7.1) and noting that V_m, V_T, ζ_m, ζ_T and f are independent of pressure (height). Integration over the layer 0 to p_s involves the following steps:

$$\frac{1}{p_s} \int_0^{p_s} \left(\frac{\partial \zeta_g}{\partial t} + V_g \cdot \nabla_p(\zeta_g + f) \right) dp = \frac{f_0}{p_s} \int_0^{p_s} \frac{\partial \omega}{\partial p} dp \qquad (7.2a)$$

or

$$\frac{1}{p_s} \int_0^{p_s} \left(\frac{\partial \zeta_m}{\partial t} + B(p) \frac{\partial \zeta_T}{\partial t} \right) dp$$

$$+ \frac{1}{p_s} \int_0^{p_s} \{ [V_m + B(p) V_T] \cdot \nabla_p[\zeta_m + \zeta_T B(p) + f] \} dp$$

$$= \frac{f_0}{p_s} \int_0^{p_s} \frac{\partial \omega}{\partial p} dp = \frac{f_0 \omega_s}{p_s} \qquad (7.2b)$$

which, since $B_m = 0$, becomes

$$\frac{\partial \zeta_m}{\partial t} = - V_m \cdot \nabla_p(\zeta_m + f) - \overline{B(p)^2}\ V_T \cdot \nabla_p \zeta_T + \frac{f_0 \omega_s}{p_s}. \qquad (7.3)$$

Equation (7.3) resembles forms of the equivalent barotropic vorticity equation presented in Chapter 6. The second term on the right-hand side, however, represents a correction to the conservation of vorticity at the mean level (p_m), and is effectively a *development term* that describes the local rate of vorticity generation at the mean level. Here, development is not expressed in terms of the divergence, as it is at 1000 mb, but by the advection of thermal vorticity by the thermal wind multiplied by a constant, $\overline{B(p)^2}$. This term implicitly relates development at 500 mb to the advection of temperature via the thermal vorticity advection. (Recall that the geostrophic vorticity is proportional to the Laplacian of the geopotential height and, therefore, thermal vorticity is proportional to the Laplacian of the thickness.) Briefly stated, where the advection of thermal vorticity is positive (negative), the local vorticity tendency due to the correction term is positive (negative).

Equation (7.3) provides a framework in which one can examine the feedback between the surface temperature and vorticity patterns and those at mid-levels. The second term on the right-hand side of (7.3) is referred to as the *500 mb development term*. According to climatological data (e.g. Table 6.1) the value of $\overline{B^2(p)}$ tends to be a relatively small fraction (typically about 0.2–0.4) at the mean level.

Perhaps the most abstract aspect of (7.3) is the concept of thermal vorticity, which is simply the relative vorticity of the thickness pattern, e.g.

$$\zeta_T = \frac{g}{f_0}\, \nabla_p^2(Z_m - Z_0) \approx \frac{g}{f_0}\, \nabla_p^2 h$$

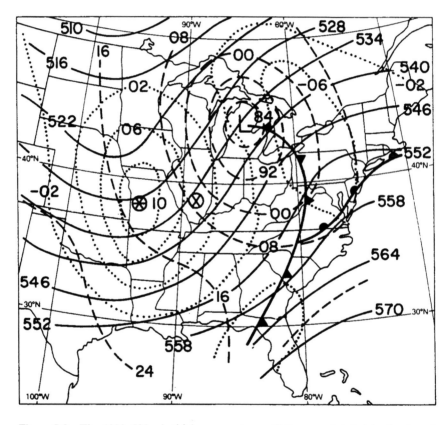

Figure 7.2 The 1000–500 mb thickness contours (full curves labeled in dam) and sea-level pressure (broken curves in mb; 1000s digit omitted) with surface fronts. Relative geostrophic thermal vorticity of the 1000–500 mb thickness pattern is shown by the dotted contours (at intervals of $4 \times 10^{-5}\,\mathrm{s}^{-1}$) for 1200 GMT 4 January 1982 (see Fig. 10.5). The location of the 1000–500 mb thermal vorticity maximum is denoted by a circled asterisk and the location of the maximum absolute vorticity at 500 mb is denoted by a circled cross.

where thickness and thermal wind are defined here with respect to differences between 1000 mb and the mean level p_m. Note that the thermal vorticity bears the same relationship to the thermal trough and to the thickness contours as does the relative vorticity to the geopotential height trough and contours; maximum thermal vorticity resides in the thermal trough, minimum thermal vorticity in the thermal ridge.

This relationship between the location of the trough and the thermal vorticity maximum is illustrated in Figure 7.2 for a mature cyclone associated with a 500 mb trough. The thermal vorticity maximum (10 units) is located in the middle of the 1000–500 mb thickness trough and just upstream from the 500 mb vorticity maximum (crossed circle). Positive thermal vorticity advection by the thermal wind is occurring over the 500 mb vorticity maximum. Solenoids of thermal vorticity advection by the thermal wind correspond closely with solenoids of cold thickness advection by the 1000 mb thermal wind. Development of the 500 mb geopotential height and vorticity patterns is therefore occurring in the region southwest of the surface cyclone. (This case is discussed in further detail in Chapter 10.)

To illustrate further the behavior of the 500 mb development term, let us consider a typical wave/cyclone pattern in Figure 7.3, in which a cyclone at the surface, a trough at 500 mb (approximately the mean level) and a thickness trough are shown. Indicated also are the locations of the maxima of absolute geostrophic and thermal vorticities, respectively ζ at point (b) and ζ_T at point (a).

In Figure 7.3, the maximum cold air advection occurs near (c). Similarly, warm thickness advection by the 1000 mb geostrophic wind correlates with negative thermal vorticity advection northeast of the surface low (near (e)), and developmental height rises at 500 mb. *Importantly, however, maximum cold thickness advection by the 1000 mb geostrophic wind corresponds closely in location to the maximum positive geostrophic thermal vorticity advection by the thermal wind and to 500 mb developmental height falls.*

In the case of a surface cyclone lying ahead of a 500 mb trough, the regions between the surface trough and the 500 mb trough, and between the surface trough and the downstream 500 mb ridge, experience temperature advections leading to developmental height changes both at the surface and at 500 mb. Negative 500 mb geopotential height tendencies and positive surface pressure tendencies occur west and southwest of the surface cyclone in association with cold air advection. Positive 500 mb height tendencies and negative geostrophic vorticity tendencies at that level occur northeast of the surface cyclone in association with warm advections.

These principles are illustrated in a three-dimensional schematic drawing (Fig. 7.4). Maximum developmental 500 mb geopotential height falls occur just ahead of the 500 mb trough and to the rear of the surface cold front, where there is large-scale descent and surface pressure rises. Near (c) positive

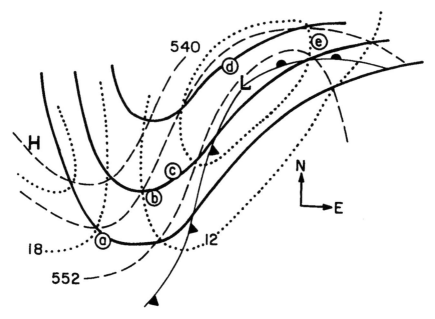

Figure 7.3 Schematic analysis showing contours of geopotential height at 500 mb (full curves), 1000–500 mb thickness (broken curves labeled in dam) and 1000 mb geopotential height (dotted curves labeled in dam) for a typical open wave cyclone. At the surface the centers of the high and low are indicated with symbols H and L. Cold and warm fronts at the surface are shown with conventional symbols. The location of the maximum absolute geostrophic vorticity at 500 mb is near point (b), that of the maximum thermal vorticity is near (a). Maximum cold air thickness advection by the 1000 mb geostrophic winds is near (c) and maximum warm thickness advection is near (e). Minimum thermal vorticity lies between (d) and (e).

absolute vorticity advection at 500 mb and cold air advection are juxtaposed, so that there is both rising motion and cold air advection. Similarly, surface pressure falls, 500 mb geopotential height rises and positive vorticity advection at 500 mb characterize the region east of the surface cyclone. A similar statement can be made regarding 500 mb height rises and absolute vorticity decreases associated with development over the region east of the surface cyclone and within the 500 mb ridge. This arrangement characterizes developing waves/cyclones.

For simplicity, the mean level is equated with 500 mb, although the former is actually closer to 600 mb as shown in Chapter 6. The approximation involved in using 500 mb as the mean level is sufficiently accurate for making a reasonable visual assessment of development at 500 mb in view of the other simplifying assumptions made in the quasi-geostrophic equations. It is the experience of this author that an optimum value to use for $\overline{B^2(p)}$ (which

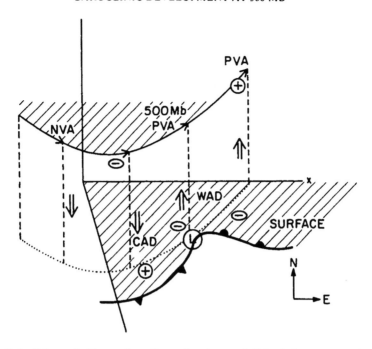

Figure 7.4 Schematic illustration of a surface low and 500 mb flow pattern (represented by a single streamline). Regions of cold and warm advection (CAD, WAD), positive and negative absolute vorticity advection (PVA, NVA) and lower-tropospheric vertical motion (double-shafted arrows) are shown. Plus and minus signs refer to the sign of developmental height changes at 1000 and 500 mb.

strictly applies at p_m) is about 0.2 when applied at the 500 mb level. The following equations for smooth terrain can be useful in evaluating the height and absolute vorticity tendencies at 500 mb:

$$\partial \zeta_{g5}/\partial t \approx -V_{g5} \cdot \nabla_p(\zeta_{g5} + f) - 0.2 V_T \cdot \nabla_p \zeta_T \qquad (7.4a)$$

which follows from (7.3), and

$$\frac{\partial Z_5}{\partial t} \approx \left(\frac{\partial Z_5}{\partial t}\right)_{\text{barotropic}} - 0.2 V_{go} \cdot \nabla_p h \qquad (7.4b)$$

which is empirical.

The second term on the right-hand sides of (7.4b) and (7.4a) describes the developmental effect of thickness and thermal vorticity advections on the 500 mb vorticity and geopotential height tendencies, respectively. Here the thermal wind or thermal vorticity refers to the 1000–500 mb layer, rather than the layer between 1000 mb and p_m. Equation (7.4b) is empirical, but it is justified by the height tendency equation (4.9), has considerable observational basis and is analogous to the more rigorously derived equation (7.4a). Equation (7.4b) is more convenient to apply than (7.4a) because conventional weather maps display thickness rather than thermal vorticity. The relatively small value of the coefficient (0.2) reflects the weak (but not insignificant) strength of the coupling between temperature advection in the lower troposphere and local height or vorticity changes with time at 500 mb. The weak coupling between 1000 and 500 mb reflects the relatively small magnitude of divergence at middle levels. Presumably, however, the value of this coefficient varies somewhat from case to case. For a unit advection solenoid in the thickness/1000 mb geopotential height field, the magnitude of the thickness advection by the 1000 mb wind is about 1200 m day^{-1}, equivalent to a developmental height tendency of about 240 m day^{-1}.

In the case where both the thermal and geopotential height troughs are in

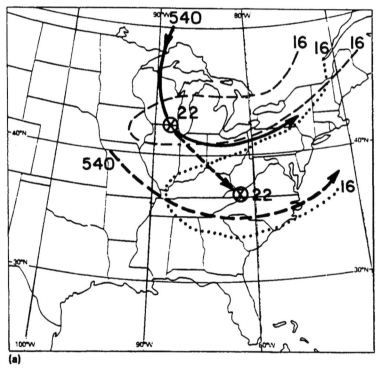

(a)

Figure 7.5 (a) Isopleths of 16 unit absolute vorticity (broken and dotted contours), location of 22 unit absolute vorticity center (circled crosses separated by broken arrow showing 12 h displacement) and 540 dam geopotential height contours (direction of 500 mb geostrophic wind velocity, the full and broken streamlines) for 0000 and 1200 GMT 9 April 1986, respectively.

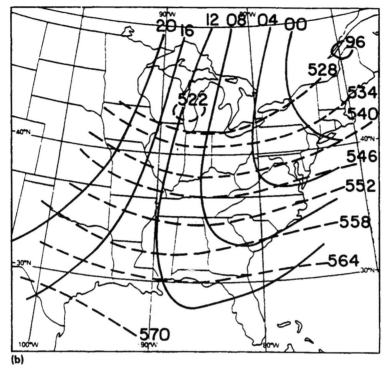

(b)

Figure 7.5 (b) Isopleths of sea-level pressure (mb; 1000s unit omitted) and 1000–500 mb thickness (broken curves labeled in dam) for 0000 GMT 9 April 1986.

phase, advection of thermal vorticity by the thermal wind at the 500 mb trough axis is negligible and there is no development of the absolute vorticity maximum. *Figures 7.3 and 7.4 illustrate that a necessary condition for 500 mb wave development is a phase lag between thickness and 500 mb waves with colder air lagging behind the 500 mb trough.*

Figure 7.4 shows that development of the 500 mb trough occurs east of the surface anticyclone in the region where there are surface pressure rises, and therefore low-level divergence. The local vorticity tendency is positive at the 500 mb absolute vorticity maximum (even though the advection of 500 mb vorticity is zero) because advection of relative thermal vorticity (cold 1000–500 mb thicknesses) is positive over that region. It follows that, given the association between 500 mb height rises (falls) and surface pressure falls (rises), the implied divergence (convergence) at 500 mb is consistent with the level of non-divergence being slightly below 500 mb. Development at 500 mb alters both the magnitude and movement of the vorticity pattern. It is not uncommon, especially in winter, to find that troughs and ridges develop rapidly equatorward of their absolute vorticity maximum, although there is

167

an eastward barotropic steering current at 500 mb. Trough or ridge development usually accompanies rapid cyclogenesis at the surface in the region of strong cold or warm air advection in the lower troposphere.

This is illustrated by the case study in Figure 7.5. Note that the 500 mb geopotential height contours show no indication of equatorward steering, as described in the previous chapter, yet the absolute vorticity maximum and the 540 dam geopotential height contour shift southward. Development is signaled by an expansion of the area enclosed by the 16 unit contour and by a deepening of the geopotential heights near to and equatorward of the vorticity maximum. The result of this development was the formation of a deep 500 mb trough along the east coast of the United States. Small temperature advection solenoids (signifying strong cold air advection) in the thickness pattern occur in Figure 7.5b south of the center of maximum absolute vorticity at 500 mb, resulting in geopotential height falls (in the developmental sense) over the southeastern part of the United States and a corresponding southward shift and expansion of the 16 unit isopleth of absolute vorticity. This development of the 500 mb trough is consistent with positive thermal vorticity advection by the thermal wind.

7.2 Large-scale developmental changes that occur at 500 mb during cyclogenesis

Let us now illustrate the large-scale changes that occur at 500 mb in the geopotential height, vorticity and wind patterns during cyclogenesis. Consider Figure 7.6, which shows two sets of geopotential height contours, one representing an initial 500 mb height pattern at the onset of cyclogenesis (full curves) and the other the 500 mb height pattern 24 h later following a period of development (broken curves). Let us isolate developmental from barotropic changes (those height changes that are not simply due to advection of absolute vorticity) by superposing the axes of the wave at 0 and 24 h. We will assume that differences in the two height fields represent changes that are largely due to non-barotropic effects, since the overall translation of effects of the wave movement are largely eliminated by the process of superposition of the waves.

It is clear that maximum 500 mb geopotential height falls take place southwest of the surface low center, which is west of the cold front in the region of strongest cold air advection. Corresponding height rises occur poleward of the low center in the region of strongest warm air advection. This evolution in the height pattern is consistent with the following changes.

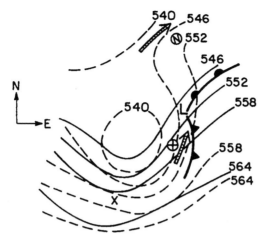

Figure 7.6 Superimposed wave patterns 24 h apart, showing characteristic changes that occur at 500 mb during cyclogenesis. Full curves are the initial 500 mb geopotential height pattern (labeled in dam) and the broken curves are the 500 mb height lines 24 h later. The surface low center (with surface fronts) is indicated. The initial location of the absolute vorticity maximum is shown by a cross; 24 h later it is indicated by a circled cross. The location of the negative relative vorticity maximum after 24 h is given by a circled N. Wind speed maxima at the later map time are shown by hatched arrows.

Shape of the wave

Trough and downstream ridges begin to tilt from the customary "positive" (northeast–southwest) to the "negative" (northwest–southeast) sense at the onset of rapid cyclogenesis. The half-wavelength between the maximum developmental height rises and height falls tends to shorten, becoming less than that of the initial half-wavelength of the 500 mb pattern. (The reason for this shorter wavelength of development is further discussed in Chapter 11.) Changes in the tilt of trough and ridge are therefore an indication of cyclogenesis at low levels in the troposphere.

Configuration of the wind field

Developmental decreases in the 500 mb heights southwest of the surface low center are accompanied by increases in the geopotential height gradient between the region of 500 mb height falls and the warm sector where developmental height changes are small. Consequently, there is an increase in the geostrophic (and real) wind speeds in this southwesterly flow southwest of the surface low center, leading to formation of a jet streak in this sector of the wave (the lower hatched arrow in Figure 7.6). Note that the wind velocity has a strong poleward component.

Corresponding changes occur in the downstream ridge, where height rises in the middle and upper troposphere lead to the formation of a westerly wind speed maximum (the upper hatched arrow) in the upper troposphere.

The vorticity pattern

With the strengthening of the gradient in the geopotential heights and in the wind speed west and equatorward of the low center, there is a corresponding increase in absolute vorticity just to the left side of the newly forming southwesterly jet (the circled cross). Similarly, there is a strengthening of the vorticity minimum near the ridge (the circled N).

Absolute vorticity and temperature advections: feedback to surface development

Positive vorticity advection becomes stronger and more meridional over the center of the surface low, leading to increased quasi-geostrophic forcing of ω_d over the cyclone and an increased poleward track of the cyclone. Where positive vorticity advection and cold air advection are juxtaposed, the cooling effects of adiabatic ascent may form an isolated pool of cold air and a closed low in the mid- and upper-tropospheric geopotential height pattern. This occurs in the region just west and equatorward of the surface low and cold front, as indicated in Figure 7.6. Having formed in the region of strong developmental height falls at 500 mb, the cold pool migrates toward the surface low pressure center. The result is that the system eventually becomes vertical and the advections cease.

The inception of a developing surface cyclone circulation signals that temperature advection has begun to influence cyclone growth. In the initial stages of cyclogenesis, the thermal advection term is necessarily small due to the weak circulation of the surface low. Once the surface low has formed, there is an increase in temperature advection, an increase in the geopotential height gradient at 1000 mb, and an increase in ascent to the east and poleward of the storm and descent on its west and south. The storm moves eastward and poleward, generally toward the region of smallest temperature and vorticity advection solenoids; this path is typically almost parallel to and on the cold side of the surface warm front. Thus, the inception of a surface cyclone in a region of baroclinicity initiates the feedback between temperature advection and 500 mb vorticity advection. *This constitutes an essential first step in cyclogenesis.*

Barotropic steering effects

At the end of development of the wave, equatorward movement of the absolute vorticity center is halted and there is a concentration of absolute

170

vorticity on the eastern side of the wave, as indicated in Figure 7.6. Accordingly, barotropic steering principles suggest that the wave and vorticity maximum migrate poleward and experience apparent (barotropic) decay, in which the relative vorticity decreases. As discussed in Chapter 5, this involves a transfer of eddy kinetic energy to the longer waves, resulting in a loss of eddy kinetic energy by the short wave.

7.3 An illustration of coupled surface and 500 mb development

An example is now presented in which the principles discussed in this chapter are illustrated. In particular, we show how there is feedback in the developmental process between the surface and at 500 mb. The case presented in Figure 7.7 is one in which a remarkably rapid and extensive cyclogenesis took place over North America at the surface and 500 mb.

At the onset of rapid surface development, an unremarkable surface pattern in Figure 7.7a contrasts with that of the thickness and 500 mb height

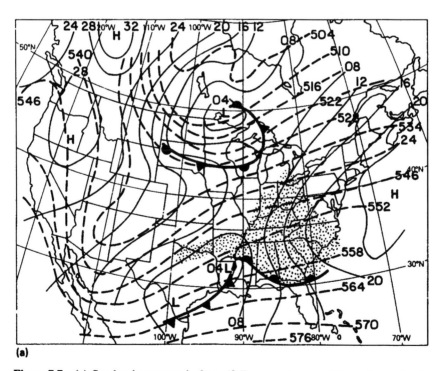

(a)

Figure 7.7 (a) Sea-level pressure isobars (full curves in mb; 1000s units omitted), fronts and the 1000–500 mb thickness contours (broken curves labeled in dam) for 0000 GMT 25 January 1978. Shading denotes the region of continuous precipitation.

171

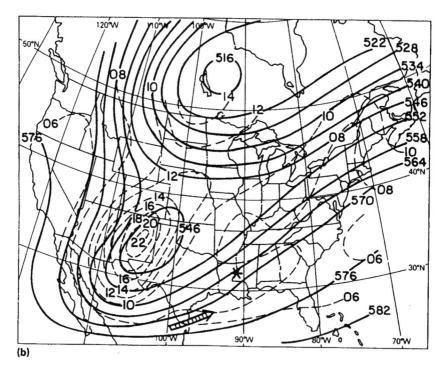

(b)

Figure 7.7 (b) The 500 mb height contours (full curves labeled in dam) and absolute vorticity isopleths (broken curves labeled at intervals of $2 \times 10^{-5}\,\mathrm{s}^{-1}$) for 0000 GMT 25 January 1978. The location of the surface low is marked by the star. Hatched arrow represents location of 500 mb wind speed maximum.

and vorticity fields (Fig. 7.7b). The latter were extraordinarily intense, with the region of strongest geostrophic wind speed and absolute vorticity being displaced far equatorward (over Mexico). Note that, even before the start of rapid surface development, there is a rather imposing vorticity maximum of about 22 units over New Mexico (NM) within the 500 mb trough axis. Associated with the intense gradients of temperature and vorticity is a 300 mb jet whose maximum speed exceeds 150 kt ($75\,\mathrm{m\,s}^{-1}$) over northern Mexico. Strong vorticity advection is occurring west of the surface cyclone but not over its center. Precipitation is largely associated with ascent forced by thickness advection over the region north of the cyclone and its warm front.

The cyclone moved northeastward in the direction of the strongest surface pressure falls, where warm air advection (small solenoids in Figure 7.7a) constitutes the primary forcing term in the omega and pressure tendency equations. Cyclogenesis was slow initially, but changes in the 500 mb height pattern took place as the result of *strong cold air advection behind the cyclone.*

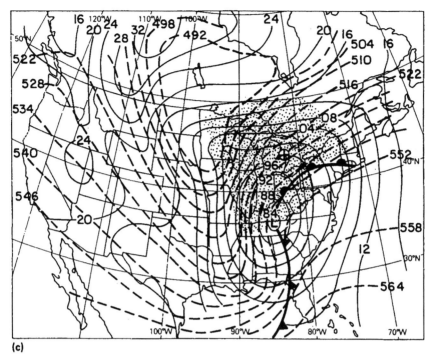

(c)

Figure 7.7 (c) Same as (a) but for 0000 GMT 26 January 1978. The location of the surface low at 0900 GMT is denoted by the circled cross.

As a result of these changes, the 500 mb vorticity field at 500 mb intensified and the center of positive vorticity moved eastward from its position in Figure 7.7b. *Evidence of the importance of the development effect at 500 mb in the early stages of rapid cyclogenesis is manifested by the change in orientation of the trough axis from northeast–southwest (positive) to northwest–southeast (negative).* This effect can be seen by comparing Figures 7.7b and d in the region south of the surface low center.

The result of this development at 500 mb is that the principal vorticity center moved to a position just south of the surface low pressure center in Figures 7.7c and d. Similarly, as the result of warm air advection, developmental geopotential height rises occurred at 500 mb northeast of the surface cyclone. The shift and development in the 500 mb absolute vorticity pattern produced an extraordinary gradient of absolute vorticity over the surface cyclone center, approximately that of a unit advection solenoid! Explosive deepening of the surface low occurred, reducing the central sea-level pressure of the cyclone to 984 mb. The cyclone continued to develop rapidly, with central pressure falling by another *26 mb* (to 958 mb) over the next 9 h.

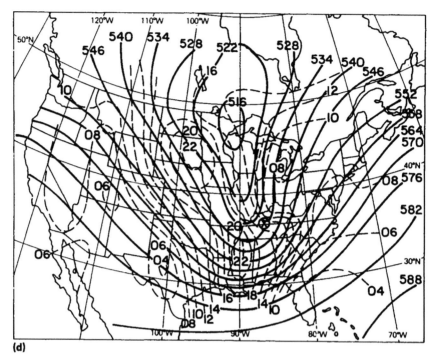

(d)

Figure 7.7 (d) Same as (b) but for 0000 GMT 26 January 1978. The surface low position is again denoted by the circled cross.

The equatorward displacement of the vorticity maximum with respect to the surface cyclone is understandable in view of the location of maximum cold air advection, which is strongest southwest of the intense surface lows. (This aspect of development is discussed further in Chapter 10.)

By 0900 GMT 26 January 1978 the center of the cyclone was situated over the tip of Lake Huron (circled cross in Figure 7.7c). *Not surprisingly, the movement of the surface cyclone center was almost due north in the direction of the strongest low-level convergence as diagnosed by the distribution of vorticity and thickness advection solenoids.* The key to rapid development in this cyclone is the intense gradient of temperature, which provided the initial baroclinic forcing that led to strong ascent and low-level convergence and to 500 mb geopotential height falls in the region of cold advection. This temperature gradient is therefore at the root of the process known as baroclinic instability, whose manifestation is the vertically coupled cyclone development illustrated in Figure 7.7.

High winds and heavy snow and rain were present to the north and west of the warm front in this case. As the system continued to intensify, the warm

174

front moved westward and subsequently *southward* from Canada, with much colder air at the surface west and south of the low center than north of it. Consequently, regions south and southwest of the cyclone received heavy snow, along with high winds, while rain and freezing rain fell north and northwest of the low center. Perhaps the most remarkable aspect of the advection pattern is that warm advection occurred *southwest* of the surface cyclone after the low reached the Canadian border later on 26 January! Evidently, large-scale ascent was taking place in a quadrant where there is normally descent.

Recall from Chapter 3 that height falls or rises at a given level must be commensurate with a decrease or an increase in column density (warming or cooling) of the total air column above that level. Equations 7.4a and b, however, make no reference to this thermodynamic consideration. Yet it is clear that the air column above the location of the 500 mb trough axis in Figure 7.7 must have undergone rapid warming during the explosive cyclogenesis phase. Similarly, the geopotential height falls at 1000 mb at the location of the surface low must be accounted for by a column warming at upper levels, as our theory does not permit geostrophic temperature advection over the surface low in the 1000–500 mb layer. This issue will be resolved in Chapter 10.

The deficiency in the barotropic forecasts

Versions of the barotropic model have been used continuously by the United States Weather Service since the inception of operational numerical weather prediction in the 1950s. Although the models are no longer in service, barotropic forecasts had been able to compete with the forecasts based on the complete equations of motion, the so-called primitive equation models, because the 500 mb barotropic model forecasts were almost as accurate and faster executing than more complex models. It is useful to consider the error pattern for the barotropic model because it underscores the developmental effects at 500 mb.

Barotropic forecasts were found to suffer from characteristic errors, most notably the ones associated with cyclogenesis. Maximum error occurs in the region of the 500 mb trough, where there is a change of tilt in the wave from northeast–southwest to northwest–southeast. Trough displacement in barotropic forecasts tended to be slow on the equatorward side of the 500 mb vorticity maximum. Poleward of the latitude of the vorticity maximum the trough is predicted to move too rapidly.

Problems

7.1 Rapid deepening of a trough at 500 mb occurred between 0000 GMT 2 October 1978 and 0000 GMT 4 October 1978. Trough development was accompanied by a rapid southward shift in the 500 mb trough and vorticity maximum with time (cf. Figs 7.8a & b). This southward displacement is not predicted by equivalent barotropic steering, as discussed in Chapter 6, but by developmental processes associated with the production of absolute vorticity at 500 mb, during the two-day period. The vorticity and height patterns as predicted by the U.S. Weather Service barotropic model were considerably in error, as can be seen in Figure 7.8c.

    (a) Outline on a blank map the region where the 500 mb development term in the two-level baroclinic model is appreciable and positive (i.e. where the contours of the relative thermal vorticity advection

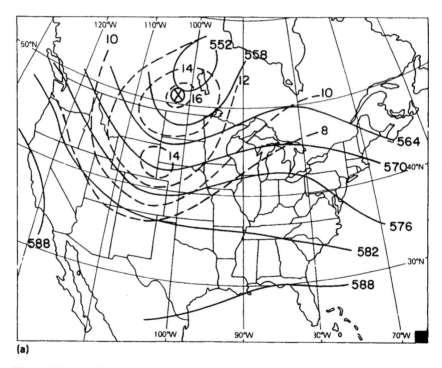

(a)

Figure 7.8 (a) The 500 mb absolute vorticity (broken curves at intervals of $2 \times 10^{-5} \, s^{-1}$, greater than or equal to $8 \times 10^{-5} \, s^{-1}$) and geopotential height contours (labeled in dam) for 1200 GMT 2 October 1978. The principal vorticity center is marked with a circled cross. The blackened square at lower right denotes the size of a unit advection solenoid.

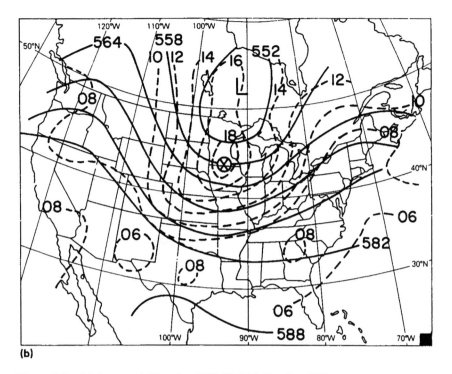

Figure 7.8 (b) Same as (a) but for 1200 GMT 3 October 1978.

by the thermal wind in Figure 7.8d form advection solenoids for positive vorticity). At the location of maximum advection compute the 500 mb vorticity tendency due to the development term (in units of s^{-1} per day; 1 day $\approx 10^5$ s). The development term is written:

$$- 0.2\, V_T \cdot \nabla_p \zeta_T = \left(\frac{\Delta \zeta_s}{\Delta t}\right)_{development}$$

(Note that a unit advection solenoidal area is provided by the blackened one-degree latitude box in the lower right-hand corner.) Examine the pattern of errors in the 24 h barotropic forecast of height and vorticity shown on Figure 7.8c. Assess whether the pattern of $- V_T \cdot \nabla_p \zeta_T$ as diagnosed from inspection of Figure 7.8d conforms *approximately* in extent to that of the 500 mb vorticity error pattern. (Realize, of course, that the errors arise over the period of a 24 h forecast and are not necessarily

177

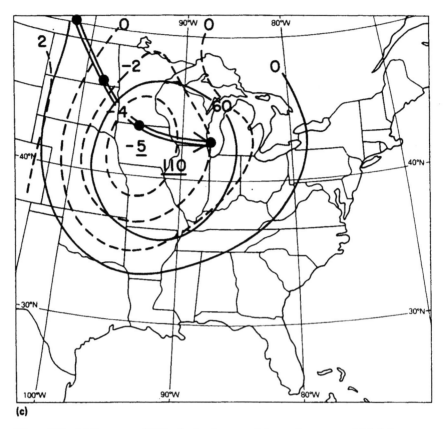

(c)

Figure 7.8 (c) Patterns of 500 mb absolute vorticity and geopotential height errors for 24 h barotropic forecast ending at 1200 GMT 3 October 1978. Contours are at intervals of $2 \times 10^{-5}s^{-1}$ and in meters, respectively, associated with maximum errors of $5 \times 10^{-5}s^{-1}$ and 110 m. Filled circles denote 12 h intervals along the path of the principal vorticity center (following the bold-faced arrow) between 1200 GMT 2 October and 0000 GMT 4 October 1978.

equivalent to those diagnosed at one instant of time.) Does the observed maximum vorticity error in the 24 h forecast (5 units in 24 h) agree approximately with your computed 24 h vorticity tendency using the above expression? Note that you can use the calculation of unit vorticity advection, referred to in Chapter 2, to determine thermal vorticity advection by the thermal wind; however, these vorticity isopleths are drawn at intervals of 4 (rather than 2) units, so that a unit advection solenoid here is equivalent to twice that for a contour interval of 2 units.

(b)   Examine the sea-level pressure and thickness chart (Fig. 7.8e) and

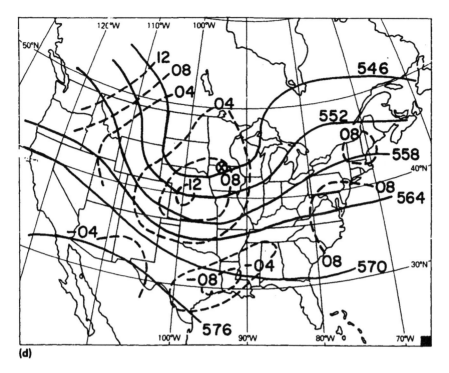

Figure 7.8 (d) The 1000–500 mb thickness contours (full curves labeled in dam) and relative geostrophic thermal vorticity of the thickness field (broken curves labeled at intervals of $4 \times 10^{-5} \mathrm{s}^{-1}$) for 3 October 1978. The location of the absolute vorticity maximum at this time is indicated by a circled cross.

outline on the blank map the area encompassing solenoids of cold advection. In the area of maximum thickness advection (smallest advection solenoid) compute the local thickness tendency (in units of meters per day) and the local developmental 500 mb height tendency associated with the thermal advection using the empirical relationship.

$$ - 0.2\,V_0 \cdot \nabla_p h = \left(\frac{\Delta Z_5}{\Delta t} \right)_{\text{development}} $$

You can use a previous value of a unit advection solenoid as in (a). Compare the area of strong thermal advection with the area of significant 500 mb barotropic height error for the 24 h forecast shown in Figure 7.8c and discuss whether the magnitude (110 m in 24 h) and extent of the barotropic forecast error pattern can be *roughly* explained by the distribution of cold thickness advection.

179

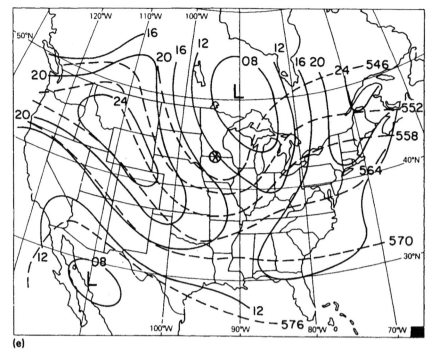

(e)

Figure 7.8 (e) Sea-level pressure (full curves are isobars labeled in mb; 1000s unit omitted) and 1000–500 mb thickness contours (broken curves labeled in dam) for 1200 GMT 3 October 1978. The location of the 500 mb absolute vorticity maximum is shown by a circled cross.

Do you see any other places over the continental United States where significant 500 mb trough development may be taking place? Why is development at 500 mb continuously taking place *south* of the absolute vorticity maximum?

Further reading

Haltiner, G. J. 1971. *Numerical weather prediction*. New York: Wiley.
Haltiner, G. J. and F. L. Martin 1957. *Dynamical and physical meteorology*. New York: McGraw-Hill.

8

Alternative expressions for vertical motion and divergence

The appeal of the quasi-geostrophic omega equation is that it provides a practical means for evaluating vertical motion and development using conventional meteorological charts. Forcing and response are neatly segregated on either side of the equation, with the former consisting of two terms that represent conceptually simple processes that can be evaluated solely from analyses of the geopotential heights.

The quasi-geostrophic omega equation has become so familiar, however, that it is sometimes forgotten that the terms in this equation are mathematical artifices. Indeed, it is possible to derive alternative relationships expressing vertical motion and divergence in terms of advections of vorticity or temperature at 500 mb or other levels. For instance, the Sutcliffe or Trenberth relationships represent further simplifications of quasi-geostrophic equations without great loss of accuracy. In addition, the so-called Q-vector formulation, which is exact in the context of quasi-geostrophic theory on the constant f plane, demonstrates that the two terms in the omega equation can be combined into one term. It is clear that vorticity and temperature advection are not inseparable quantities but are framework-dependent. We will show in Chapter 12 that vertical motion can be expressed solely in terms of temperature advection in relative-wind isentropic coordinates because the (implicit) vorticity advection vanishes in the transformation from absolute to relative-wind coordinates.

All of these formulations, including the quasi-geostrophic omega equation, represent expressions of atmospheric processes with simplifications. We now consider some of these other approaches to diagnosing vertical motion, not only because an awareness of these other formulations allows one to gain a better appreciation for the broader aspects of quasi-geostrophic theory, but because these other expressions provide additional useful insights into how weather patterns move and develop. The purpose of this chapter is partly historical: the source of modern development theory, the

Sutcliffe model (and more modern variations), is examined within the context of models presented in previous chapters.

8.1 Sutcliffe development theorem

R. Sutcliffe was the first to describe development in terms of a vertical distribution of divergence, specifically that at the surface, and readily measurable quantities at 1000 and 500 mb. He realized that cyclogenesis had to be accompanied by low-level convergence (slightly overcompensated by upper-level divergence). The 500 mb pressure surface represented a level of non-divergence. Sutcliffe, after some justification, arrived at the simple quasi-geostrophic version of the vorticity equation given by equation (3.6a). He applied this equation at two levels, 1000 and 500 mb, subtracting them to get

$$-f_0(V_p \cdot V_s - V_p \cdot V_0) = \partial(\zeta_s - \zeta_0)/\partial t + V_s \cdot V_p(\zeta_s + f) - V_0 \cdot V_p(\zeta_0 + f) \quad (8.1)$$

where subscript g, for geostrophic quantities, is understood in this derivation. The Laplacian of the height is equated with relative vorticity as in (2.8), so that

$$(\zeta_s - \zeta_0) = \frac{g}{f_0} \, V_p^2(Z_s - Z_0) = \frac{g}{f_0} \, V_p^2 h. \quad (8.2)$$

Next, Sutcliffe assumed that thickness changes only as the result of advection and he proposed a truncated version of the thickness tendency equation (3.16b) in which the vertical motion and diabatic terms are omitted, leaving

$$\partial h/\partial t \approx -\bar{V} \cdot V_p h \quad (8.3)$$

where the overbar refers to a mean quantity over the 1000–500 mb layer. Combining (8.2) and (8.3) and expanding the Laplacian operator in terms of the thickness advection yields

$$\frac{g}{f_0} \, V_p^2 \frac{\partial h}{\partial t} = -\frac{g}{f_0} \, V_p^2(\bar{V} \cdot V_p h) = -\left(\frac{\partial^2}{\partial x^2} + \frac{\partial^2}{\partial y^2}\right)(\bar{u} v_T - \bar{v} u_T) \quad (8.4a)$$

where

$$u_T = -\frac{g}{f_0} \frac{\partial h}{\partial y} \qquad \text{and} \qquad v_T = -\frac{g}{f_0} \frac{\partial h}{\partial x}.$$

Sutcliffe contended that terms representing the geostrophic deformation (those which contain products of derivatives, e.g. $(\partial \bar{u}/\partial x)/(\partial v_T/\partial x))$ are likely to be small except in the vicinity of fronts, where they may even be dominant. Therefore, given (8.4a) and (8.2),

$$\frac{g}{f_0} \nabla_p^2(\bar{V} \cdot \nabla_p h) = \left(\bar{u}\frac{\partial}{\partial x} + \bar{v}\frac{\partial}{\partial y}\right)\zeta_T + \left(\bar{u}\frac{\partial}{\partial y} - \bar{v}\frac{\partial}{\partial x}\right)(\nabla_p \cdot V_T)$$

$$- \left(u_T\frac{\partial}{\partial x} + v_T\frac{\partial}{\partial y}\right)\bar{\zeta} - \left(u_T\frac{\partial}{\partial y} - v_T\frac{\partial}{\partial x}\right)\nabla_p \cdot \bar{V}. \qquad (8.4b)$$

Sutcliffe neglected the divergence terms ($\nabla_p \cdot V_T$; $\nabla_p \cdot \bar{V}$) in (8.4b) in comparison with the vorticity terms (ζ_T; ζ), consistent with scale analysis. He also allows the advecting wind, which is the mean in the layer, to be that at 1000 mb. Recall that, for parallel isotherms at all levels within a layer (e.g. 1000–500 mb), the thickness advection by one geostrophic wind is identical to the thickness advection by any other wind velocity in the layer, including the vertically averaged wind in the layer. This assumption is consistent with the development of the linearized omega equation (Ch. 4) and the model of baroclinic development at 500 mb (Ch. 7). Given this simplification, we find that

$$-\frac{\partial}{\partial t}(\zeta_5 - \zeta_0) = \frac{-g}{f_0}\nabla_p^2\frac{\partial h}{\partial t} = V_0 \cdot \nabla_p \zeta_5 - V_5 \cdot \nabla_p \zeta_0. \qquad (8.5)$$

Sutcliffe then substituted this expression in the equation for divergence (8.1) to obtain

$$f_0(\nabla_p \cdot V_5 - \nabla_p \cdot V_0) = -[V_5 \cdot \nabla_p(\zeta_0 + \zeta_5 + f) - V_0 \cdot \nabla_p(\zeta_0 + \zeta_5 + f)]$$

$$= -[(V_5 - V_0) \cdot \nabla_p(\zeta_0 + \zeta_5 + f)]. \qquad (8.6)$$

Letting $V_T = V_5 - V_0$ (the thermal wind) and $\zeta_T = \zeta_5 - \zeta_0$ (the thermal vorticity), and assuming that the divergence at 500 mb is nearly zero (being close to the level of non-divergence), Sutcliffe arrives at the following diagnostic expression for the 1000 mb divergence:

$$-\nabla_p \cdot V_0 = -\frac{2}{f_0}V_T \cdot \nabla_p \zeta_0 - \frac{1}{f_0}V_T \cdot \nabla_p \zeta_T - \frac{1}{f_0}V_T \cdot \nabla_p f. \qquad (8.7)$$

The first term on the right-hand side of (8.7) represents a steering effect. It means that vertical wind shear (horizontal temperature gradient) over a surface maximum of vorticity (manifested by thickness lines crossing the feature) is associated with cyclonic development ahead of the feature and anticyclonic development behind it, resulting in a displacement of the feature in the direction of V_T. Thus, in a broad sense, cyclones move at any given instant in the direction of the thickness lines over

183

the center of the low (warm air to the right of the motion vector). This point is made in Chapter 4 with regard to Figure 4.3. Although the thickness lines are also changing with time, observations do suggest that cyclones tend to migrate along the locus of maximum thickness gradients. For example, cyclones tend to move in a direction somewhat parallel to the thickness lines along the warm front. Sutcliffe felt that the first term on the right-hand side of (8.7) controls the storm's motion, but the second term controls the development of the vorticity over the center of the cyclone, while also influencing the cyclone's motion. The form of the second term is precisely that which emerges in the two-parameter baroclinic model and can be interpreted in the same manner as discussed in the first part of Chapter 7. The second term operates to create low-level convergence and spin up of cyclonic vorticity over the surface low if the thermal trough lags behind the surface low. The last term, the beta effect, is relatively small, and Sutcliffe could not isolate any development solely arising from the configuration of this term. The Sutcliffe development equation is satisfactory for diagnosing surface development, although it neglects the vertical motion term ($\omega \tilde{s}$) (and the diabatic term) in the thickness tendency equation (8.3).

Steering has been an important concept in synoptic meteorology and the Sutcliffe development equation (published just after the end of the Second World War) was a revolutionary advancement in understanding cyclogenesis. Implicit in (8.7) is the idea that a cyclone is not carried like a bubble in a stream, but that the pressure field is reconstituted continuously by the divergence field. At the same time, the divergence field is being fashioned by forcing associated with vorticity and thermal advections. Therefore, lows move from one point to another because the surface pressure pattern is being continuously reconstituted due to forcing. Thus, pressure falls occur ahead of the low and rises behind it. Although it appears that an entity is moving in the direction of the 500 mb winds and is thus being steered by the upper flow, surface pressure features move not as entities but are more properly conceived as holes or hills in the pressure pattern; these features move as the result of local variations in time imposed by upper forcing. Thus, a surface low moves because it is continuously being filled up behind and deepened ahead of its center due to the divergence and convergence patterns associated with the geostrophic advections at higher levels.

Recall that confluence or diffluence is the squeezing together or pulling apart of streamlines without a decrease or increase in the area enclosed by a ring of fluid, such as occurs in convergence or divergence. Sutcliffe examined a variety of different configurations in the geopotential height and thickness patterns including those of diffluent and confluent ridges and troughs. He was able to show that some patterns at 500 mb are much more favorable for development than others. In a previous chapter it was shown that diffluent troughs tend to maximize the gradients of vorticity along the direction of flow and therefore are more favorable for cyclogenesis. Sutcliffe showed that the larger the development term in (8.7), the more diffluent the trough. In practice, however, unless the vorticity and thickness fields are unavailable, it is more efficient to examine the solenoidal fields in determining patterns of synoptic-scale forcing, rather than to assess the degree of diffluence in the geopotential height contours at 500 mb.

8.2 Petterssen's development equation

Sutcliffe's pioneering work during the late 1940s established the foundation for quantitative weather prediction and provided a framework for interpreting the behavior of weather patterns with the aid of conventional maps. Following Sutcliffe, a variety of similar interpretive and predictive models were formulated for use at both 500 and 1000 mb. One of these, which governs the vorticity tendency at 1000 mb, was derived by S. Petterssen in the early 1950s. Petterssen made the same assumptions that have been made in this text: that the divergence at 500 mb is negligible compared to that at the surface, and the surface vorticity advection is negligible compared to that at 500 mb. Petterssen begins by letting

$$\frac{\partial \zeta_s}{\partial t} = - V_s \cdot \nabla_p (\zeta_s + f) = \frac{\partial \zeta_0}{\partial t} + \frac{\partial \zeta_T}{\partial t} \tag{8.8}$$

where ζ_T is the thermal vorticity for the 1000–500 mb layer. The next step is to use the relationship

$$\frac{\partial \zeta_T}{\partial t} = \frac{g}{f_0} \nabla_p^2 \frac{\partial h}{\partial t} \tag{8.9}$$

where h is the 1000–500 mb thickness, and substitute for $\partial h / \partial t$ the vertically averaged 1000–500 mb thickness equation (3.16b). Thus,

$$\frac{\partial \zeta_0}{\partial t} = - V_0 \cdot \nabla_p (\zeta_0 + f) - \frac{g}{f_0} [\nabla_p^2 (\bar{A}_T + \bar{S} + \bar{H})] \tag{8.10}$$

where

$$\bar{A}_T \equiv - V_0 \cdot \nabla_p h \qquad \bar{S} \equiv \frac{R_d \bar{s} \Delta p_s}{g \bar{p}} \bar{\omega} \qquad \bar{H} \equiv \frac{R_d \bar{s} \Delta p_s}{g c_p \bar{p}} \dot{\bar{Q}}_{db}.$$

As in the Sutcliffe notation, terms with the overbar signify an average over the 1000–500 mb layer. (As in Chapter 4, we let $\overline{\omega s} = \bar{\omega} \bar{s}$.) Here, all wind

185

quantities are assumed to be geostrophic (subscript g omitted). Evaluation of this expression involves the Laplacian of three terms: the vertical motion, thickness advection and diabatic heating fields. The equation is not very convenient to apply to conventional weather maps, but it is of some historical interest in that it has been used extensively as a classroom tool.

8.3   The Trenberth approximation

Some models of 1000 mb vorticity tendency developed during the 1950s were fairly elaborate, employing approximations and empirically derived coefficients, as in the case of the equivalent barotropic and two-parameter models discussed in Chapters 6 and 7. The purpose of such models was to make operational predictions in situations where there was no access to high-speed computers or to the output of Weather Service prediction models. These models could be solved graphically by manipulation of the 500 and 1000 mb height fields. More recently, alternative forms of the quasi-geostrophic omega equation have been derived, which combine the forcing advections on the right-hand side of the omega equation (4.3) in such a way as to permit a more rigorous (but nevertheless straightforward) diagnosis of the vertical motions using conventional weather charts. Two of these derivations are now considered, that of Trenberth and (in Ch. 14) that of Hoskins and his co-workers.

Interpretation of the omega equation in the form presented in Chapter 4 (equation (4.3)) requires knowledge of two forcing functions and several map fields; two are for geopotential height, at least one for vorticity and one for thickness. Since geopotential height and temperature are related hydrostatically, and since the geostrophic advecting winds are also transformations of the geopotential height or thickness patterns, it seems self-evident that terms on the right-hand side of (4.3) can be combined into fewer terms or at least arranged differently, as has been done by K. Trenberth.

Consider (4.3) expressed as

$$\left(\frac{R_d}{pg} \tilde{s} \nabla_p^2 + \frac{f_0^2}{g} \frac{\partial^2}{\partial p^2} \right) \omega = F_1 + F_2 \tag{8.11a}$$

where

$$F_1 = f_0 \frac{\partial}{\partial p} \left(u_g \frac{\partial \zeta_g}{\partial x} + v_g \frac{\partial \zeta_g}{\partial y} + v_g \frac{\partial f}{\partial y} \right) \tag{8.11b}$$

and

$$F_2 = - \nabla_p^2 \left(u_g \frac{\partial}{\partial x} \frac{\partial Z}{\partial p} + v_g \frac{\partial}{\partial y} \frac{\partial Z}{\partial p} \right). \tag{8.11c}$$

Substitution of the Laplacian relationship (2.8) for the relative vorticity in F_1 yields

$$F_1 = f_0 \frac{\partial}{\partial p} \left[\frac{g}{f_0} \left(u_g \nabla_p^2 \frac{\partial Z}{\partial x} + v_g \nabla_p^2 \frac{\partial Z}{\partial y} \right) + v_g \frac{\partial f}{\partial y} \right]. \quad (8.11d)$$

Employing the geostrophic relationships $u_g = -(g/f_0)(\partial Z/\partial y)$ and $v_g = (g/f_0)(\partial Z/\partial x)$ for terms containing $\partial Z/\partial x$ and $\partial Z/\partial y$, the term F_1 becomes

$$F_1 = f_0 \overbrace{\left(\frac{\partial u_g}{\partial p} \nabla_p^2 v_g - \frac{\partial v_g}{\partial p} \nabla_p^2 u_g \right)}^{A} + f_0 \overbrace{\left(\frac{\partial v_g}{\partial p} \frac{\partial f}{\partial y} \right)}^{C}$$

$$+ f_0 \underbrace{\left(u_g \nabla_p^2 \frac{\partial v_g}{\partial p} - v_g \nabla_p^2 \frac{\partial u_g}{\partial p} \right)}_{B} \quad (8.11e)$$

$$\equiv A + C + B \quad (8.11f)$$

For convenience the three groups of terms in (8.11e) are labeled A, B and C. In F_2 the expansion involves the Laplacian of products. Noting that

$$\nabla^2(AB) = B \nabla^2 A + A \nabla^2 B + 2 \left(\frac{\partial A}{\partial x} \frac{\partial B}{\partial x} + \frac{\partial A}{\partial y} \frac{\partial A}{\partial y} \right)$$

F_2 can be expanded, after first substituting the geostrophic wind relationships, to yield

$$F_2 = f_0 \overbrace{\left(\nabla_p^2 v_g \frac{\partial u_g}{\partial p} - \nabla_p^2 u_g \frac{\partial v_g}{\partial p} \right)}^{A} - f_0 \overbrace{\left(u_g \nabla_p^2 \frac{\partial v_g}{\partial p} - v_g \nabla_p^2 \frac{\partial u_g}{\partial p} \right)}^{B}$$

$$- 2 f_0 \underbrace{\left(\frac{\partial u_g}{\partial x} \frac{\partial}{\partial x} \frac{\partial v_g}{\partial p} + \frac{\partial u_g}{\partial y} \frac{\partial}{\partial y} \frac{\partial v_g}{\partial p} - \frac{\partial v_g}{\partial x} \frac{\partial}{\partial x} \frac{\partial u_g}{\partial p} - \frac{\partial v_g}{\partial y} \frac{\partial}{\partial y} \frac{\partial u_g}{\partial p} \right)}_{\Lambda} \quad (8.11g)$$

$$\equiv A - B - 2\Lambda. \quad (8.11h)$$

The term F_2, therefore, can be separated into its components; using continuity

$$\frac{\partial u_g}{\partial x} + \frac{\partial v_g}{\partial y} = 0.$$

We can see that two of the three bracketed terms in (8.11g), terms A and B, are identical to those in (8.11e). Summing F_1 and F_2, term B cancels to yield the following expression for the right-hand side of the omega equation:

$$F_1 + F_2 = 2f_0 \left(\nabla_p^2 v_g \frac{\partial u_g}{\partial p} - \frac{\partial v_g}{\partial p} \nabla_p^2 u_g \right) + f_0 \frac{\partial v_g}{\partial p} \beta - 2\Lambda. \qquad (8.12a)$$

Here Λ, the deformation term, is defined as

$$2\Lambda = 2f_0 \left(\frac{\partial u_g}{\partial x} \frac{\partial}{\partial x} \frac{\partial v_g}{\partial p} + \frac{\partial u_g}{\partial y} \frac{\partial}{\partial y} \frac{\partial v_g}{\partial p} - \frac{\partial v_g}{\partial x} \frac{\partial}{\partial x} \frac{\partial u_g}{\partial p} - \frac{\partial v_g}{\partial y} \frac{\partial}{\partial y} \frac{\partial u_g}{\partial p} \right). \qquad (8.12b)$$

Given that the divergence of the geostrophic wind is zero, then

$$\Lambda = \tfrac{1}{2} \left(X \frac{\partial Y}{\partial p} - Y \frac{\partial X}{\partial p} \right) \qquad (8.12c)$$

where X and Y are deformation terms defined as

$$X = \frac{\partial u_g}{\partial x} - \frac{\partial v_g}{\partial y} \qquad \text{and} \qquad Y = \frac{\partial v_g}{\partial x} + \frac{\partial u_g}{\partial y}. \qquad (8.12d)$$

Cancellation allows the omega equation to reduce to a more compact expression upon summing F_1 and F_2. As in the Sutcliffe derivation, the deformation terms arise when the Laplacian of products is taken. Trenberth similarly argues that these terms tend to be large only in the vicinity of strong jets or frontal systems and, therefore, are unlikely to be important below 500 mb except on subsynoptic scales and near frontal boundaries. The term $f_0 \beta \, \partial v_g / \partial p$ corresponds to the beta effect and is relatively small.

Without much loss of accuracy it is, therefore, possible to neglect the deformation term and largely to ignore the beta term. Arbitrarily adding the beta term to the right-hand side of (8.12a) and setting $\Lambda = 0$ allows one to obtain a closed expression for $F_1 + F_2$, which is

$$F_1 + F_2 = 2f_0 \left(\frac{\partial \zeta_g}{\partial x} \frac{\partial u_g}{\partial p} + \frac{\partial \zeta_g}{\partial y} \frac{\partial v_g}{\partial p} + \beta \frac{\partial v_g}{\partial p} \right)$$

$$= -\frac{2f_0}{\Delta p_5} V_T \cdot \vec{\nabla}_p (\zeta_g + f) \qquad (8.13)$$

where $\Delta p_5 = 500$ mb.

The latter form of (8.13) is in keeping with previous simplifications in which finite differences are substituted for vertical derivatives. Here, the thermal wind can represent the advecting wind in the layer between 1000 and 500 mb (the geostrophic wind of the thickness field) and the vorticity advected by the thermal wind is at a mean level between 1000 and 500 mb, say 700 mb. Equation (8.13) is a crude approximation to the Trenberth formulation that is convenient to use in conjunction with conventional products; one needs only the 700 mb absolute vorticity and the 1000–500 mb thickness patterns to deduce the approximate sign and magnitude of $-\omega_{d7}$. The latter is assessed using the advection solenoids formed by the advection of 700 mb geostrophic vorticity with the 1000–500 mb thermal wind field, upward motion being associated with positive vorticity advection.

The result (8.13) is similar to the development equation of Sutcliffe (8.7). Sutcliffe found the thermal wind appropriate for assessing steering of surface systems. Equation (8.13) allows one to diagnose forcing in ambiguous situations where positive (negative) vorticity advection is being offset by negative (positive) thermal advections. The equation is also suitable for evaluating the sign and relative magnitude of the vertical motions by inspection of the solenoids of geostrophic vorticity and thickness.

In this approximation to the omega equation, the evaluation of vertical motion is analogous to the evaluation of the vorticity advection term in the original omega equation. In the original version, positive vorticity advection by the thermal wind corresponds to upward motion, and vice versa for sinking motion. This can be seen more simply in terms of the linearized omega equation (4.6) and (4.7). Thus, the right-hand side of (8.13) can be written in the notation of (4.7), assuming sinusoidal patterns of thickness and vorticity, as

$$-\omega_{d7} = -2a[V_T \cdot \nabla_p(\zeta_{g7} + f)] \qquad (8.14)$$

where the coefficient a, defined in Chapter 4, is

$$a = f_0/(g\Delta p_s D).$$

The ambiguity in the omega equation resulting in F_1 and F_2 having opposing signs can be resolved by the use of the Trenberth approximation (8.14), as shown in Figure 8.1. This case is identical to that of Figure 4.1 in which positive vorticity advection and cold air advection coexist just to the west of the surface cold front (the location of the circled cross in Figures 4.1 and 8.1). The conventional omega equation (4.3) presents us with an ambiguity in determining the sign of ω_d. It can be seen that the advection of 700 mb geostrophic vorticity by the 1000–500 mb thermal wind is clearly negative in the region where the signs of the quasi-geostrophic forcing terms (F_1 and F_2) possess opposite signs. Moreover, the disturbed weather east of the low, and the pressure falls along the warm front and in the warm sector, are suggested by positive advections of ($\zeta_{g7} + f$) by V_T.

Since negative (positive) thickness advection and positive (negative) vorticity advection sometimes cancel, it is useful to combine F_1 and F_2 in a more unambiguous form. Nevertheless, situations where the two forcing terms in

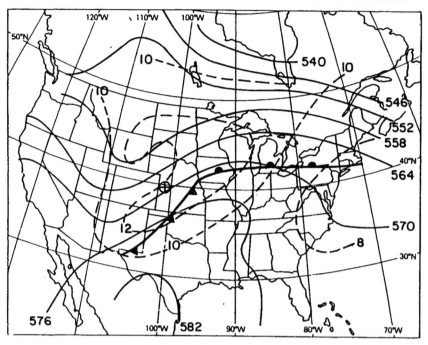

Figure 8.1 The 1000–500 mb thickness lines (full contours labeled in dam) and 700 mb absolute geostrophic vorticity (broken contours labeled at intervals of 2×10^{-5} s^{-1}) for 0000 GMT 11 September 1986. The circled cross is the location of strong cold air advections and a region where the two quasi-geostrophic forcing terms in (4.3) possess opposite signs (see Fig. 4.1). Surface fronts are shown in conventional notation.

the omega equation possess opposite signs are relatively infrequent. In (8.11), term B cancels regardless of the magnitude of the advections in the omega equation. In general, however, terms A and B are of approximately equal magnitude, but the latter possesses opposite signs in F_1 and F_2.

The infrequent occurrence of opposing contributions by the geostrophic forcing terms to the vertical motion indicates that cold air advection is usually coupled with descent and warm air advection with ascent. Indirectly, this association reflects a correlation between warm air advection and ascent, and cold air advection and descent. It also means that thickness contours generally move slower than the advecting wind, C_* being less than 1.0 in the empirical equation (3.19). Exceptions to this rule may be important, however, as in the case of weakening cold fronts in which broad ascent begins to take place over the cold air in association with secondary development along the front. In such instances cold advection is accompanied by a broad area of cyclonic vorticity advection, ascent and cloudiness on the cold side of the front.

8.4 Mathematical unity of quasi-geostrophic forcing

While it is tempting to attach differing physical processes to the two quasi-geostrophic forcing terms, their mathematical separation is a matter of convenience. Sutcliffe arranged the forcing of low-level divergence in terms of three different terms, one of which includes the surface (1000 mb) relative vorticity field. Trenberth demonstrates that thermal and vorticity effects can be incorporated in a single forcing term. Separation between vorticity advection and temperature advection is not independent of the coordinate system. Chapter 12 shows that vertical motion can be ascribed entirely to temperature advection in a relative coordinate system moving with the particular feature of interest, the wave, cyclone or front.

A compact but mathematically rigorous derivation of the omega equation has been derived by B. Hoskins and M. Pedder in terms of Q-vectors. In the Q-vector notation the two forcing terms in the quasi-geostrophic omega equation are combined in a single quantity, a two-dimensional vector, which makes no explicit reference to advections. A complete description of the Q-vector, its derivation and applications, is presented in Chapters 14 and 15.

Problems

8.1 In terms of the Sutcliffe development equation (8.7), what is the sign of the vertical motion implied by the surface vorticity advection by the thermal wind term over the center of a developing cyclone (ζ_0 maximum)? How about with regard to the thermal wind advection of absolute 700 mb vorticity in the Trenberth approximation (8.14) for the use of surface cyclone development with a phase lag between surface and 700 mb?

8.2 (a) Use the Sutcliffe development equation (8.7) to justify the rule of thumb that surface cyclones tend to move along the warm front just poleward of its surface position.

(b) Using the Trenberth approximation, determine the sign of the vertical motion and the surface pressure tendency in the warm sector of the cyclone shown in Figure 8.1.

Further reading

Hoskins, B. J., I. Draghici and H. C. Davies 1978. A new look at the ω-equation. *Q. J. R. Met. Soc.* **104**, 31–8.
Hoskins, B. J. and M. A. Pedder 1980. The diagnosis of middle latitude synoptic development. *Q. J. R. Met. Soc.* **106**, 707–19.

Petterssen, S. 1956. *Weather analysis and forecasting. II.* New York: McGraw-Hill.
Sutcliffe, R. C. 1947. A contribution to the problem of development. *Q. J. R. Met. Soc.* **73**, 370–83.
Trenberth, K. E. 1978. On the interpretation of the diagnostic quasi-geostrophic omega equation. *Mon. Wea. Rev.* **106**, 131–7.

9

Some additional dynamic aspects of the baroclinic wave/cyclone: effects of friction, terrain and diabatic heating

Until now we have been concerned primarily with dynamically forced vertical motion (ω_d). Recall that we define ω_d as the vertical motion that would be obtained from a solution of (4.3), the quasi-geostrophic omega equation with no lower boundary effects (friction and orography) or diabatic heating. In the absence of these effects, the synoptic-scale vertical motion profile tends to resemble the bowstring model discussed in Chapter 2. We now consider these three non-quasi-geostrophic components of the vertical motion profile not treated in the diagnosis of the omega equation in Chapter 4. Although they represent deviations from the quasi-geostrophic formulations, friction, orography and diabatic heating constitute important influences on cyclogenesis.

9.1 The role of friction in cyclogenesis and cloud formation

Even a casual observer is struck by the fact that surface winds do not blow parallel to the surface isobars but across isobars toward lower pressure (Fig. 3.1a). As we showed in Chapter 3, frictionally induced convergence in areas where there is positive relative vorticity in the planetary boundary layer (generally regions of low pressure) contributes to a positive vorticity tendency that tends to cancel the opposing effect of frictional dissipation.

Condensation occurs where frictionally induced ascent within a moist planetary boundary layer leads to saturation of the air. In regions where the atmosphere is conditionally and latently unstable, this cloud takes the form of cumulus convection.

193

At middle latitudes, however, frictional convergence and divergence usually degrade the existing vorticity pattern and sometimes form low-level stratiform cloud. A balance between friction, pressure gradient and Coriolis forces necessitates a cross-isobaric flow. Figure 9.1 shows an air parcel moving with wind speed V, and experiencing a Coriolis force (COR), a pressure gradient force (PGF) and a frictional force (F) in the planetary boundary layer. Friction acts opposite to the direction of motion; the Coriolis and pressure gradient forces act, respectively, to the right (left) of the wind motion in the Northern (Southern) Hemisphere and toward lower pressure. The frictional component opposite to the direction of the wind velocity exactly balances the component of pressure gradient force along the direction of motion, while the Coriolis force is exactly balanced by the component of pressure gradient force normal to the wind velocity. In the absence of accelerations of forces other than these three, the velocity vector is at some angle α with the isobars, about 30° at the surface (anemometer) level for typical surface wind speeds.

The result of frictional convergence or divergence is the production of a vertical motion component, which we call $\omega_F(p)$ and whose maximum value, at the top of the planetary friction layer ($p = p_f$), is ω_f. Figure 9.2 shows that frictional convergence or divergence leads to a vertical motion profile with a maximum at the top of the friction layer (a distance Δp_f above ground) and a decay of the vertical motion above. Convergence in the boundary layer is almost exactly balanced by divergence in the layer above p_f, a form of Dines' compensation.

The change in sign of the divergence at p_f implies that the vorticity

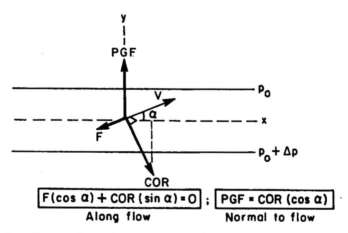

Figure 9.1 Force balance in unaccelerated flow between the pressure gradient (PGF), Coriolis (COR) and friction (F) forces (bold-faced vectors) for an air parcel moving with speed V at an angle α with the isobars.

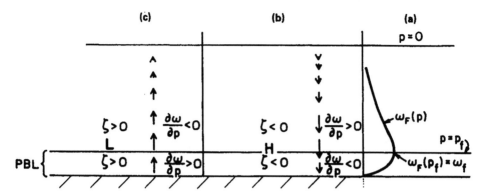

Figure 9.2 (a) Schematic vertical motion profile for frictionally induced vertical motion ($\omega_F(p)$). (b) Descent (shown by arrows proportional to magnitude of vertical motion and by vertical profile for $\omega_F(p) > 0$) occurs with negative relative vorticity in the planetary boundary layer. (c) Ascent occurs with positive relative vorticity in the planetary boundary layer. The depth of the friction layer is denoted by the symbol PBL whose top is at p_f where $\omega_F(p_f) = \omega_f$.

tendency due to frictional divergence changes sign at this level. Positive relative vorticity in the planetary boundary layer (a surface low) leads to a negative vorticity tendency due to the compensating divergence above p_f in the region of indirectly forced ascent. The result of this "secondary spin down" is a degradation of the vorticity field above the friction layer (see Ch. 3), which damps rapidly with height, however. Near the level of p_f, the frictionally forced vertical motion may constitute a dominant component in the total vertical motion and differ in sign from the dynamically forced vertical motion (ω_d).

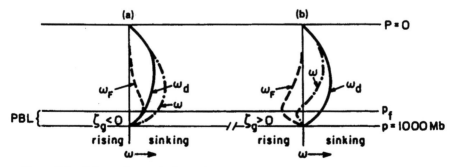

Figure 9.3 Schematic vertical motion profiles for dynamically forced vertical descent (ω_d), that forced by friction (ω_F) and the total vertical motion (ω): (a) negative relative vorticity in the planetary boundary layer ($\zeta_0 < 0$); (b) positive relative vorticity in the planetary boundary layer ($\zeta_0 > 0$).

Let us consider some schematic examples. The two cases illustrated in Figure 9.3 show frictionally forced vertical motion ($\omega_F(p)$) superposed on a profile of dynamically forced vertical motion (ω_d). The total vertical motion $\omega(p)$ is

$$\omega(p) = \omega_d(p) + \omega_F(p)$$

which consists of dynamic and frictional vertical motion components.

Figure 9.3a pertains to relative anticyclonic vorticity and Figure 9.3b to cyclonic vorticity in the planetary boundary layer (PBL). In the former, synoptic-scale descent is reinforced by frictionally forced descent. Since the maximum of frictional descent is at the level p_f, the effect of the friction in this example is to increase the anticyclonic vorticity tendency within the boundary layer. Conversely, positive vorticity in the boundary layer results in a shallow layer of upward motion near the top of the boundary layer

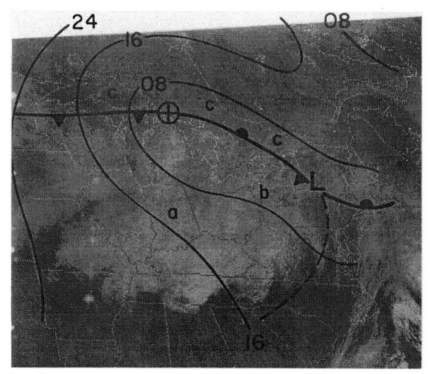

Figure 9.4 Visible-wavelength GOES satellite image for 1631 GMT 7 January 1985, showing a mass of dense cloud associated with sustained precipitation (labeled c) and low stratiform cloud associated with frictionally forced ascent. This cloud is rather shallow near a, but at b there are light showers and radar echoes with tops reaching 4.5 km.

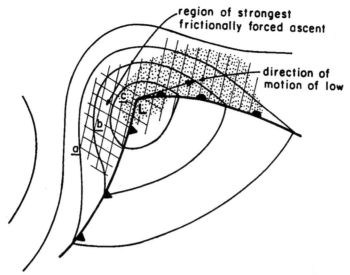

Figure 9.5 Schematic illustration of cyclone with surface isobars and fronts. Shading denotes continuous precipitation, vertical hatching is low-level cloud produced primarily by dynamically forced ascent, and cross-hatching is low-level cloud produced by frictionally forced ascent or both dynamically and frictionally forced cloud. The arrow indicates the direction of motion of the low. The letters a, b and c are referred to in the text.

where the magnitude of dynamically forced subsidence is less than that of ascent due to frictional forcing. Convergence occurs within the PBL and frictional divergence (contributing to the dynamic component) in the layer above. The example provided in Figure 9.3b suggests the reason why layers of stratus tend to be confined to the region of cyclonically curved sea-level pressure isobars and is most prevalent to the west of developing cyclones, since that is where surface relative vorticity is maximized and the surface wind speeds are strongest.

Frictionally forced ascent is often marked by an extensive layer of low, shallow stratiform clouds, whose top appears on satellite infrared photographs to be rather uniform and at a level typically below 700 mb (3 km). Frictionally forced cloud is visible on the satellite photograph in Figure 9.4. Highly reflecting cloud is present southwest of a large low-pressure system in this figure. In the vicinity of point a in the figure, the cloud consisted of stratus and stratocumulus with a base of about 500 m. There were no radar echoes in this region and only scattered reports of light precipitation. Near point b there were very weak radar echoes reaching 4.5 km and a small area of continuous precipitation. North of the warm front there was continuous precipitation (mostly in the form of snow) along a cloud band that was being sustained by dynamically forced vertical motion. Infrared images, however, show little contrast between the ground and stratus because the top of the stratus was at a temperature not much lower than that of the surrounding ground surface.

Figure 9.5 shows a combination of dynamically and frictionally forced low cloud cover near point c; at b the dynamically forced vertical motion is downward and there is no middle- and upper-layer cloud, which is a visible indicator of large-scale dynamically forced descent marked by overcast stratiform cloud in the boundary layer. Precipitation in this area, except in certain circumstances involving terrain effects or adjacent bodies of water, tends to be showery and its duration is brief. Rain or snow amounts are usually light, except in isolated squalls that occur where the air is convectively unstable. Near a, the cloud consists of stratocumulus and fair-weather cumulus, revealing patches of bright, cloudless sky through breaks in the stratus. This figure illustrates the importance of frictional convergence, which offsets the dynamically forced divergence at low levels. The stratiform cloud vanishes where either the frictionally forced ascent is absent or the dynamically forced descent offsets the effect of frictionally forced ascent and produces a net descending motion within the planetary boundary layer.

The frictionally forced cloud west of the cyclone center tends to expand with increasing size of the developing cyclone. This cloud formation is aided by the vertical flux and turbulent mixing of moisture that normally occur to the west of a cyclone. For a layer of cloud to form as the result of friction, turbulent vertical moisture flux from the surface to the level of saturation is also necessary. Upward flux of mixture is enhanced by the presence of relatively warm, moist ground below colder, drier air, which is advected to the west of a cyclone. This process is illustrated in Figure 9.6. The strength of

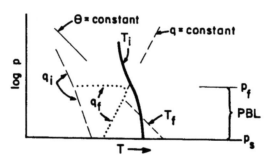

Figure 9.6 Schematic temperature and dew-point distribution on a skew T–$\log p$ diagram for the case when an initially dry and descending air stream becomes modified by moving over a warmer and relatively moist surface. Sensible heat flux from the surface mixes the air over the depth of the planetary mixing layer (PBL) whose top is at the pressure of p_f. The initial and final (after mixing) temperature (T) and specific humidity (q) distributions are labeled with the subscripts i (initial) and f (final).

the frictionally forced vertical motion is maximized in the cross-hatched area of Figure 9.5.

Other types of friction

There are other types of frictional effects not expressed by the type of formulae presented in Chapters 1 and 3. These equations govern the fluxes of momentum through the planetary surface layer to the ground, where its kinetic energy is dissipated. The wind stress (τ) constitutes the vertical flux of momentum; τ/ρ is defined as the covariance of the turbulent u or v component of the wind and the turbulent component of the vertical velocity (w). (This is $-\overline{u'w'}$ and $-\overline{v'w'}$, where overbars represent time averages (Ch. 1).)

In general we may consider that τ represents a loss of kinetic energy by the synoptic scales to the subsynoptic scales, the smallest of which is the molecular scale in the form of heat energy. In the quasi-geostrophic system, friction effects a loss of kinetic energy from the synoptic scales, which are governed by quasi-geostrophic theory (i.e. cyclones and waves), to that of smaller scales, which are governed by turbulent exchanges.

Above the planetary boundary layer, surface-layer similarity laws are inapplicable, although τ may still be regarded as a measure of the local dissipation of synoptic-scale kinetic energy to smaller scales. Thus, τ can be non-zero although its vertical derivative may be either positive or negative. Some studies show that the kinetic energy dissipation term (5.37b), which expresses the rate of dissipation of synoptic-scale kinetic energy, sometimes acts to create kinetic energy locally, particularly at upper levels in the atmosphere.

Turbulent dissipation due to cumulus convection is referred to as "cumulus friction". Cumulus friction may be rather important in tropical disturbances, although at middle latitudes, where cumulus convection plays a more peripheral role in atmospheric energetics, it is not a major factor in wave/cyclone development. To date, there have been very few studies to illuminate the role of cumulus friction in mid-latitude disturbances.

Besides cumulus convection, a cause of turbulent dissipation above the planetary boundary layer is due to the breakdown of strongly sheared flows. Since this type of turbulence may occur in the absence of cloud, it is given the name "clear-cut turbulence". Clear-air turbulence tends to occur where the mechanical effects of shear exceed the restoring effects of static stability (buoyancy) in a statically stable environment. More precisely, the breakdown of vertically sheared flows into turbulence can take place when the Richardson number $Ri = (g/\theta)(\partial\theta/\partial z)/(\partial V/\partial z)^2$ is less than or equal to about 0.25. The dominating effect is the vertical wind shear, which is squared in the denominator of this expression. In practice, however, the critical value of Ri

is larger than 0.25, because the fine-scale structure of the vertical wind shear is not captured by conventional rawinsonde measurements. A more realistic threshold seems to be $Ri \approx 1.0$. Regions of strong vertical wind shear are often those in which intermittent episodes of clear-air turbulence may occur. It is unlikely that this source of synoptic-scale energy dissipation figures importantly in the growth or decay of waves and cyclones, but it is certainly a primary reason why conservative properties such as potential temperature or potential vorticity do not remain constant along material trajectories, even in clear air.

Differential friction

Under some circumstances, the horizontal variation of frictional drag can also create cyclonic or anticyclonic vorticity over a limited region. Consider Figure 9.7, which is a schematic illustration of a high-pressure system located along a coastline. Normally there is less surface drag over the ocean than over the land, and parcels moving westward cross isobars toward lower pressure at a sharper angle (see Fig. 9.1) over land than over the ocean. The result is a wind shift and a cyclonically curved trajectory along the coastline. In this case, friction acts to produce a zone of relative cyclonic vorticity (and therefore frictional convergence) along the coastline and to advect cooler continental air more rapidly equatorward over the land, thereby enhancing the land–sea contrast and the frontal nature of the air along the land–sea boundary. Differential friction advection is of some importance in certain

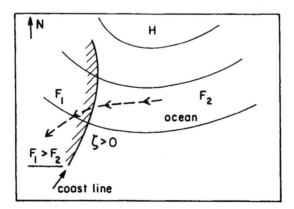

Figure 9.7 Schematic illustration of differential frictional effect on air parcel (broken trajectory) moving from ocean to land. The frictional drag force over the ocean (F_2) is less than that (F_1) over land, so that a region of relative cyclonic vorticity ($\zeta > 0$) is created at the surface along the coast. Full contours are sea-level isobars.

areas, such as along the east coast of the United States where so-called "cold air damming" by the mountains constitutes an important modifying factor in coastal cyclogenesis. The regional consequences of such mesoscale modifications to the large-scale flow can be important for cyclogenesis, but they lie outside the scope of this text.

9.2 Terrain-forced vertical motions: the effects of orography

Troughs and cyclones frequently form in the lee of large mountain chains when there is strong downslope motion on the synoptic scale. Synoptic-scale, downslope winds produce the spin up of the vorticity that is responsible for the lee-side trough or cyclone. Locally, downslope motion is associated with dry warm winds, which are referred to in the United States as "snow eaters", chinook or foehn winds. At synoptic scales, the warm downslope motion is consistent with development of a temperature ridge in the lee of the mountains. The amplitude of this ridge tends to decrease with height, and is normally very weak at 500 mb.

In the presence of sloping terrain, the vertical motion at the surface (ω_s) may no longer be negligible; specifically, $w_s (\partial p / \partial z)_s$ is non-zero in (2.26a). As a result, the lower boundary condition for the quasi-geostrophic omega equation is non-zero and the solution to (4.3) must contain an additional term involving w_s.

Theory suggests that, if the width of the mountain range has synoptic-scale dimensions and the Rossby number for the flow is less than about 1, the *effective* vertical motion at the surface (w_s) can be expressed approximately as $V_s \cdot \nabla_p H_s$, where H_s is the height of a smoothed terrain. Let us define the vertical profile of synoptic-scale orographic vertical motion as $\omega_M(p)$ and that at the surface as ω_m. In the equivalent barotropic model, the decay with height of the terrain-forced vertical motion, expressed by the function $N(p)$ in Chapter 6, is relatively gradual and monotonic. Although the actual profile is likely to be more complex, especially on the smaller scales, the large-scale vertical motion profile tends to resemble that in Figure 9.8a.

This simplified view of the flow over sloping terrain is shown by the schematic picture of a column of air moving up a sloping surface. In the quasi-geostrophic system, the vertical motion at the surface over flat terrain is given by $\omega_s \approx (\partial p / \partial t)_s$. Over sloping terrain

$$\omega_s = (\partial p / \partial t)_s - \rho g V_s \cdot \nabla_p H_s$$

where the last term is simply ω_m.

Because the top of the column remains at $p = 0$, the column shrinks vertically from below and expands horizontally while ascending; con-

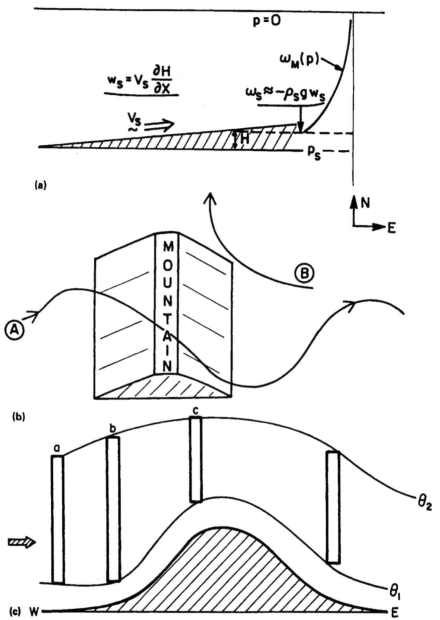

Figure 9.8 (a) Illustration of vertical shrinking of air column as it moves up a mountain of slope $\partial H/\partial x$ with horizontal velocity V_s. A schematic vertical motion profile is shown whose surface value is $\omega_s = \omega_m\,(= -\rho_s g w_s)$, the vertical component of V_s. (b) Airflow over a synoptic-scale mountain ridge for westerly (A) and easterly (B) trajectories (isobaric projection). (c) Schematic illustration of air columns (vertical rectangles) moving with a westerly wind (bold-faced arrow to left) over an idealized mountain between two potential temperature surfaces (θ_1 and θ_2; thin full curves). The letters a, b and c refer to locations of air columns (see text).

sequently $\partial\omega/\partial p$ within the column is negative. In the absence of dynamically forced vertical motion, the shrinking column implies, through the development term in the vorticity equation, a negative vorticity tendency. A solution to the omega equation in the presence of terrain-induced vertical motions is obtained by including

$$(\partial p/\partial t)_s - r^2\rho g V_0 \cdot \nabla_p H_s = \omega_s$$

as a lower boundary condition. (Recall that the magnitude of the surface pressure tendency, expressed by $(\partial p/\partial t)_s$, is generally quite small compared to that of ω at 700 mb.)

In general, the total vertical motion due to dynamic and terrain effects is

$$\omega(p) = \omega_d(p) + \omega_M(p).$$

If we consider the effect of terrain slope alone on the generation of vorticity, it is evident that the effect of vertically stretching and shrinking vertical columns results in a spin up or down of vorticity. This stretching or shrinking can be stated in terms of a *potential vorticity* conservation theorem for adiabatic flow, which is

$$(\zeta + f)\frac{\partial\theta}{\partial p} = \text{constant} = (\zeta + f)\frac{\Delta\theta}{\Delta p} \tag{9.1}$$

where Δp is the depth of the column between a pair of isentropic surfaces. Equation (9.1) follows directly from the quasi-geostrophic vorticity equation (3.6a).

Imagine a column of air approaching a north–south mountain chain from the west. In the simplest sense, the movement of a column of air would resemble Figure 9.8a, ascending the mountain slope, shrinking vertically and expanding laterally (diverging). From (3.6a) or (9.1) we conclude that this air column would acquire anticyclonic relative vorticity and therefore begin a clockwise rotation in ascending the mountain. There is some evidence to suggest that isentropic surfaces along the cross-mountain flow do not exactly follow the terrain; rather, they broaden with altitude, while the amplitude of the perturbation diminishes. Consequently, as shown in Figure 9.8c, a column of air ascending the mountain between two potential temperature surfaces would first expand vertically and contract laterally (converge) between locations a and b.

This model of how columns expand slightly before ascending the mountain seems to avoid the issue of what happens to the entire column between the surface and some upper, undisturbed isentropic surface. Provided that the wavelength of the potential temperature perturbation continues to broaden with height, however, all air columns would first experience some vertical stretching before undergoing vertical compression while crossing the mountain. As the results of this lateral convergence, the air parcels would acquire cyclonic rotation and briefly follow a poleward path, as indicated in Figure 9.8b.

Subsequent to this initial vertical expansion and horizontal convergence, the air column contracts vertically and diverges horizontally, as illustrated in Figure 9.8c between locations b and c. Accordingly, the column first acquires cyclonic vorticity and starts to execute a cyclonic path, as shown in Figure 9.8b. The ascending (diverging) column then acquires anticyclonic vorticity and executes an anticyclonic curve such that the trajectory forms a ridge on the windward side of the mountain crest (Fig. 9.8b). In gaining anticyclonic rotation, the air turns equatorward in the manner shown in the figure. Subsequently, as the air descends the leeward slope, the column expands vertically, contracts horizontally (converges) and begins to gain cyclonic vorticity.

As it moves equatorward, the air parcel experiences a decrease in its Coriolis parameter and an increase in its relative vorticity. The result is that the column arrives equatorward of its initial latitude on the lee-side of the mountain, thereby acquiring a net gain in relative cyclonic vorticity as the result of crossing the mountain. Because of the acquired cyclonic curvature, the parcel turns poleward (analogous to the constant absolute vorticity trajectory described in Chapter 6), loses relative cyclonic vorticity and curvature and arrives at its original latitude. The parcel continues to move poleward, gaining relative anticyclonic vorticity, and eventually moves equatorward. In principle, the parcel can execute a series of wave-like undulations of alternating troughs and ridges downstream. In the absence of the beta effect, no real wave trough would be formed along the mountain. Successive perturbations can propagate downstream from the site of the initial standing wave and resemble ordinary baroclinic waves. Usually only one lee trough is readily observed in conventional synoptic-scale data, but the trailing perturbations may nevertheless influence baroclinic development and precipitation some distance east of the lee slopes. Development of lee cyclones into major storms occurs after the lee trough has moved away from the mountains, rather than within the region of downslope motion.

For easterly flow ascending a mountain slope, the shrinking or diverging column develops an anticyclonic curvature, in accordance with (9.1), causing the column to move poleward. In so doing, the Coriolis parameter increases and the relative vorticity decreases still further, so the trajectory executes an anticyclonic path poleward on the east side of the mountain and may fail to cross it. In such instances a closed high-pressure cell may form on the east side of the mountain. Easterly upslope motion leading to the formation of an anticyclone on the windward side may contribute to a

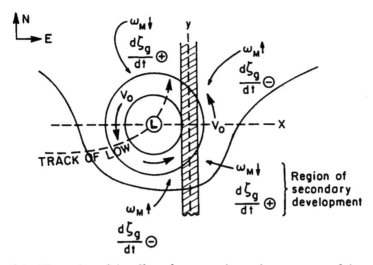

Figure 9.9 Illustration of the effect of a mountain on the movement of the parent cyclone and the development of a secondary cyclone in the lee of a mountain. The signs of the vertical motions and the vorticity tendencies are shown, respectively, by arrows next to the symbol ω_M and by positive and negative signs beside the symbols for vorticity tendency.

mesoscale distribution of winds and pressure referred to earlier as cold air damming. Variations of the large-scale perturbations described above occur with differing mountain configurations.

For a low- or high-pressure system approaching the mountain, the situation is slightly more complex than for uniform upslope flow. Consider the case where a low-pressure cell moves eastward toward a north–south mountain ridge, and assume the flow throughout the lower troposphere is that of the lower-tropospheric geostrophic wind. According to Figure 9.9 there are two quadrants where the air is ascending and two where it is descending. The vorticity equation indicates that the southeast and northwest quadrants (of a coordinate system with its y axis along the crest of the mountain and its x axis through the center of the low) experience descent and a positive vorticity tendency ($\omega_m > 0$ and $d(\zeta + f)/dt > 0$). Similarly, the northeast and southwest quadrants experience ascent and a negative vorticity tendency. Cyclonic vorticity tendency is created poleward of the low center on the west side of the mountains and also on the east side of the mountains equatorward of the low center. The effect of this arrangement on the pressure pattern and on the motion of cyclones (in the quasi-geostrophic framework) is that the parent low center migrates poleward along the windward side of the mountain ridge toward the region of terrain-forced spin up of vorticity, while secondary cyclone development begins southeast of the cyclone center in the lee of the mountains where the local vorticity tendency is positive. This secondary development is called *lee-side cyclogenesis*. As a result of the vorticity tendencies imposed by the secondary cyclone, the lee cyclone begins to migrate equatorward.

Descent in the lee of mountains may deviate greatly from the smoothed terrain, such as when cold air abuts the lee-side of mountains, causing descending air to flow over the wedge of stable air. This effect is particularly

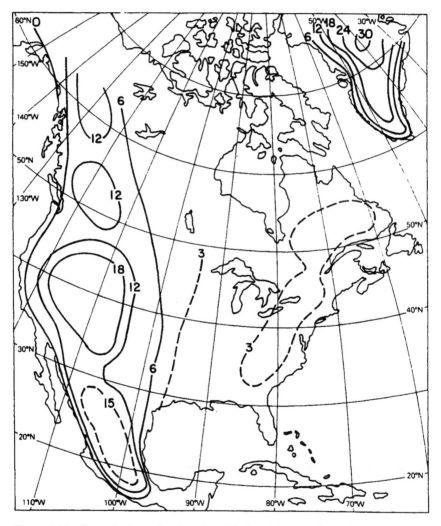

Figure 9.10 Smoothed terrain elevation (National Meteorological Center, United States Weather Service). (Elevation in hundreds of meters.)

noticeable in narrow valleys between mountain ridges, where cold air trapped in the valley may cause warmer air to flow over both mountain and valley. The trapped cold air effectively smoothes the terrain for the path of the warmer air, serving as a kind of plateau.

It is typical for intense cyclones to approach the Rocky Mountains from the west and move northward just west of the Pacific Ocean coastline without ever crossing the mountains. At the same time, a separate cyclone development (or trough formation) takes place east of the mountains and further south than the latitude of the approaching disturbance. This type of lee-side cyclogenesis involves a vigorous cross-mountain flow in the lower atmosphere (e.g. at 700 mb). Many case studies exist describing lee-side cyclogenesis over the Western Plains of the United States, over Europe in the lee of the Alps, over the Gulf of Genoa, east of the Appalachian Mountains of the United States and in the lee of the Palmer Peninsula of Antarctica.

Over North America, the western cordillera extends almost due north–south, as illustrated by the smoothed terrain map in Figure 9.10. Synoptic-scale quasi-geostrophic perturbations forced by flow over mountains are more closely related to such smoothed terrain than to the smaller-scale features. Thus, as a first approximation the Rocky Mountains resemble the idealized barrier in Figure 9.9.

Let us now consider an example of lee-side cyclogenesis. A deep cyclone had approached the west coast of North America (Fig. 9.11a). A weak lee-side trough is almost imperceptible at this time (1200 GMT 15 December 1981) just east of the parent cyclone on the lee-side of the mountain ridge. At 700 mb, a geopotential height ridge was centered over the mountains with strongest cross-mountain flow occurring just to the east and south of the parent cyclone. Maximum pressure falls (exceeding 3 mb $(3\,h)^{-1}$) were taking place in the lee of the mountains (Figure 9.10).

During the subsequent 24 h, the lee-side trough developed a well-defined center ESE of the parent low and moved (or skipped) rapidly southward in the lee of the mountains. Maximum pressure falls during the formative state in the lee cyclone coincided with the location of the strongest lee-side descent and cross-mountain component at 700 mb. (Note that this strong cross-mountain flow is also reflected in the 700 mb geostrophic winds.) By 1200 GMT 16 December (Fig. 9.11b), the central pressure of the lee-side low fell to 1006 mb, a change of about 12 mb in 24 h. Twelve hours earlier, surface pressure falls along the lee of the mountains were very large, approximately 12 mb in 12 h at the location of the filled circle in Figure 9.11b. No precipitation was reported in the vicinity of this secondary cyclone during this first 24 h of its lifetime.

Subsequently, the secondary low moved northeastward very rapidly (without much further deepening) in the direction of dynamically forced pressure falls, which were in excess of 3 mb $(3\,h)^{-1}$ ahead of the storm

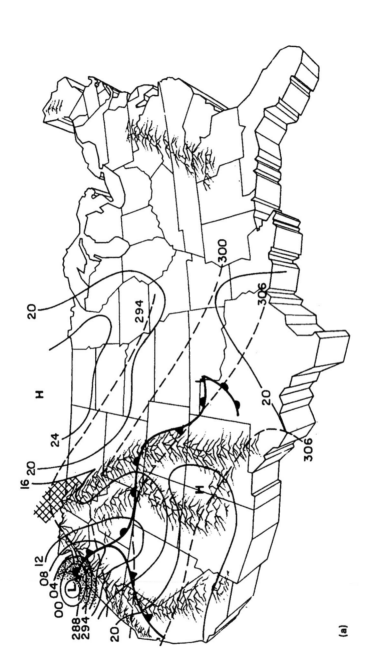

Figure 9.11 (a) Sea-level isobars (full curves labeled in mb above 1000 mb), 700 mb height contours (broken curves labeled in dam) and conventional frontal symbols for 1200 GMT 15 December 1981. Shading denotes regions of continuous precipitation and cross-hatching is where the local rate of pressure falls exceeds 3 mb $(3\ h)^{-1}$.

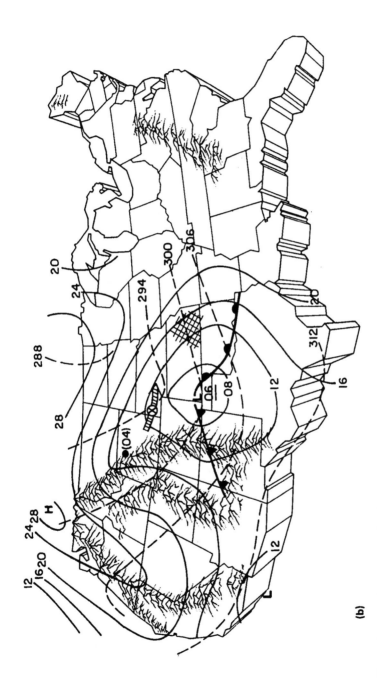

Figure 9.11 (b) Same as (a) but for 1200 GMT 16 December 1981. The filled circle east of the mountains marks the position and central pressure of the lee-side cyclone 12 h earlier (1004 mb at 000 GMT 16 December). The arrow through the circled cross denotes the location of the 500 mb absolute vorticity maximum and the direction of the 500 mb geostrophic winds.

(b)

(Fig. 9.11b). At the same time, the parent cyclone moved northward along the coast, remaining west of the mountains.

Such rapid southward movement of a lee cyclone during its formative stages is a very common occurrence over the lee slopes of the Rocky Mountains, especially during the winter and spring seasons. Figure 9.11 illustrates the sequence of lee-side cyclogenesis when a Pacific Ocean cyclone approaches the mountains from the west. (Frequently, however, weak, elongated troughs form in the lee of the Rocky Mountains in the absence of a parent cyclone to the west. These troughs tend to occur where the cross-mountain component of the 700 mb winds is relatively strong.) After a lee cyclone forms, the circulation around the secondary low modifies the geo-strophic flow over the mountains, favoring increased downslope motion on its southern side. Accordingly, a positive vorticity tendency occurs south of the developing low; in the absence of synoptic-scale vorticity or thermal advections, the track of the low is toward the south. The increase of the downslope motion south of the cyclone is reflected in Figure 9.11 by the southward shift in the strongest cross-mountain component at 700 mb. Note, however, that the center positive vorticity advection at 500 mb is not located near the cyclone, but well to its north in Figure 9.11b. Thus, initial formation and movement of the cyclone was dictated more by the configuration of downslope motion over the Rocky Mountains than by dynamically forced ascent associated with vorticity and temperature advections.

Many such dramatic examples of lee-side development take place east of the Rocky Mountains in the absence of appreciable dynamic forcing. A strong cross-mountain wind component alone, however, is not a sufficient cause of cyclogenesis. In general, cyclogenesis is favored where there is a strong cross-mountain component in the upper-tropospheric flow, as shown by the location of the 300 mb jet axis in Figure 9.12. The jet is situated where the atmosphere is most baroclinic (in the sense of providing solenoids of temperature and vorticity advections), but the distribution of cyclogenesis is strongly modulated by the static stability in the lee of the Rocky Mountains. Therefore, baroclinic forcing, cross-mountain winds and low static stability are necessary for cyclogenesis in that region. Typically, the lee-side cyclone migrates from the lee of the mountains with the arrival of a mobile short-wave trough and its attendant vorticity advection, which provides the forcing once the lee cyclone has been generated.

Lee cyclones function in a similar manner to ordinary cyclones, deriving their kinetic energy from the available potential energy of the baroclinic atmosphere. The wavelength of the amplifying lee disturbance is, according to some theories, proportional to the size of the mountain (height and width). The preferred wavelength for further cyclogenesis beyond the initial trough is dictated by baroclinic instability criteria and is independent of the mountain size. As suggested in Figure 9.12, low static stability, brought

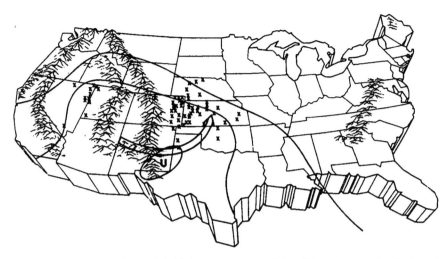

Figure 9.12 Location of initial occurrences of lee-side cyclogenesis during the springs of 1964–1968 (small crosses) and the mean position of the 300 mb wind speed maximum (bold-faced streamline). Thin full curves are contours of the mean spring-time static stability over the middle troposphere; the most unstable area is marked by the letter U. (After Hovanec and Horn, 1975.)

about by intense heating throughout a deep layer, such as that which occurs over the arid regions of the American southwest during spring, is also an important factor in the development of lee cyclones. (The influence of static stability on cyclogenesis is treated more thoroughly in other sections of this text.) The mountain's role is to reinforce the development potential of baroclinic instability in the lee of the mountains.

9.3 Diabatic forcing and convective heating

A desirable goal in making simple models of the atmosphere is to diagnose, using conventional weather maps, large-scale forcing. Diabatic forcing, however, is not easily assessed on the basis of quasi-geostrophic advection fields, although it can exert a profound effect on the vertical motions and pressure tendency. We have emphasized that a quasi-geostrophic approach excludes an explicit description of the complex and subsynoptic processes, such as cumulus convection. However, at present there is no single method for introducing the effects of diabatic heating in atmospheric prediction or simulation models. The subject remains one for vigorous debate.

Nevertheless, it is clear that many diabatic processes at middle latitudes are forced by (or intimately related to) synoptic-scale motions. In fact, diabatic heating constitutes an important forcing of development in mid-latitude cyclogenesis, although the latter is dominated by synoptic-scale baroclinic processes.

Let us first consider diabatic heating as an adjustment to the thermodynamic and quasi-geostrophic omega equations. In the temperature equation (3.7), the heating term is mathematically separated into two components: one is due to moist adiabatic ascent, accounted for by the moist static stability, and the other is due to all non-adiabatic processes. The latter consists of heating due to a variety of factors:

(a) heating due indirectly or directly to cumulus convection;
(b) heating due to sensible heat flux at the surface;
(c) heating due to solar and long-wave radiation absorption; and
(d) heating due to turbulent mixing.

Although some radiative warming is due to the absorption by atmospheric constituents of solar and long-wave radiation, the net radiation effect, *internally* in the atmosphere, is generally one of cooling. Radiational cooling is typically more pronounced in clear rather than in cloudy regions, varying between 1 and 2°C per day for deep layers. The effects of cumulus convection, turbulent mixing and surface sensible heating are extremely variable in time and space, however.

Retaining the diabatic heating term containing $\dot{Q}_{db}$ in (3.14) and deriving omega equation (4.3), one obtains

$$\left(\frac{R_d}{gp} \tilde{s} \nabla_p^2 \omega + \frac{f_0^2}{g} \frac{\partial^2 \omega}{\partial p^2} \right) = F_1 + F_2 - \frac{R_d}{gpc_p} \nabla_p^2 \dot{Q}_{db}$$

$$= F_1 + F_2 + F_3 \tag{9.2}$$

where F_1 and F_2 on the right-hand side represent the quasi-geostrophic advection terms discussed in Chapter 4, and F_3 is the forcing due to diabatic heating. In the absence of quasi-geostrophic forcing, the solution to (9.2) is $\omega(p) = \omega_{db}(p)$, where $\omega_{db}(p)$ is the vertical motion forced by all forms of diabatic heating. Thus, the general solution for the omega equation for forcing by quasi-geostrophic advection and by diabatic heating takes the form

$$\omega(p) = \omega_d(p) + \omega_{db}(p).$$

According to (9.2), where the Laplacian of the diabatic heating is positive, the vertical motion due to diabatic heating is upward. Assuming that the heating is introduced in some sinusoidal manner (as in (2.16)) and the wavelength of the heating is L_H, the simple arguments for linearizing the Laplacian lead to the result that

212

$$-(\omega_{db})_7 = \frac{R_d}{g\bar{p}c_p} \frac{8\pi^2}{L_H^2} \frac{\overline{\dot{Q}_{db}}}{D} \tag{9.3}$$

where D is defined in (4.7c) and $\overline{\dot{Q}_{db}}$ is the mean diabatic heating rate between 1000 and 500 mb. As in the case of quasi-geostrophic forcing expressed by (4.7), diabatic heating causes the magnitude of the vertical motion to be greater where the static stability is smaller. Recall from Chapter 5 (e.g. equation (5.10)) that the generation of eddy available potential energy is more efficient where diabatic heating (cooling) occurs in a relatively warm (cool) region.

Unlike the vertical profiles of dynamically forced vertical motion, or even those due to friction and orography for synoptic-scale motion, there is an infinite variety of possible vertical heating profiles for any of the four principal diabatic heating processes. Consider three vertical profiles of $\dot{Q}_{db}$ and of the vertical motion due to diabatic heating ($\omega_{db}(p)$) as shown in Figure 9.13; the three profiles represent (1) that due to sensible heat flux from the ground, (2) that due to large-scale moist adiabatic ascent or shallow rain-producing cumulus clouds, and (3) that due to large cumulonimbi. The

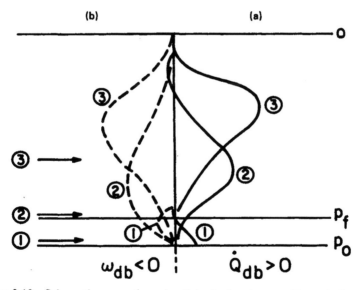

Figure 9.13 Schematic synoptic-scale diabatic heating profiles (a) for surface sensible heating in the boundary layer (below p_f), for latent-heat release by shallow convection or synoptic-scale moist adiabatic ascent and for deep convection, respectively labeled 1, 2 and 3. The corresponding vertical motion profiles induced by the heating are shown in (b); levels at which $\partial\omega/\partial p$ is a maximum are indicated at the left by arrows.

vertical motion profiles corresponding to the three heating distributions are shown on the left-hand side of the figure. Thus, vertical motion maxima can occur in the lower or higher troposphere as the result of $\dot{Q}_{db}$, thereby distorting the simple bowstring profile (Fig. 2.8), which pertains to the case of dynamically forced vertical motion by the quasi-geostrophic advections. In so doing, diabatic heating can profoundly alter the vertical derivative of ω (the divergence) and thereby affect the rate of vorticity change as expressed by the divergence term in the vorticity equation. In the absence of precipitation reaching the ground, there is no net diabatic heating and the convection constitutes a form of turbulent mixing in which total static energy in a column is redistributed without undergoing a net gain, although heating or cooling can occur at a given level as the result of the vertical mixing of potential temperature.

The role of cumulus convection in diabatic warming

Before discussing the consequences of diabatic heating and the profiles shown in Figure 9.13, it is instructive to examine the manner in which cumuli heat their surroundings. At first glance it is obvious that latent-heat release in convective clouds serves to warm the surroundings, at least in precipitating clouds. Most of the water vapor in the atmosphere is contained in the lowest 200 mb, there being very little moisture in the high troposphere where the air is relatively cold. Energy budget studies of convective systems show, however, that the effective warming by deep cumulus tends to resemble the deep heating profile of Figure 9.13, in which the maximum diabatic warming and its resultant vertical motion are maximized in the upper troposphere. One tends to develop a false impression from such profiles that the convective elements act as local radiators in which the heat is dispersed from the core of the cloud into the surroundings.

In fact, the warming produced by convection operates in a two-step process, which will now be described. The first step involves ascent in convective clouds. Cumulus clouds with active updrafts typically occupy a very small fraction of any region, no more than a few percent. Within the cores of large convective elements, air rises almost moist adiabatically from the lower to the upper troposphere with relatively little lateral mixing in so-called protected cores or "hot towers".

Let us imagine an array of deep cumulus towers in the warm sector of an extra-tropical cyclone, such as shown in Figure 9.14. Parts of the warm sector may be convectively and latently unstable and are therefore favorable for the deep cumulus convection, as is often the case within the boundary layer at low latitudes or ahead of cold fronts in baroclinic disturbances. Buoyancy drives the air upward within the clouds, which serve as efficient conductors of mass from lower to higher levels. The temperature within the cumulus, though no more than a few degrees above that of the environment,

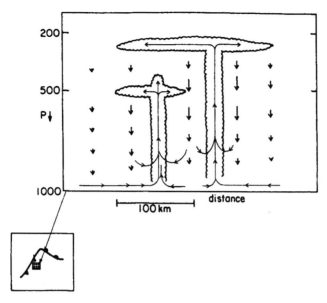

Figure 9.14 Diabatically forced vertical, lateral and convective motions (arrows and thin streamlines) in a convection region within the warm sector of a wave-cyclone (insert lower left).

constitutes sufficient buoyancy to produce rather large updraft speeds, particularly in the upper part of the cloud. The paradox is that ascending buoyant parcels within the cumulus, even those which follow saturated adiabats, do not come to rest and mix with the cloud environment until they reach their level of equilibrium; when they do, the virtual temperature of the cloud core is equal to that of the surroundings. Alternatively stated, $\tilde{s}_m$ is small in moist adiabatic ascent within deep cumulus clouds. Subsequently the cloud elements, after overshooting the equilibrium level, collapse, spread laterally and mix with their surroundings, whose temperature is approximately equal to that of the cloud.

Precipitation from convective clouds originates primarily at levels below 500 mb. Indirectly, diabatic warming, due to latent heating however, can occur over a deep layer above the level of maximum condensation rate. We can resolve this paradox by considering continuity within a region affected by the convection. Continuity requires that the ascent within the convective clouds be balanced by an equal amount of descent, which occurs in the cloud surroundings beneath the laterally expanding cirrus outflow (the "anvils") as depicted in Figure 9.14. The second step in the heating process is accomplished by adiabatic descent between clouds. It is therefore necessary in view of Figure 9.14 to realize that actively rising convective cells do not occupy more than a few percent of the area in disturbances, even those in the tropics. Since the profile of forced descent is similar to that of ascent,

215

the rate of compressional warming surrounding the active cloud towers tends to maximize in the middle and upper part of the cloud. Ultimately, however, the net warming due to convection is produced by condensation.

Convective clouds do not always achieve their maximum possible growth. Rather, there is a spectrum of cloud heights, which results in an arbitrary and complex diabatic heating profile over the depth of the convection layer. We have seen that $\partial \omega / \partial p$ is important in cyclogenesis. The vertical distribution of convective warming is crucial in producing the correct vertical motions and therefore the vorticity tendencies. Where convective warming produces ascent, a spin up of cyclonic vorticity is maximized below the level of maximum ascent, with a negative vorticity tendency above it. Thus, Figure 9.14 shows low-level inflow into the convection region, suggesting a positive vorticity tendency.

Synoptic-scale latent heating

Latent-heat release by deep cumulus clouds sometimes constitutes an important source of energy for developing mid-latitude cyclones, especially those that form in regions of relatively high θ_w, such as over the ocean.

Development of explosive cyclones and polar cyclones is intimately dependent on diabatic heating. In tropical systems such as hurricanes, convective cloud clusters (called mesoscale convective cloud complexes when they occur at middle latitudes) are agents of massive diabatic heating, which can be the dominant forcing. At mid-latitudes, convection occurs intermittently and serves as an adjunct to baroclinic development, although convective cloud clusters are sometimes found in weakly baroclinic summertime patterns. In fact, the weaker the baroclinicity, the more important is diabatic forcing relative to baroclinic forcing.

Conversely, sensible heating at the surface appears to exert a very minor direct influence on cyclogenesis, although the upward flux of latent and sensible heating near the surface, particularly over the ocean, is necessary in the long term in priming the system by raising the latent instability in the boundary-layer air. Where the radar coverage is complete, one finds that the distribution of deep radar echoes in vigorous cyclones is often confined to the warm sector and a region just poleward of the warm front. Diabatic warming, by increasing the synoptic-scale temperature gradients, notably in regions of existing baroclinicity, enhances baroclinically forced vertical motions by warming regions that are already relatively warm. *An important effect of diabatic heating is to create or increase the baroclinicity between the region warmed by diabatic heating and the surroundings not affected by the heating.* Alternatively stated, $\dot{Q}_{db}$ not only directly affects the profile of $\omega_{db}(p)$ via the forcing term F_3 in (9.2), but indirectly augments ω_d via the quasi-geostrophic forcing terms, F_1 and F_2. The importance of diabatic heating due to cumulus convection is that it can concentrate heating and temperature

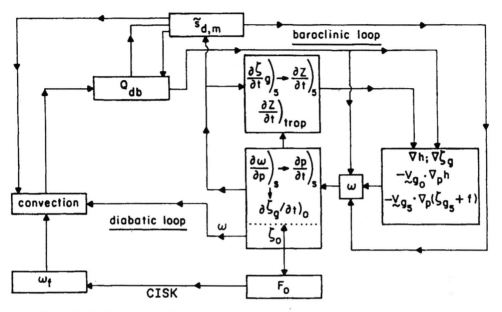

Figure 9.15 Flowchart illustrating a coupled baroclinc–diabatic system with various feedback circuits (see text).

gradients rapidly and on a relatively small scale.

Diabatic heating can be driven by frictional convergence in the planetary boundary layer or by dynamically forced vertical motions. Boundary-layer forcing of convective heating is coupled to the larger-scale circulation by a process referred to as the "conditional instability of the second kind" (CISK) mechanism. In CISK, latent heating responds to the vertical mass flux in the boundary layer and is driven by frictional convergence, i.e. by ω_f (Fig. 9.15). This feedback will only occur in an atmosphere that is potentially unstable. CISK differs from baroclinic instability in that it involves a feedback via frictional convergence (in regions of cyclonic relative vorticity in the boundary layer) and diabatically forced vertical motion, diabatic heating and surface pressure tendency.

A simple baroclinic model of convective latent heating

Modification of the quasi-geostrophic vertical motion by diabatic heating depends on static stability and on large-scale forcing. Conditionally unstable (lapse rate between dry and moist adiabatic) saturated ascent implies that the moist static stability is negative, violating the condition for solution of the quasi-geostrophic version of the omega equation. If one considers only deep layers (e.g. 1000–500 mb), however, with spatially averaged static stability appropriate to a synoptic-scale average, the moist static stability at middle

217

latitudes will almost always be positive. Its value, however, may approach zero, enabling both baroclinic and diabatic effects to be much more efficient in forcing synoptic-scale vertical motions than in a highly stable atmosphere.

We have shown that, for moist adiabatic ascent, the moist static stability ($\tilde{s}_m$) is the appropriate stability to use in the thermodynamic and omega equations, whereas $\tilde{s}_d$ governs a dry ascending or descending atmosphere. Quasi-geostrophic theory demands, however, a spatially constant value of static stability in the omega equation. An acceptable expedient for our discussion is to imagine that the atmosphere consists of subregions within which the static stability, whether moist or dry, does not vary horizontally. Since moist static stability is generally less than dry static stability, cyclogenesis development should take place much more rapidly in regions of saturation or where the efficiency of forcing is rather more than in dry regions.

In a statically stable atmosphere, convection occurs in small-scale regions where the atmosphere is potentially unstable. The driving mechanism at middle latitudes for diabatic heating is the large-scale forcing, which will generally produce heating in regions of upward motion and cooling in regions of clear skies and descent.

Let us now consider a simple model in which the diabatic heating is periodic and forced by large-scale dynamically forced vertical motion. In this framework $\dot{Q}_{db}$ is positive where ω_d is upward and negative where ω_d is downward. Thus, cloudy regions correspond to diabatic heating by convection and clear regions to a net radiative cooling. If the vertical motion (mass flux) due to synoptic-scale forcing is directly proportional to the diabatic heating rate, then we are free to suggest that

$$\dot{Q}_{db} = -\mu c_p \tilde{s} \omega_d(x, y, p) \tag{9.4}$$

where μ is an arbitrary proportionality factor, which lies between 0 and 1.0, and $\tilde{s}$ is the dry or moist static stability. The dynamically forced vertical motion in (9.4) is prescribed by the sinusoidal function as in Chapter 4. The parameter μ depends on the effectiveness of the diabatic heating but probably not directly on the magnitude of the quasi-geostrophic forcing.

There are two problems with this type of formulation: besides the arbitrary nature of the proportionality factor (1), equation (9.4) specified that $\dot{Q}_{db}$ increases with increasing static stability, which goes somewhat counter to observation. Another compromise in the physical reasoning is that the periodicity is symmetric, i.e. the size of the ascent region is equal to that of descent. Chapter 12 will show that ascent in saturated areas is more intense and occurs over a much smaller area than descent in clear areas. Nevertheless, this type of model has been used in some diagnostic studies, with the result that the physical ambiguities in the condition imposed by (9.4) do not seriously affect the conclusions. In any case, it is not our intention to treat (9.4) and its consequence as exact solutions.

It is simple to show that inclusion of (9.4) in the temperature tendency equation (3.14) results in the diabatic heating term combining with $\tilde{s}_d\omega$, such that, with the inclusion of diabatic heating, $\tilde{s}$ is replaced in (4.3) (and in the linearized forms (4.7) and (4.11)) by a scaled static stability $\tilde{s}'$, where $\tilde{s}' = \tilde{s}(1 - \mu)$.

The concept of a static stability scaled by $1 - \mu$ illustrates that systematic diabatic heating driven by large-scale dynamically forced vertical motion tends to generate a response similar to a reduction in the static stability of the system. The smaller the value of s', the more intense the vertical motions for a given advection pattern. Thus, diabatic heating acts to destabilize the atmosphere with regard to large-scale baroclinic processes. Baroclinic and adiabatic processes, therefore, tend to work in consort at middle latitudes, with one enhancing the other. Simultaneous operation and mutual feedback between the two processes is illustrated in Figure 9.15.

Problems

9.1 Assuming that the sea-surface temperature is equal to the surface air temperature and that the latter is equal to the mean potential temperature in the surface mixing layer of depth Z, calculates the horizontal variation in 1000–500 mb thickness that would be implied by a horizontal variation of 10°C in sea-surface temperature. Determine this horizontal thickness range for a variation between a deep mixing layer ($Z = 3$ km) and a shallow mixing layer ($Z = 1$ km). Assume that the mixing layer lapse rate is isentropic and that the temperature profile is continuous in the vertical. Note that 1 km equals approximately 100 mb in depth and that the lower boundary is at 1000 mb; ignore the virtual temperature correction.

9.2 Air rises over a mountain of elevation 1 km along a slope 1000 km in length. The horizontal air speed is 10 m s^{-1}. What is the value of ω_s along this slope. Referring to Table 6.1, what is the value of $\omega_M(p)$ at 500 mb in the barotropic model? Using the values of ω_s and ω_s, what are the average values of divergence and of vorticity tendency over the bottom 500 mb of the atmosphere due to orographic motion?

Further reading

Buzzi, A. and A. Speranza 1986. A theory of deep cyclogenesis in the lee of the Alps. Part II: Effects of finite topographic slope and height. *J. Atmos. Sci* **43**, 2826–37.

Buzzi, A. and S. Tibaldi 1978. Cyclogenesis in the lee of the Alps: a case study. *Q. J. R. Met. Soc.* **104**, 271–87.

Chen, S.-J. and L. Dell'Osso 1987. A numerical case study of east Asian coastal cyclogenesis. *Mon. Wea. Rev.* **115**, 477–87.

Chung, Y.-S., K. D. Hage and E. R. Reinhelt 1976. On lee-side cyclogenesis in the Canadian Rocky Mountains and the East Asian mountains. *Mon. Wea. Rev.* **104**, 879–91.

Danard, M. B. 1964. On the influence of released latent heat on cyclone development. *J. Appl. Meteor.* **3**, 27–37.

Danard, M. B. 1966. On the contribution of released latent heat to changes in available potential energy. *J. Appl. Meteor.* **5**, 81–4.

Holton, J. R. 1979. *An introduction to dynamic meteorology*, 2nd edn (Int. Geophys. Ser. Vol. 23). New York: Academic Press.

Hovanec, R. D. and L. H. Horn 1975. Static stability and the 300 mb isotach field in the Colorado cyclogenetic area. *Mon. Wea. Rev.* **103**, 628–38.

McClain, E. P. 1960. Some effects of the western cordillera of North America on cyclone activity. *J. Meteor.* **17**, 104–15.

Newton, C. W. 1956. Mechanism of circulation change during a lee cyclogenesis. *J. Meteor* **13**, 528–39.

Petterssen, S. 1956. *Weather analysis and forecasting II*. New York: McGraw-Hill.

Smith, R. B. 1984. A theory of lee cyclogenesis. *J. Atmos. Sci* **41**, 1159–68.

Smith, R. B. 1986. Further development of a theory of lee cyclogenesis. *J. Atmos. Sci.* **43**, 1582–602.

Tibaldi, S., A. Buzzi and P. Malguzzi 1980. Orographically induced cyclogenesis: analysis of numerical experiments. *Mon. Wea. Rev.* **108**, 1302–14.

Tracton, M. S. 1973. The role of cumulus convection in the development of extra tropical cyclones. *Mon. Wea. Rev.* **101**, 573–93.

10

The evolution of cyclones

Let us now consider the evolution of a wave/cyclone from the inception of a weak perturbation at the surface until the disturbance begins to decay. We begin with a schematic portrayal of the disturbance in four successive stages of development, not unlike the sequence of events portrayed in the classic Norwegian cyclone model. A unique aspect of this chapter is its restatement of the Norwegian cyclone model in terms of changes occurring at all levels in the troposphere and in light of the quantitative principles discussed in earlier chapters. A brief historical review of the Norwegian cyclone model now follows. (For a more detailed discussion of the history of cyclone theory, the reader is referred to an excellent treatise on the subject by Kutzbach (1979).)

In 1921, J. Bjerknes (and later J. Bjerknes and H. Solberg 1926) published one of the first modern, extensive accounts of the structure and evolution of cyclones. Without the advantage of upper-air soundings, but with brilliant intuition, they were able to present an account of cyclogenesis that remains remarkably unaltered to this day. Their results are contained in two well-known figures, reproduced here in Figures 10.1 and 10.2 (Bjerknes and Solberg, 1926).

The map view in Figure 10.1 depicts the surface airflow around the cyclone, the location of frontal boundaries, the precipitation (shading) and the warm sector. The lower part of Figure 10.1 shows a cross section of a vertical slice through the cold and warm fronts and the various cloud forms: nimbus or nimbostratus (Ni and Ns) at lower levels in association with the precipitation, altostratus and altocumulus (As and Acu) at middle levels, and cirrus (Ci) at high levels. Ascent is shown in the warm air and descent in the cold air, with the principal ascending air stream originating in the warm sector of the storm and rising above the retreating cold air along the warm front. The upper part of Figure 10.1 presents a vertical cross section poleward of the low center, showing the elevated warm air stream and precipitating clouds. Below, the cross section intersects the surface warm and cold fronts.

The central idea of the Bjerknes model, shown in Figure 10.2, is that precipitation occurs on the cold side of frontal boundaries, which are depicted as advancing or retreating wedges of cold air. The most extensive precipitation occurs ahead of the warm front, where warm air broadly

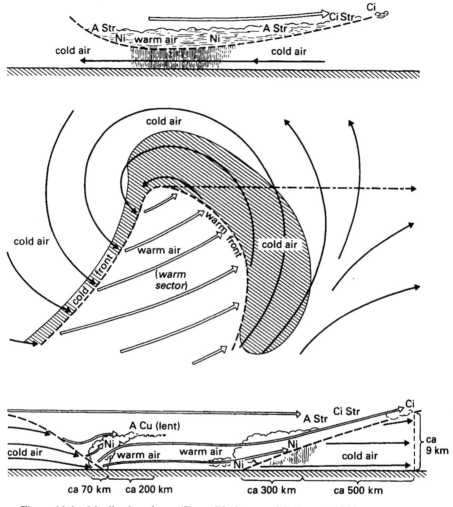

Figure 10.1 Idealized cyclone. (From Bjerkens and Solberg, 1926.)

ascends over a shallow wedge of retreating cold air. Along the cold front, precipitation is in the form of showers as the warm air is lifted more abruptly along the steeper frontal boundary. The source of the warm air is the warm sector, which is formed by the intersection of the cold and warm fronts.

In this Norwegian cyclone model, the cyclone is viewed as a perturbation on an existing frontal boundary, which gradually becomes distorted by the opposing warm and cold air streams until the advancing cold air overtakes the retreating warm air and pinches off the warm sector. Finally, the cyclone decays, the area of rainfall diminishes and the intersection of frontal boundaries is no longer located near the center of the dying storm. The main idea here is that cyclogenesis originates with a frontal boundary, grows as a result of an instability on the front and finally dissipates because the cold front overtakes the warm front. This latter stage is called "occlusion" because the fronts have occluded a portion of the warm sector nearest the cyclone center. That part of the frontal boundary where the cold and warm fronts have merged is called the "occluded front". A cross section through an occluded front would resemble the top cross section in Figure 10.1. Occlusion, therefore, involves the replacement of warm air near the surface by cold air and the isolation of the warm air aloft. The horizontal temperature gradient at the surface along the occluded front is relatively weak and of approximately equal strength on both sides of the front. Occluded fronts are recognizable not as the edges of strong temperature gradients but as features that lie along a ridge in the isotherms.

Although there are infinite variations of the Norwegian cyclone on actual weather maps, this model has proved to be a highly valuable tool for understanding and diagnosing the evolution of cyclones. Today we realize that cyclones do not represent instabilities along a frontal boundary, but occur as a consequence of large-scale forcing associated with horizontal temperature gradients. This chapter builds on the ideas of the Norwegian school within the framework of quasi-geostrophic ideas presented earlier in this text.

Modern ideas have not replaced the Norwegian cyclone model, but a great deal more has been learned about the structure and evolution of mid-latitude cyclones as the result of upper-air and satellite information. Quasi-geostrophic theory is able to provide us with a more powerful tool for understanding development, structure and evolution of weather systems such as cyclones. The behavior of cyclones, as discussed by Bjerknes and Solberg, can be expressed in terms of dynamic forcing of vertical motion and pressure tendencies, as outlined in previous chapters.

The four stages of cyclone development referred to in this chapter are loosely defined as: incipient disturbance, rapid development, maturity and decay. In the first stage the presence of a weak disturbance at the surface serves as a focus for subsequent development of that relative vorticity maximum (associated with the surface low). In the second stage, characterized by rapid development (*self-development*), changes brought about by development serve to accelerate that development. There is a positive feedback (instability) in this stage of growth. Eventually, a stage of *self-limitation* is reached in which the development processes begin to retard. The storm alters wind and temperature fields in such a way as to limit the vorticity and temperature advection and therefore to limit the availability of eddy available potential energy. Finally, there is a period of slow decay of the disturbance. This sequence of four stages corresponds roughly to the

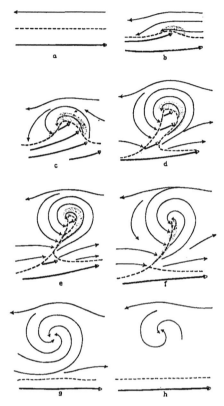

Figure 10.2   Sequence of events in cyclone development. (As shown by Bjerknes and Sol-berg, 1926.)

four pairs of drawings (a–b, c–d, e–f, g–h) in Figure 10.2. A fifth stage, that of total occlusion, is not discussed at this point in the text.

10.1   Cyclone climatology

Cyclone development does not occur uniformly over the Earth, but tends to favor certain geographical locations such as the lee of mountains, coastal regions and, in general, areas where there are strong horizontal temperature gradients in the lower troposphere. The non-uniform distribution of cyclogenesis annually and in summer over North America and the northern Atlantic and Pacific Oceans is shown in Figure 10.3. In all seasons, the principal sites for cyclone development occur over the western side of the Atlantic Ocean, in the lee of two North American mountain chains (the

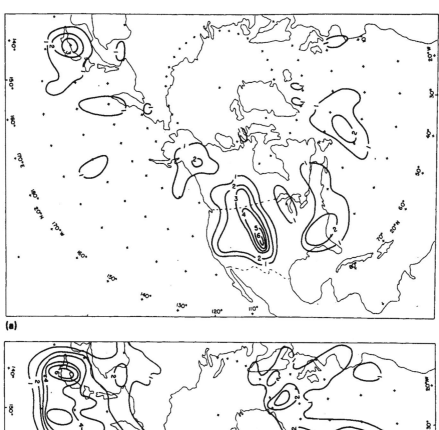

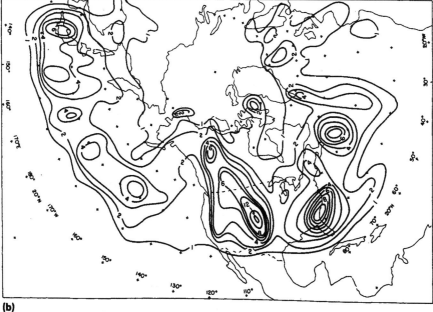

Figure 10.3 (a) Geographic distribution of formation positions of cyclones (1976–1982) for the warm season, smoothed from a 5° latitude/longitude grid of raw data. (b) Same as (a) but for annual cyclogenesis. (From Roebber, 1984.)

Rocky Mountains and the Appalachian Mountains) and over the Great Plains (see also Fig. 9.12). There are also several other prominent regions of extra-tropical cyclone development in the Western Hemisphere; these are associated with strong sea-surface temperature gradients, which occur south of Greenland over the Atlantic Ocean (40 to 60°N), and over the northern Pacific Ocean south of Alaska (55°N, 140°W). Cyclogenesis is much more frequent and intense during the colder months, which is why the annual distribution reflects that of wintertime.

Cyclogenesis tends to occur where the topography is favorable for producing low-level convergence and where there is already a local absolute vorticity maximum at the surface. Spin up of surface cyclones is likely to occur more rapidly in regions where topography favors a front or pre-existing disturbance, and low static stability, significant lower-tropospheric gradients of temperature and mid-level gradients of vorticity. Given these precursors, cyclogenesis can take place quite rapidly, even when starting with a relatively weak disturbance.

10.2   Evolution of the wave/cyclone during cyclogenesis

For purposes of simplification, let us consider the 500 mb surface as being representative of the middle troposphere. Further, let us restrict our attention to the changing patterns at 1000 and 500 mb during four stages of cyclone development, which are defined arbitrarily without intending to suggest discontinuities in the progression of events.

The start of cyclone development (stage 1) involves a weak surface trough, low pressure center or front, and an upper-level wave. Figures 10.4a and b (equivalent to part b of Bjerknes and Solberg's drawings in Figure 10.2) show a shallow surface depression some distance downstream of a diffluent, short-wave trough in the westerlies. There is a gradient of absolute vorticity along the direction of flow, as is characteristic of diffluent troughs. Positive values of vorticity advection over the surface cyclone are revealed by advection solen-oids over the surface cyclone in Figure 10.4b. The classical dynamic picture is as follows: east of the trough line (generally between the trough and the down-stream ridge at 500 mb) and over the surface disturbance the vertical motion is upward, in accordance with the positive vorticity advection. Near the surface cyclone there is low-level convergence, ascending motion and surface pressure falls. Sinking motion occurs west of the trough axis, which is oriented in typical fashion northeast–southwest. For purposes of discussion, let us place a wind speed maximum (jet streak) initially at a location upstream from the trough axis, as shown in Figure 10.4b. This wind speed maximum is associated with the principal vorticity maximum at upper levels, which is situated on the left-hand side of the jet (with respect to the direction of flow).

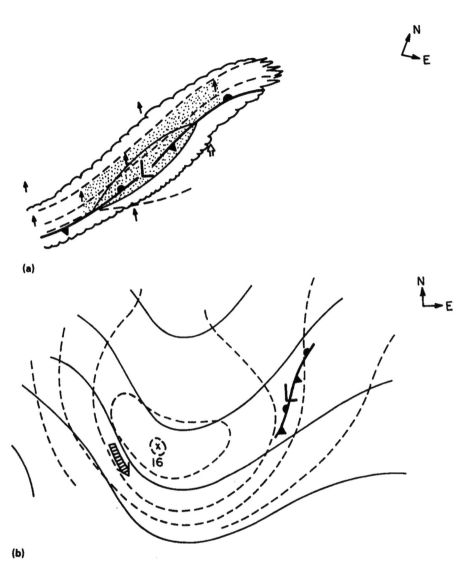

(a)

(b)

Figure 10.4 (a) Stage 1 (incipient surface cyclone development), at intervals of 4 mb, cyclone development at the surface. Full curves represent sea-level isobars and broken curves the 1000–500 mb thickness contours (conventional map intervals and units). Shading represents continuous precipitation and arrows denote the direction and magnitude of the vertical motion at 700 mb. Conventional frontal symbols are also used. Scalloping represents overcast cloud cover, as would be viewed from a satellite. (b) The 500 mb height (full curves at intervals of 120 dam) and vorticity contours (broken curves at intervals of 2 units) coinciding with the pattern at the surface shown in (a), which is represented on this figure by the configuration of surface fronts. The maximum vorticity is indicated by a cross labeled in vorticity units. The bold-faced arrow indicates the geostrophic wind speed maximum.

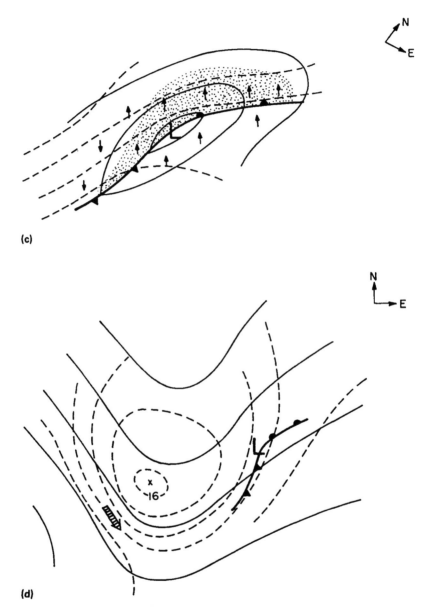

Figure 10.4 (c) Same convention as in (a) except for stage 2 (open wave stage) in cyclone development. (d) Same convention as in (b) except for stage 2 in cyclone development.

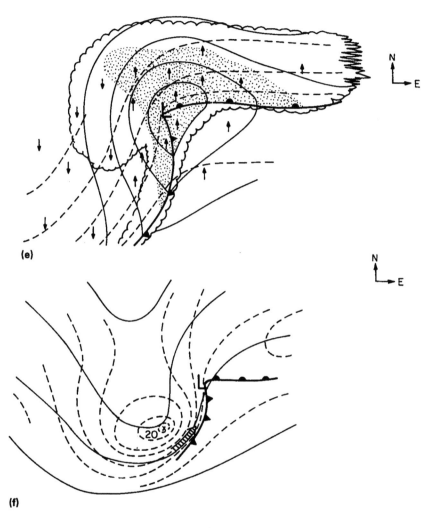

(e)

(f)

Figure 10.4 (e) Same convention as in (a) but for stage 3 (mature cyclone, incipient occlusion stage) in cyclone development. (f) Same convention as in (b) but for stage 3 in cyclone development.

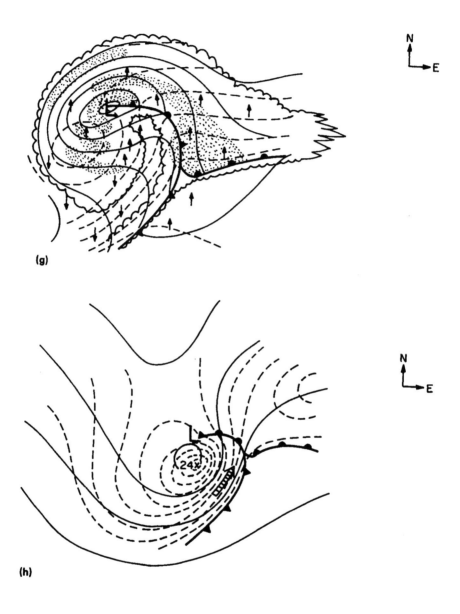

(g)

(h)

Figure 10.4 (g) Same convention as in (a) but for stage 4 (occluded, maximum intensity stage) in cyclone development. (h) Same convention as in (b) but for stage 4 in cyclone development.

Intensification of the surface low occurs as the result of forcing at upper levels, and this leads to the formation of a cyclonic circulation along the surface front. In stage 2, shown in Figures 10.4c and d (equivalent to part c in Figure 10.2), the surface pressure falls occur preferentially along the front and the surface vorticity maximum (the low), where the development term in the vorticity tendency equation is maximized.

Establishment of a cyclonic circulation near the surface initiates a pattern of 1000–500 mb thickness advection by the 1000 mb geostrophic winds. Warm advection now starts to occur east of the cyclone and cold advection to the west, with strongest thickness advection on the cold sides of the fronts. This temperature advection modifies the initial distribution of vertical motion imposed by vorticity advection from one in which there is a general area of ascent and low-level convergence downstream from the trough axis to a dipole pattern in which ascent takes place mainly east and poleward of the surface low and descent west and equatorward of the low. Now, thermal advection begins to assume greater importance as the circulation of the low intensifies within low-level temperature gradients.

The distribution of cloudiness also changes from a band situated along the surface front (a baroclinic "cloud leaf") to one in which the area of cloudiness is wider poleward of the warm front and narrower along the cold front ("comma cloud"). Cloud cover evolves in tandem with the evolution in the vertical motion pattern, in which there is increased sinking motion behind the cold front (*katabatic motion*) and increased ascent on the cold side of the warm front (*anabatic motion*).

In accordance with the distribution of pressure changes, the surface cyclone begins to migrate along the thickness contours in the direction of the maximum pressure falls, which occur poleward of the warm front. The circulation associated with the developing cyclone causes the cold front to move eastward and equatorward and the warm front to move eastward and poleward. This state of development (represented by Figures 10.4c and d) is manifested by a bend in the frontal zone and an open wave pattern with a warm sector between the cold and warm fronts. By similar reasoning, the surface high located west of the upper trough (not shown) moves eastward and equatorward toward the region of greatest positive pressure tendency.

Now, with stronger cold air advection and large-scale sinking motion west of the cold front and equatorward of the 500 mb vorticity maximum, 500 mb geopotential heights will fall in the region west of the cold front and rise poleward of the warm front since these areas experience the strongest temperature advections in the lower troposphere (see Chapter 7). Although the thermal vorticity fields are not depicted in the figure, areas of trough and ridge building at 500 mb are also ones in which thermal advections (and therefore $- V_0 \cdot \nabla_p h$ and $- V_T \nabla_p \zeta_T$) are significant.

In the third stage of cyclone development (a mature open wave cyclone shown in Figures 10.4e and f and equivalent to part d in Figure 10.2), feedback between surface and mid-tropospheric dynamics has begun noticeably to affect the mid-tropospheric flow. As discussed in Chapter 7, mid-tropospheric geopotential height falls produce an increase in the geopotential height gradient between the region of height falls and the warm sector of the cyclone. Consequently, at 500 mb, the southwesterly geostrophic flow intensifies southwest of the surface cyclone, as does the geostrophic vorticity adjacent to the region of greatest height falls; the absolute vorticity maximum tends to lie just to the left-hand side of the wind speed maximum. This region of enhanced wind speed and its attendant absolute vorticity maximum migrate into the region equatorward and west of the surface cyclone.

Until stage 4 of the storm development, changes occur at upper levels in such a way as to accelerate the development processes. Changes in the 500 mb pattern lead to an increase in the gradient of absolute vorticity over the surface cyclone. Further cyclone development, in turn, enhances the thickness advection and causes changes in the upper-level geopotential height pattern. This accelerated feedback between processes occurring at all levels is the manifestation of baroclinic instability. Large gradients of vorticity over the cyclone center in Figure 10.4f (which is similar to part e in Fig. 10.2) are due to development of the vorticity maximum at 500 mb and its migration into the region just equatorward of the surface cyclone.

These events described above are manifested by an expansion of the cloud area poleward of the warm front and west of the surface cyclone center, and a decrease in the width of the cloud band along the cold front. Some of the cloud area along the western side of the cyclone consists of the shallow, frictionally induced, stratiform clouds with light precipitation, as discussed in Chapter 9. The amplification of the 500 mb trough–ridge pattern during the development period (geopotential height falls equatorward and west of the surface cyclone and rises east and poleward of the cyclone) leads to the formation of a negative tilt (discussed in Chapter 7) in the upper trough and ridge, imparting to the 500 mb wave a more northwest–southeast orientation (Fig. 10.4f).

In this stage of cyclone development the storm has reached its maximum intensity (Figs 10.4g & h) and has begun to occlude, which is to say that its center is cut off from the warm sector. Cold air is now present at all levels over the cyclone center up to the tropopause. The principal vorticity maximum at 500 mb has migrated toward the surface cyclone center and there is a shorter-wavelength trough–ridge bulge in the geopotential contours east of the parent trough. Eventually the cyclone center may become vertical, thereby causing the advections to diminish. *This movement of the surface cyclone into the cold air and the development of an upper low and cold pool over the center is not the result of frontal occlusion but is produced by a rearrangement of mass brought about by dynamic processes.* Thus the surface cold front *appears* to overrun a portion of the retreating warm front, forming an occluded front and cutting off the interior of the cyclone from the warm sector. Here, there is a conflict between the dynamic viewpoint and the Norwegian model because the former does not require occlusion for decay to begin, nor does the occlusion process seem to deal with processes occurring at upper levels.

We have already described the increase in the upper-level wind speed that occurs west and equatorward of the low center, east of the vorticity maximum at 500 mb. A second wind speed maximum, however, forms in the upper troposphere near the ridge, poleward of the vorticity minimum and the surface of the warm front (see Ch. 12). Clouds expand to the west and southwest of the surface cyclone, especially the frictionally induced stratiform cloud in the region of cyclonically curved surface isobars. By the end of stage 4, the poleward side of the comma cloud has become extensive, bulging toward the west and equatorward, while the tail narrows to vanishing.

Secondary development can occur, however, even after the initial occlusion takes place and the parent storm ceases to develop. One area favorable for redevelopment is along the warm front where the thermal and 500 mb vorticity advections may not vanish. Such secondary development seems to be inhibited in cases where the primary vortex becomes very intense. In some cases, a family of cyclones develop in a series along the same baroclinic zone, with successive cyclone development at earlier stages upstream. This is commonly observed along the eastern coast of the United States where there may be local cyclogenetic factors, such as diabatic heating and low-level baroclinicity, imposed by land–sea contrasts.

In a fifth stage (not shown) advections over the storm center almost cease. Absolute vorticity at 500 mb may achieve a rather large value (perhaps as much as 26–34 units on the synoptic scale), but the advections, nevertheless, tend to vanish as the temperature and geopotential height fields become superimposed at all levels. The system undergoes decay due to surface friction, internal dissipation and a barotropic transfer of kinetic energy to other scales, including that of the mean flow as discussed in Chapter 5. It should be emphasized, however, that most cyclones do not develop to this stage, perhaps because of the absence of a strong horizontal temperature gradient, weak diabatic heating or high static stability (see Ch. 11).

10.3 Cyclone movement

It is sometimes said that the upper air steers the surface features. In terms of the dynamics, this is an incorrect view. As pointed out in earlier chapters, cyclone movement and development are characterized by a continuously evolving pattern of forcing manifested by the motion of the surface cyclone toward the region of greatest low-level convergence, which tends to occur where the positive quasi-geostrophic thickness advections are maximized, as suggested by the steering term in the Sutcliffe equation (8.7).

These advections are maximized along the direction of the strongest 1000–500 mb thickness gradient, which tends to lie parallel to the 500 mb geostrophic winds. Storms will give the impression of being steered by the mid-level winds, although they do not move at any fixed ratio of the 500 mb winds but may sometimes appear to leap across the map at speeds exceeding those of the winds at 500 mb. (A useful rule of thumb, however, is that surface cyclones move along the 500 mb wind direction, but with about half the speed of the 500 mb winds.)

Eventually the cyclone moves poleward following the region of strongest ascent, which shifts toward the cold air during its evolution toward occlusion. Similarly, descent shifts equatorward of the cyclone center.

The result of these changes may be a highly contorted pattern in which the cyclone, at both 500 mb and the surface, ceases to move eastward entirely, as in the case depicted in Figure 7.7. Note, however, that the concentration of absolute vorticity on the east side of the parent 500 mb trough results in an asymmetry in the flow that will cause the wave to move poleward and subsequently undergo apparent weakening, as discussed in Chapter 6.

10.4 The mature cyclone: a satellite view

Consider the cloud pattern for a mature cyclone, as viewed on a satellite photograph in the thermal infrared. Figure 10.5a is an "enhanced" infrared image, so-called because the gray scale is arranged to bring out certain ranges of temperature. The black and white shading in the image constitutes a temperature scale, which can be converted to altitude, provided that one has access to an appropriate vertical temperature sounding. Here, darkest shading represents the surface, white within the surrounding dark areas represents mid-level cloud tops, and darker masses within the light high cloud represent upper-level cloud temperatures. Small dark regions within the latter correspond to cloud near or above the tropopause and therefore most probably to the tops of convective cells.

Because of its shape, this type of cloud pattern is sometimes referred to as the comma cloud. The comma cloud possesses a "comma tail", narrowing equatorward, and a "comma head", which bulges toward the west, similar to the cloud pattern in Figure 10.5b. In stage 1, the cloud pattern tends to lie parallel to the upper wind flow in a band that is sometimes referred to as a baroclinic cloud leaf, because it is associated with strong tropospheric horizontal temperature gradients. In the developmental stage, the cloud narrows along its equatorward "tail", but expands on the poleward side, eventually developing a bulge westward on the poleward side of the surface cyclone (the comma head). Figure 10.5a shows that the western part of the comma head is surrounded by a circular area of lower cloud, constituting the frictionally forced stratus cloud with tops below 700 mb. A more specific discussion of the comma cloud structure is reserved for Chapter 12.

The clear zone between the comma tail and the cloud to the west is the "dry tongue". Dry tongues are typical of developing cyclones. They tend to coincide with a region of strong cold air advection and positive vorticity advection along its poleward side. Negative vorticity advection is found

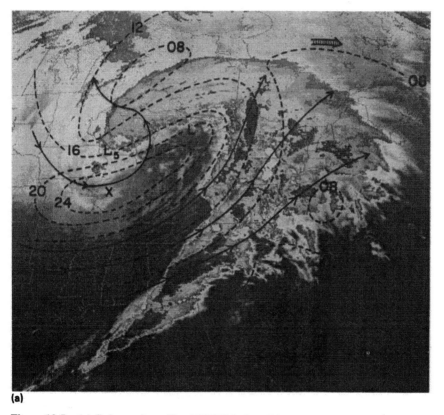

(a)

Figure 10.5 (a) Enhanced satellite (GOES) infrared image at 1200 GMT 4 January 1982 with superimposed absolute vorticity isopleths (broken curves at intervals of $2 \times 10^{-5} \, \text{s}^{-1}$ with center located by a cross) and geopotential height contours (represented as streamlines and at intervals of 120 m). The location of the surface and 500 mb low centers are indicated by the letters L and L_5, respectively. Wind speed maxima at 300 mb are shown by hatched arrows.

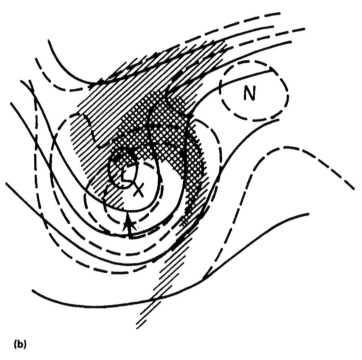

(b)

Figure 10.5 (b) Schematic illustration of comma cloud pattern and absolute vorticity isopleths (at intervals of $2 \times 10^{-5}\,\mathrm{s}^{-1}$; maximum value located at X and minimum at N). L denotes center of 500 mb low. Shading denotes region of overcast cloud and heavy shading the region of densest cloud and heaviest precipitation. Arrow denotes cloud cusp referred to in text.

equatorward and west of the 500 mb absolute vorticity maximum. The latter is situated equatorward and east of closed contours in the geopotential height field at 500 mb. Consequently, air is descending over the equatorward side of the dry tongue and ascending on the poleward side of the vorticity maximum. Because of the origins of this air entering the dry tongue (the high troposphere west of the trough), it is extremely dry at mid-levels.

Figure 10.5b, an idealization of Figure 10.5a, shows that closed geopotential height contours at 500 mb are accompanied by the formation of a cusp of cloud at the end of the comma head (arrow). This cusp tends to be located very close to the 500 mb low center and vorticity maximum, as shown in Figure 10.5b. The surface low, however, tends to be located at the point where the comma head joins the comma tail. Strong absolute vorticity advection occurs over the comma head region, except along the western edge where there is only frictionally forced ascent. Clouds are generally much higher over the eastern part of the comma head, with the dividing line

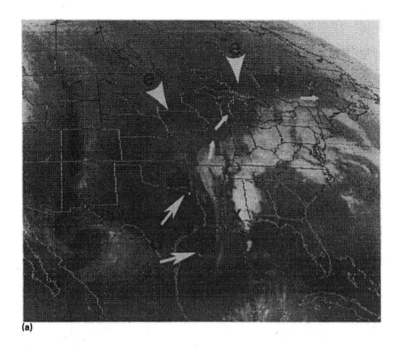

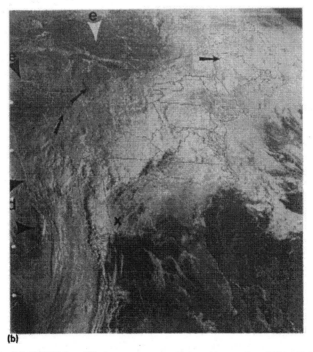

Figure 10.6 (a) GOES (satellite) water vapor channel image for 1601 GMT 5 March 1989, showing dry tongues (arrows labeled d and e), 300 mb jet axis in region of strongest winds (thin arrows) and surface low position (X). (b) Goes (satellite) visible image for 1431 GMT 5 March 1989. Symbols have same meaning as in (a), which covers approximately the same surface area.

between very high- and medium-level cloud tops lying close to the 500 mb wind speed maximum, which crosses the comma head. Warm air advection occurs over the comma head except where there is only frictionally driven low cloud.

Another aspect of cyclone development, which was pointed out earlier in this text, is the formation of a vorticity minimum along the poleward side of the comma head. Note that in both Figures 10.5a and b there is a marked intrusion of lower absolute vorticity over the poleward edge of the comma head from the direction of the ridge. Accordingly, the poleward side of the comma head corresponds in this case to the limit of positive vorticity advection and to a region of developmental geopotential height *rises* in this area. Separate wind speed maxima tend to form in regions of enhanced geopotential height gradients between the maximum or minimum vorticity centers, as shown in Figure 10.5a.

Evidence for the intrusion of dry air from the west of the trough west of the surface low can be seen on the water vapor channel images, such as the 6.7 μm channel from the GOES (satellite), shown in Figure 10.6a, which is an image from the Visible Infrared Spin-Scan Radiometer (VAS) aboard the GOES. This figure represents the total water vapor in the column by shades of gray, lighter shades indicating more precipitable water. The situation depicted was a slowly developing cyclone east of a very deep trough in the westerlies. A tongue of very dry air, indicated by arrows (labelled d in both the VAS and visible images), extends poleward on the western side of a surface low (marked with a cross).

Interestingly, the core of dry air (arrows labelled d and e) was situated some distance west of the visible cloud mass and the axis of maximum 300 mb winds (thin arrows). Note that the VAS image shows a comma-shaped moisture distribution, with a very sharp poleward and western edge. This feature does not coincide precisely with the 300 mb jet axis, which lies along the poleward edge of the visible cloud comma (Fig. 10.6b). The small width of this clear zone (arrows labelled e) suggests the possibility that the dryness is maintained by mesoscale descent forced by transverse/vertical circulations associated with the jet (see Ch. 14).

Occlusions: real or artifact?

Occlusion of the cyclone is viewed in the Norwegian cyclone model as an overrunning of the warm front by the approaching cold front, thereby trapping the parent low center in the cold air. This is the classical picture of the frontal evolution depicted by Bjerknes and Solberg. Indeed, the concept of fronts advancing and overtaking each other was undoubtedly inspired by the then very recent First World War, in which battles were fought along fronts that advanced and retreated. Today, it is not uncommon to hear explanations of violent weather given in the popular media in terms of warring cold and warm air masses. We prefer to view cyclone evolution mainly as an inevitable end result of the quasi-geostrophic forcing. The movement of the occluding cyclone into the cold air to the west of the principal polar front jet, which gives the impression of a roll-up of the warm front, is a result of increasing ascent north and west of the cyclone (in association with the westward and poleward expansion of the cloud mass and its response to increased vorticity and temperature advections in that region). Where cold advection and ascent are juxtaposed, the air experiences a cooling rate greater than that given by temperature advection alone (3.14). Consequently, adiabatic cooling produces a thickness minimum that is lower than the surrounding thicknesses and perhaps colder than any value poleward of the region. This thickness minimum eventually coincides with the center of the cyclone. Therefore, occlusion is not so much an overrunning of retreating cold air by advancing cold air, as in the classical Norwegian cyclone model, but the result of an evolution of the pattern involving the migration of the cyclone into the cold air, an interleafing of moist and dry air streams and a cessation of forcing (advection of vorticity and temperature). This process involves a lowering of the center of gravity of the system and a decrease in potential energy of the cyclone; the latter is associated with increase in the kinetic energy.

It is certain that occlusions are more complicated than portrayed in the early cyclone models. Rather than exhibiting a pool of warm air above the occluded part of the front, occluded lows tend to contain spiral bands of precipitating clouds, alternating with bands of broken cloud cover or nonprecipitating clouds. Examination of satellite pictures in sequence suggest that some of the dry air from the dry tongue region has become interleaved with the cloud, and the original comma cloud mass is detached from the cyclone. Such an example of an occluded low is shown in Figure 10.7. Note that the center of the cyclone consists of a band of cloud interleaved with a narrow band of dry air. Farther east, the pattern more closely resembles the typical comma cloud embedded in a baroclinic zone. (At present, there are few studies of the large-scale nature of occlusions, partly because of their enormous size and complexity.)

By the time the storm reaches occlusion, surface winds may have attained a relatively destructive force, notably in the case of explosive cyclogenesis, a topic to be addressed later in this chapter. High winds, accompanied by blowing and drifting snow in the wintertime, may give the correct impression in winter storms that the occluding cyclone is powerful. The rate of intensification diminishes and the storm weakens with time in accordance with a reduction in the number of advection solenoids and a corresponding reduction in vertical motion and precipitation amounts in all quadrants of the cyclone. Thus, the eddy kinetic energy of a storm may be large, but the rate

Figure 10.7 Visible channel satellite image from GOES at 2031 GMT 2 April 1988, showing an occluding cyclone over the central United States.

of energy conversion and vertical motions may be very small at this stage in cyclone development. Further discussion concerning the large-scale structure of occlusions is presented in Chapter 12.

10.5 Changes occurring at the tropopause

Although it is convenient to reduce the processes of cyclone development to events occurring at the surface and at 500 mb, one must recognize that cyclogenesis involves changes that occur throughout the troposphere and lower stratosphere. Recall from Chapter 2 that the magnitude of the divergence, which is intimately related to the local geopotential height tendency, tends to maximize in both the lower and upper troposphere. Emphasis on the lower troposphere avoids the fact that pressure falls at the surface not only necessitate a net excess of divergence at upper levels over convergence in the lower troposphere, but also require that the *average column virtual temperature between the surface and some undisturbed pressure surface near the top of the atmosphere (say 10 mb) must rise in order for the surface pressures to fall at the center of a low*. According to our simple treatment of the omega equation in Chapter 4, surface lows can deepen in response to the advection of positive absolute vorticity over the center, but not directly in response to advection of warm 1000–500 mb thicknesses by the 1000 mb geostrophic wind, which is zero at the center of a low. Consequently, one must look to the upper troposphere or lower stratosphere for that warming. There, the reason for the deepening of a surface low is apparent in terms of the thermodynamics of column warming despite the absence of geostrophic warm air advection in the 1000–500 mb layer over the low center.

Consider the evolution of the tropopause, the distribution of isotherms near the tropopause and the 200 mb geopotential height pattern during cyclogenesis, as represented by a conceptual model proposed by P. Hirschberg and shown in Figure 10.8. At the start of cyclogenesis (top of Fig. 10.8I) the axis of an upper trough lies west of a weak surface disturbance (a maximum in 1000 mb absolute vorticity). Horizontal variation in temperature across the tropopause resembles that of a front. Warm temperatures and high static stability are found in the lower stratosphere where the tropopause is lowest. The downward bulge in the troposphere is associated with descent at all levels. Although static stability considerations require the vertical motion to damp out rapidly with height above the tropopause, vertical motion exerts a profound influence in the lower stratosphere by causing the isentropes to descend or rise according to the vertical circulations. Because of the high static stability above the tropopause, even weak descent can produce a large adiabatic warming. Unlike the surface, which constitutes a hard boundary across which vertical air motions do not occur, the tropo-

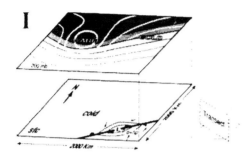

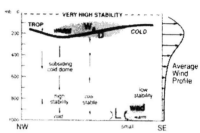

I

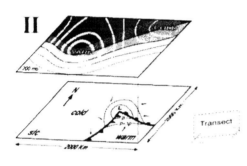

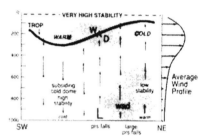

II

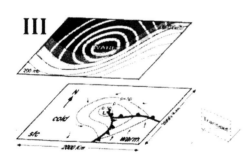

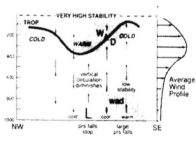

III

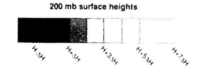

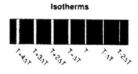

200 mb surface heights

Isotherms

pause is still free to deform in response to air motion. When it rises or sinks, the local rate of adiabatic cooling or warming can be very large.

If the wind direction is generally from the west to east (left to right in the figure), warm air advection occurs east of the low tropopause. Because of adiabatic descent, horizontal temperature gradients along the sloping tropopause tend to be quite large, and temperature advections at this level can account for large hydrostatic changes in pressure in the lower troposphere. Such pressure changes must be compatible with those brought about by the dynamics, which we have expressed in terms of geostrophic vorticity and temperature advections in the lower and middle troposphere. *Therefore, because of warm air advection at 200 mb, a surface low center situated between low and high tropopause levels can experience hydrostatic surface pressure falls due to the warming of the column by horizontal advection of warmer temperatures.*

This argument for deepening of the surface low is consistent with pressure falls associated with positive vorticity advection at 500 mb over the surface cyclone center. Further, the cold air at lower levels near the cyclone center is consistent with adiabatic cooling due to ascending motion. One practical aspect of the correlation between positive vorticity advection over the low center at 500 mb and warm air advection near the tropopause is that one can use the geostrophic temperature advection solenoids at 200 or 300 mb in place of the 500 mb absolute vorticity advection to diagnose cyclone development, should the 500 mb chart be unavailable.

For example, the warm 200 mb temperatures in the case of explosive cyclogenesis (see Fig. 10.12f) correspond to a region of low tropopause heights and relatively cold air in the lower troposphere. Strong warm air advection is occurring at tropopause level over the center of the surface low, as noted by the shaded advection solenoid. This high-troposphere, lower-stratosphere warm air advection correlates closely with positive vorticity advection solenoids at 500 mb (see Fig. 10.12e).

Figure 10.8 The 200 mb schematic geopotential height and temperature contours above a sea-level pressure chart (left) and a vertical cross section through that three-dimensional volume (right) at the location by the dotted transect (center). The sequence is one associated with a developing cyclone and tropopause undulation, chronologically from stage I to stage III. The 200 mb geopotential heights (dotted) are shaded to indicate height gradient (scale below on left) and the isotherms (white contours) denote the temperature values by their thickness (scale below on right). The lower plane of the volume depicts the sea-level isobars, the direction of the surface winds and the fronts in the vicinity of a surface low. Arrows in the cross section at the right denote the direction of vertical motion and the shading shows regions of warm air advection (WAD). Static stability, location of surface pressure falls and temperature (cool or warm) are indicated in words. The vertical profile of the wind speed is shown at the extreme right. Figure courtesy of Paul Hirschberg (Hirschberg and Fritsch 1991).

243

It is evident that temperature advections near the tropopause are required if the argument posed in Chapter 4 is to hold, namely that the warm (cold) air advection in the lower troposphere tends to diminish with height and change sign near the tropopause. Note that the warm temperatures at 200 mb are associated with cold temperature advection in the lower troposphere, descent at all levels and a lower tropopause.

In the last phase of cyclogenesis (Fig. 10.8III) the tropopause undulation is most pronounced, as is the amplitude of the temperature, height and vertical motion perturbations. Where warm air advection at tropopause level over the surface low is beginning to cease at the tropopause minimum, the 200 mb temperature maximum, the 200 mb geopotential height minimum and the surface low become superimposed. In extreme cases, the undulation in the tropopause descends into a frontal zone and can reach the middle or even lower troposphere; this feature is known as a "tropopause fold". Tropopause folding is an extreme outcome of the amplification in the tropopause undulation, which is associated with troughs and ridges in the troposphere. Accompanying tropopause folding is a downward extrusion of stratospheric air containing high amounts of potential and absolute vorticity to quite low levels. For this reason, tropopause folding is thought to be important for the development of surface cyclones. (This phenomenon is discussed in more detail in Chapter 15 in relation to upper-tropospheric fronts.)

10.6 Explosive cyclogenesis: coastal storms

Figure 10.3 shows a major site of cyclone development near the east coast of North America. In explosively developing cyclones, pressure falls of 10–20 mb over 12 h are not uncommon. Central pressures occasionally decrease to 960–980 mb, with hurricane-force winds over a large area. Such storms have crippling effects not only on maritime facilities but also over coastal inland areas. Indeed, snow accumulations of up to 60–90 cm have been recorded in major metropolitan areas of the east coast. The size, frequency and intensity of these explosive cyclones over the North Atlantic Ocean between America and Europe make them potentially more dangerous and destructive than many tropical hurricanes.

Because of their explosive deepening and destructive capability, the most rapidly developing maritime cyclones have been separately categorized. Explosive cyclogenesis has been defined as a deepening of the central sea-level pressure by the geostrophic equivalent (at 45° latitude) of 12 mb $(12 \text{ h})^{-1}$ or more. Explosive cyclogenesis occurs mainly over maritime areas (an apparent exception being the rapidly deepening continental cyclone demonstrated in Figure 7.7), is almost exclusively confined to the cold season, tends to be located about 400 km downstream from a 500 mb trough, is situated on the cold side of the belt of strongest westerlies, forms along strong air and sea-surface temperature gradients, and moves along those gradients. At the present time, they are almost always underforecasted in regard to intensity. The climatology of explosive cyclogenesis is shown in Figure 10.9. Here the axis of maximum frequency of explosive cyclogenesis is seen to lie slightly east of the axis of the Gulf Stream in the Atlantic Ocean and along the Kuroshio Current in the Pacific.

Storms along the east coast of the United States have been grouped into two categories by J. Miller: type A and type B. Figure 10.10a is an illustration of the tracks taken by some of these coastal cyclones during a 20 year period. These storms all tended to form along the southeastern part of the

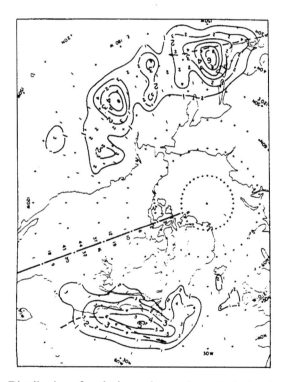

Figure 10.9 Distribution of explosive cyclogenesis events during three cold seasons. Raw non-zero values appear in each 5 × 5° quadrilateral of latitude and longitude. Isopleths represent smoothed frequencies. The column of numbers to the left and right of the heavy line along longitude 90 °W represent, respectively, the normalized frequencies for each 5° latitude belt in the Pacific and Atlantic regions. Heavy broken curves represent the mean winter positions of the Kuroshio Current and Gulf Stream. (From Sanders and Gyakum, 1980.)

245

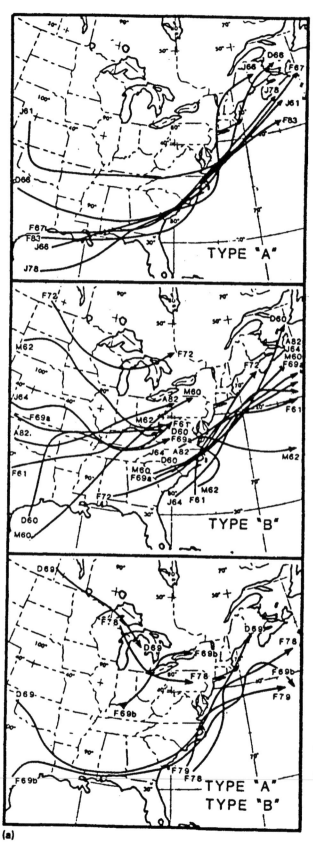

Figure 10.10 (a) Paths of surface low pressure centers grouped according to Miller's (1946) definitions.

(a)

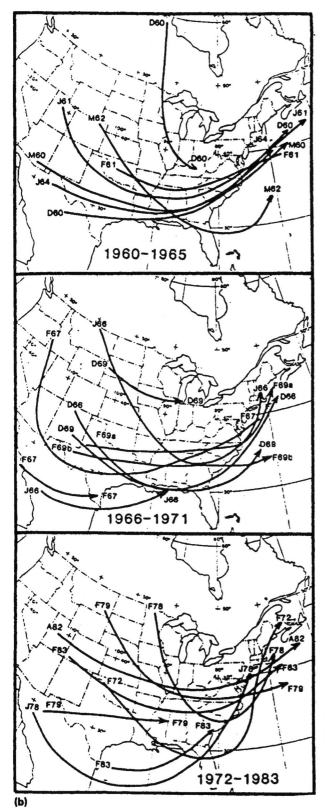

Figure 10.10 (b) Paths of inferred 500 mb geostrophic vorticity maxima associated with cyclone-producing troughs. (From Kocin and Uccellini, 1984.)

(b)

United States coastline and generally followed the coastline. Type A cyclones form along a frontal boundary separating cold continental air from warmer maritime air. Type B cyclones form along the land–ocean front but are initiated by a pre-existing cyclone, often occluding, which traveled through the Ohio Valley. Miller found that a secondary cyclone usually forms along the boundary between a shallow wedge of cold air east of the Appalachian Mountains and the warm maritime air. In such cases, the secondary development is located typically southeast of the parent storm, suggesting a topographic influence. In these cases, the parent storm vanishes or slowly dies without crossing the mountains, as in the case of the lee-side cyclone generated in association with the Pacific cyclone in Figure 9.11. Cases in which both type A and type B characteristics are present are also shown in Figure 10.10a.

Overall, the synoptic pattern in which many of these cyclones form is equatorward of a large cold high at the surface situated north of the incipient cyclone center. This configuration allows easterly winds north of the storm center to move along an extended fetch from the warmest ocean temperatures toward the coast. The location of the Gulf Stream, the warmest sea surface temperatures and the largest sea surface temperature gradient is situated along a line approximately parallel to the coastline but some distance from the land (Fig. 10.11). An even larger horizontal air temperature gradient is found along the boundary between land and water. Typically, explosive coastal cyclones form along a weak trough that lies approximately parallel to the coastline, the so-called "coastal front". The coastal front is closely related to the confluence of easterly flow from the ocean and the northerly or northwesterly land flow, which establishes a zone between the warm maritime and cold continental air streams. Although the North American coastal front is a regional mesoscale phenomenon, the presence of an existing absolute vorticity maximum near the surface is of general significance. (The effect of confluence and lateral shear on frontogenesis is discussed in more detail in Chapters 13 and 14.)

An example of rapid coastal cyclogenesis is shown in Figure 10.12. The initial map at 1200 GMT 14 December 1981 (Fig. 10.12a) shows a weak low and stationary front within a weak surface trough situated along the coast. A day later (Fig. 10.12b) the storm is better defined. The principal low center has begun to develop rapidly along the southern part of the front near 30°N. Over the next 24 h, the cyclone center moved northeastward directly along the coast following the stationary front, deepening by 7 mb in 12 h by 0000 GMT (Fig. 10.12c) and by 12 mb (to 986 mb) over the 12 h period ending at 1200 GMT 16 December (Fig. 10.12d). The storm's central pressure decreased by an additional 6 mb during the following 12 h to a central pressure just above 980 mb on 17 December. By 1200 GMT 17 December (Fig. 10.12g) the cyclone had split into two centers along the coast of eastern

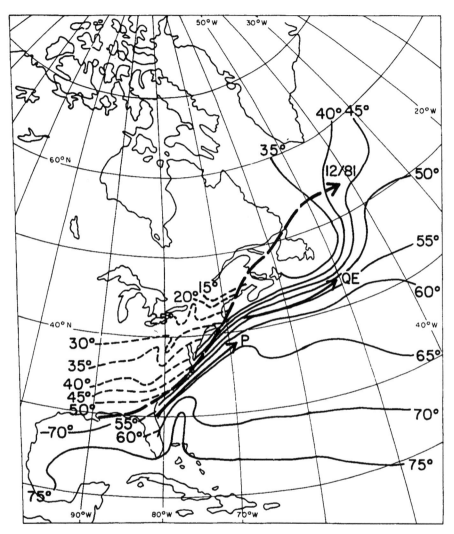

Figure 10.11 Mean January sea surface temperatures (U.S. Air Force atlas) in degrees Fahrenheit (°F). The tracks of three coastal storms are indicated by the arrows: the so-called Queen Elizabeth II storm (labeled QE), the President's Day storm (labeled P) and the example presented in Figure 10.12 (heavy broken curve). Thin broken curves are mean surface air temperature (°F) for January over the land.

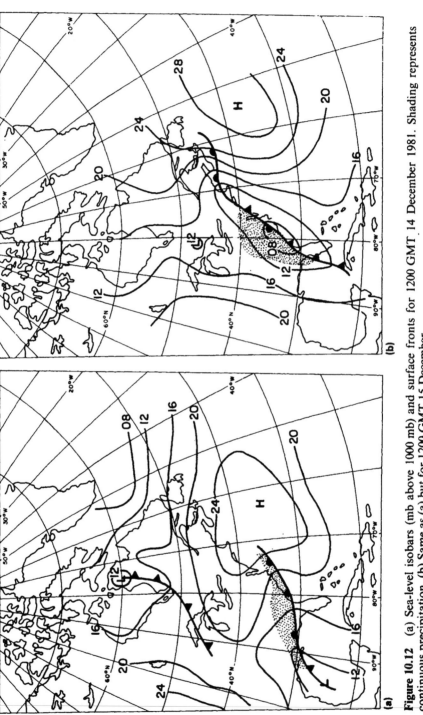

Figure 10.12 (a) Sea-level isobars (mb above 1000 mb) and surface fronts for 1200 GMT 14 December 1981. Shading represents continuous precipitation. (b) Same as (a) but for 1200 GMT 15 December.

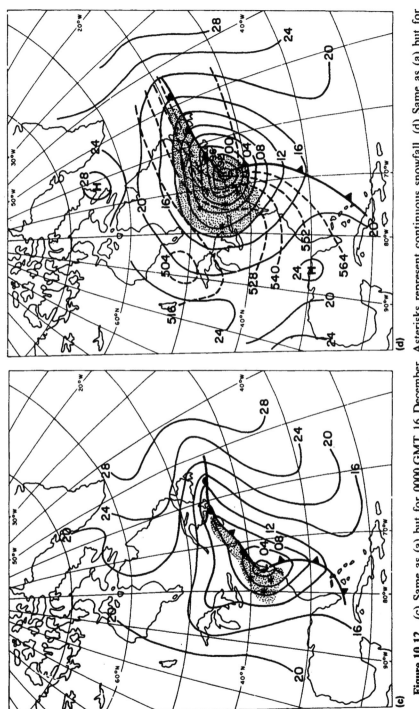

Figure 10.12 (c) Same as (a) but for 0000 GMT 16 December. Asterisks represent continuous snowfall. (d) Same as (a) but for 1200 GMT 16 December. Broken curves represent isopleths of 1000–500 mb thickness (in dam) at 120 m intervals. Asterisks represent continuous snowfall.

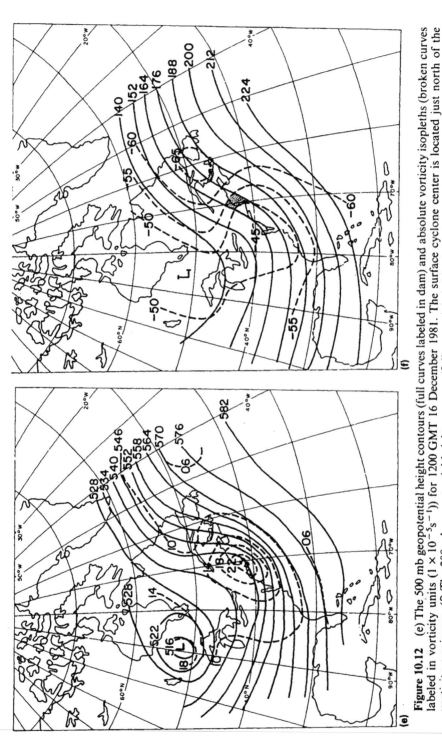

Figure 10.12 (e) The 500 mb geopotential height contours (full curves labeled in dam) and absolute vorticity isopleths (broken curves labeled in vorticity units ($1 \times 10^{-5} s^{-1}$)) for 1200 GMT 16 December 1981. The surface cyclone center is located just north of the vorticity maximum. (f) The 200 mb geopotential height contours (full curves labeled in dam above 10 000) and temperature isopleths (broken curves labeled in °C) at 1200 GMT 16 December 1981. The shaded solenoid denotes the region of strongest warm air advection at 200 mb.

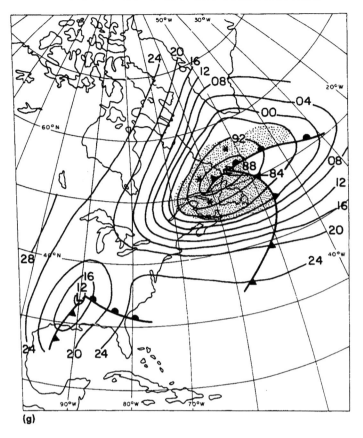

(g)

Figure 10.12 (g) Same as (a) but for 1200 GMT 17 December. Asterisks denote continuous snowfall.

Canada and was beginning to occlude. Subsequent motion of the storm was toward the east with a further slight decrease in central pressure over the following day or two.

Development of this cyclone occurred in association with the approach of a 500 mb trough from the west, which resulted in positive vorticity advection at 500 mb over the cyclone center. Simultaneously, the absolute vorticity maximum at 500 mb intensified and moved to a location just south of the cyclone center by 1200 GMT 16 December (Fig. 10.12e), thereby imparting a pronounced negative tilt to the wave. Although positive vorticity advection over coastal cyclones is crucial for their development and is associated with all the storms shown in Figure 10.10, one must look further than 500 mb absolute vorticity advection to understand the reasons for explosive cyclogenesis of coastal storms.

253

First, consider that explosive maritime cyclogenesis occurs typically about 400 km downwind from an upstream trough and on the poleward side of the belt of strongest winds, thereby allowing the storm to be located in the region of both strongest positive vorticity advection at 500 mb and low static stability.

Second, coastal storms tend to develop and remain along the boundaries of strong sea surface and air temperature gradients suggesting the importance of temperature advection, particularly warm advection in the lower troposphere. In the sequence of figures (Fig. 10.12), the warm front remains very close to the land–sea boundary where a surface cyclone was creating the most intense surface pressure falls due to warm advection on the north side of the storm. Cold advection occurs on the equatorward south side of the storm very close to the coastline. As a result, the cyclone tends to move parallel to the coastline and Gulf Stream.

Over the western North Atlantic, this temperature gradient is enhanced when the surface-level winds move over a large fetch from the region of warmest sea surface temperatures, i.e. from over the Gulf Stream, which lies to the east and southeast. An easterly air trajectory assures that warm air reaching the warm sector of the cyclone possesses a relatively high wet-bulb potential temperature, whose upper value is bounded by the temperature of the warmest ocean surface. The air also, by virtue of its warmth, enables a low static stability to develop over the lower troposphere.

Numerical simulations of storms along the eastern coast of the United States show that the presence of surface-sensible and latent-heat fluxes are not essential for the short-term intensification of the cyclone. Rather, the crucial factors in coastal development appear to be the high surface wet-bulb temperatures and corresponding low-static stability. The static stability over the 1000–500 mb layer has an approximate lower limit that is the difference between the sea surface temperature and the 500 mb potential temperature. As we show in various examples, low static stability is crucial for explosive cyclogenesis. These two factors, low static stability and high surface θ_w, promote condensational warming (through either convective or large-scale moist adiabatic ascent), which feeds back to the system to maintain the low-level convergence (see Fig. 9.15).

The role of surface-sensible and latent-heat fluxes in rapid cyclone development, even that over the ocean, is more passive. Surface fluxes of heat and moisture serve to maintain the high wet-bulb potential temperature especially when the fetch extends over the Gulf Stream, thereby enhancing condensation and producing a relatively unstable lapse rate over the lower troposphere. The large contrast in temperature augmented by confluence of this warm and moist air stream with cold air over land maintains or intensifies the coastal trough and lower-tropospheric temperature gradient. Consequently, cyclones tend to form in this confluence zone and thereafter migrate along the maximum temperature gradient, which is along the direction of the strong low-level temperature gradient.

The Gulf Stream current is clearly evident in Figure 10.11, which also shows part of the path of the 18–20 February 1979 storm (labeled P) and that of 9–10 September 1978 (labeled QE), the latter having inflicted damage on the liner Queen Elizabeth II and caused another vessel to sink. Both storms closely adhered for a time to the mean location of the sea surface temperature gradient, which lies along the edge of the Gulf Stream. The tendency of cyclones to move along sea surface temperature gradients is related to temperature gradients within which are found the strongest warm and cold advections. (Note that the warm front in Figure 10.12 is located along the coastline, more or less along the path of the cyclone.)

Movement and development of a rapidly developing coastal storm are strongly affected by the low-level temperature gradient, the moisture supply in the form of high θ_w and the static stability. Figure 10.10 shows that the 500 mb vorticity pattern is modified during coastal development, such that the motion of the vorticity maximum tends to follow the coastline. Since the coastal cyclone is advecting warm (cold) air north (south) of the center, geopotential height falls at the surface and rises at 500 mb tend to be concentrated north of the storm along the coastline, and geopotential height rises at the surface and falls at 500 mb tend to be concentrated along the coast south of the surface cyclone.

Explosive cyclogenesis is nevertheless responsive to the large-scale dynamics. The President's Day storm track deviated markedly from the sea surface temperature gradient after first moving along the path indicated in Figure 10.11. In the case shown in Figure 10.12, the cyclone first moved along the coast but then changed direction for the open sea. Such deviation from surface temperature gradients emphasizes the importance of large-scale forcing in modulating the effects of strong low-level temperature contrast.

Extremely large boundary-layer temperature gradients imposed by land–sea or oceanic temperature contrasts translate into gradients in the 1000–500 mb thickness pattern. Absence of abundant data over the ocean results in a degradation of operational analyses, which fail to capture the correct location and intensity of the temperature gradients. Thus, the conventional 500 mb analyses, such as those shown in this text, probably mute the true nature of the lower-tropospheric temperature contrast along the coastline.

Figure 10.13 shows an infrared satellite photograph of the eastern coastline of the United States during autumn. The Gulf Stream appears to meander, forming eddies (arrow labeled D), and creating rather complex patterns of sea surface temperature. The Gulf Stream temperature gradient extends from Florida, where it adjoins the land, to the North Atlantic several hundred kilometers east of Newfoundland. The warmth of the Gulf Stream

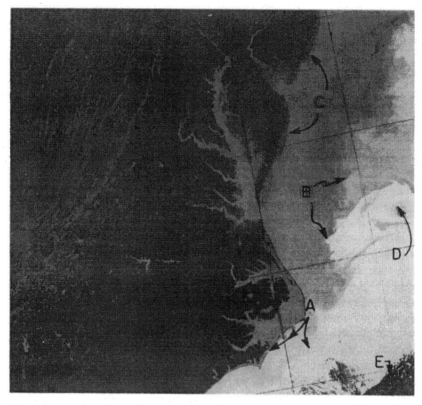

Figure 10.13 Thermal infrared satellite image taken by Heat Capacity Mapping Mission (HCMM) satellite at approximately 0230 local time on 22 October 1978, for a portion of the east coast of North America. White represents warm temperatures and black cold. The dark region over the extreme southeast corner is cloud. Arrows denote significant features (see text).

over the southeastern quadrant of the figure contrasts with the cooler land areas to the west (arrows labeled C). The sharpest horizontal temperature gradient is found between the ocean in the southeastern corner and the adjacent land (arrows labeled A). There is also a sharp gradient between the Gulf Stream in the southeastern sector and the cooler (but still relatively warm) water to the north (arrows labeled B), and between the ocean and the land along the coast to the north (arrows labeled C). Over the extreme southeast portion of the picture there is a mass of clouds situated along the axis of the Gulf Stream (arrow labeled E).

Clearly, if cyclones tend to migrate along strong sea surface temperature gradients, they may follow either the land–sea or Gulf Stream temperature gradients, including those which may be transient. Some cyclones appear to remain very close to land while others appear to move along the edge of the Gulf Stream some distance from land. The cyclone illustrated in Figure 10.12 moved along the coast from the Gulf of Mexico (30°N) to the Canadian maritime provinces (55°N). A similar dependence of cyclogenesis on surface-level baroclinicity occurs elsewhere, e.g. the South China Sea.

10.7 Polar lows

A special case of marine cyclogenesis is the polar low. Polar lows constitute an interesting deviation from the Norwegian cyclone model. The polar low is usually a non-frontal low that occurs over the ocean to the rear of the cold front, often a considerable distance from the belt of strongest westerlies and poleward of the 500 mb vorticity center. Polar lows are usually of relatively small scale (1000 km or less) and are often quite shallow, extending only to 400–500 mb in some cases. Like ordinary coastal cyclones, they form near strong sea and air temperature gradients and can become very intense. They are often associated with lines of cumulus convection or comma-shaped cumulus aggregates situated along the axis of the upper trough. Polar lows most frequently occur over the North Atlantic between Iceland and Great Britain and over the North Pacific Ocean.

Regardless of their geographical location, polar lows have three features in common: they have conditionally unstable lapse rates in the lower troposphere, they are associated with strong low-level baroclinicity, and they occur in regions where relatively cold air is advected over a warm surface. Over the Atlantic Ocean, genesis typically occurs near 60°N as compared with about 40°N over the Pacific Ocean. This latitudinal difference is due to the difference in sea surface temperature distribution, which is related to the distribution of sea ice in the northern part of the Northern Hemisphere. In the Pacific the ice extends south of the Bering Strait, approximately to 58°N, whereas the ice extends only to about 65°N over the Atlantic Ocean. Consequently, there is warm water in a latitude belt in the eastern Atlantic south of Iceland, whereas over the Pacific at that latitude there is ice. Another difference between the Atlantic and Pacific lows is that the latter generally form just poleward of relatively strong, deep baroclinic regions, while Atlantic lows form far to the north of the mid-latitude baroclinic zones.

Over each ocean, two types of polar lows have been identified. The more common polar lows form in a strong surface polar air stream on the cyclonic side of an upper-level wind maximum and are strongly influenced by diabatic heating due to convection. This is illustrated in Figure 10.14, which shows that the polar trough and polar low are not frontal in nature, although they may display a convective comma-shaped cloud pattern. The less common

257

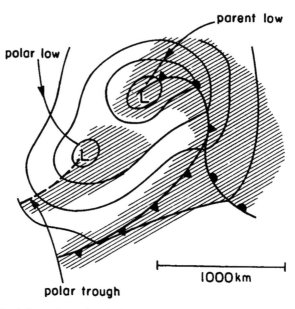

Figure 10.14 Schematic surface isobaric chart showing polar low and polar trough embedded in a larger-scale cyclonic system. Hatching denotes dense cloud. Surface fronts are shown in conventional symbols. (Based on a figure by Reed, 1979.)

type of low occurs on the anticyclonic side of an upper jet streak and appears to be initiated by baroclinic processes (vorticity and temperature advection) in regions of relatively weak winds but along strong surface temperature gradients, such as those imposed by the proximity of an ice shelf to warmer ocean. Low-level temperature advection is a prominent mechanism for development of this disturbance.

Atlantic polar lows tend to form in a northerly or northwesterly air stream, which is relatively homogeneous except for a shallow layer of strong baroclinicity near the surface (Fig. 10.15a). Precipitation forms near the depression center and along a trough extending equatorward from the center. The disturbances are generally shallow, but are accompanied by a weak perturbation at mid- and upper levels. Over the Atlantic they tend to form near Iceland, move southward and affect the weather of Europe. In one such instance a polar cyclone formed northeast of Great Britain and struck Europe following rapid intensification over the ocean. Although the central pressure was not very deep, the horizontal pressure gradient was large. Analysis of radiosonde data showed that the lapse rates near the center of the storm were nearly dry adiabatic, and that the cyclone was relatively warm over the lowest 300 mb.

Much of the heavy snowfall over Britain and northwestern Europe during the winter is associated with the passage of polar lows. Snowfall occurs just east and north of the center. Conditionally unstable air extends from the surface to about 500 mb. Extensive low stratiform cloud is found around the cyclone, while cumulus clouds are confined to the west and north of the center. A tongue of warm, moist air oriented in a southwest–northeast direction is situated south of the center in a stream of air that moves generally northwestward.

Pacific polar lows or troughs resemble Atlantic ones, but with some notable differences. They are generally larger (1000 vs. 600 km) than their Atlantic counterparts, they are deeper (the baroclinicity extends throughout the troposphere) and they form farther south within the colder air, just north of the main mid-latitude baroclinic zone but not within the very cold air farther north. The conditionally unstable air extends up to 500–700 mb.

The surface pressure pattern is often referred to as a polar trough, since there are usually no closed contours around the low. In such instances, comma-shaped cloud bands form along surface pressure troughs on the

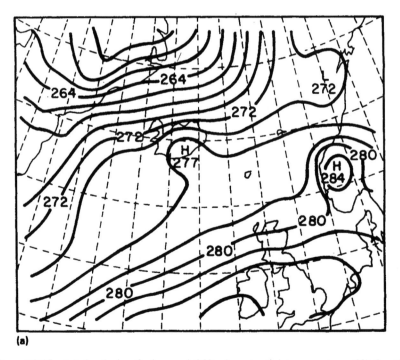

(a)

Figure 10.15 (a) Analysis of observed 950 mb potential temperature (K) for the Atlantic domain at 1200 GMT 23 November 1978.

259

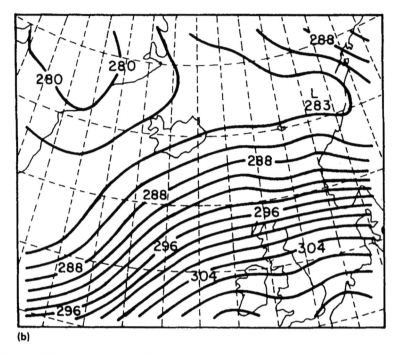

(b)

Figure 10.15   (b) Analysis of observed 500 mb potential temperature (K) for the Atlantic domain at 1200 GMT 23 November 1978. (After Sardie, 1984.)

poleward side of the zone of maximum winds at 500 mb and not far north of the main front. (Conversely, Atlantic polar lows seem to be divorced from middle-latitude frontal activity.) The comma cloud area corresponds to a region of enhanced convection, especially along the comma tail, and to locally large values of baroclinicity and vertical and horizontal wind shear.

For Atlantic polar cyclones, the shallow baroclinic zone along which the disturbances form is closely related to the topography of Greenland and the horizontal variation in the amount of surface ice, which imposes gradients of surface temperature between the ice and water (Fig. 10.15a). These surface temperature gradients are reflected as intense baroclinic zones in the lower troposphere and play the same role in the dynamics of the cyclone as do the horizontal surface temperatures in the coastal storm shown in Figure 10.12.

A crucial influence on polar low development is the distribution of sea surface temperature. The region of strong horizontal temperature gradients near the surface is shifted poleward of the most intense temperature gradients at 500 mb (Fig. 10.15b). This shift results in a lowering of static stability in the zone between the 500 mb and surface temperature gradients, thereby providing a corridor for explosive cyclogenesis by polar lows.

Radar observations indicate that cyclones originate at upper levels and then extend to the surface, thereby hampering operational forecasting of polar low formation. Penetration to the surface occurs only after the initial disturbance is advected over the warmer ocean. As in the case of coastal storms, the low-level temperature gradient is very important in the early stages of the polar low development.

When cold air, originally over the ice-covered regions, flows southward over the warmer waters, a very large sea–air temperature difference ensues, favoring large vertical eddy fluxes of sensible and latent heat from the surface throughout a layer 4–5 km deep. Not only does the presence of a semipermanent oceanic baroclinic zone provide for the growth of a deep layer of conditional instability and of low static stability in the lower troposphere when cold air flows over warm water, but the gradient of air temperature constitutes a source of large-scale baroclinicity and therefore of available potential energy.

Latent-heat release, especially in the form of convection, may play a vital role in the intensification of the disturbances into mature vortices. Therefore polar lows present another example of cyclone growth by the dual (baroclinic–convection) instability feedback loop illustrated in Figure 9.15, but with somewhat less emphasis on the existence of pre-existing forcing at 500 mb and more on the role of lower-tropospheric baroclinicity and convective heating. For Atlantic lows, the diabatic loop is essential for early development of the cyclone, whereas Pacific lows are able to profit by a deeper baroclinic zone and therefore depend less on convection and more on baroclinicity for development.

Some Atlantic polar lows that affect northwestern Europe are quite small and intense (several hundred kilometers across, with destructive winds and very low central pressures) and are characterized by near-neutral static stability throughout a very deep layer. Such a storm sometimes exhibits little convection, an indication that diabatic heating by cumulus is not very important during its mature stages. Diabatic processes are, however, responsible for the initial formation of weak vortices and their associated comma clouds, although subsequent development into an intense vortex depends on the presence of strong lower-tropospheric baroclinicity for both Atlantic and Pacific lows.

Although polar lows are thought to be an exclusively maritime phenomenon, weak troughs resembling the polar lows are often observed over continents in winter, in the vicinity of well-developed upper troughs and behind vigorous surface cold fronts. Such systems, sometimes referred to as "secondary troughs", are of interest in that they exhibit some of the features of polar lows and may originate as the result of the same array of circumstances. Secondary troughs over land, while often accompanied by squally convection and comma-shaped cloud patterns, seldom intensify beyond the stage of a weak depression, probably because strong lower-troposphere baroclinicity is absent, while high static stability and dry air exist in the boundary layer.

Progress in improving the prediction of maritime storms has been limited by the sparseness of data over the oceans and by the current inability to represent satisfactorily the effects of cumulus convection in forecast models. At present, satellite temperature and moisture soundings can provide some resolution of the basic large-scale temperature and wind structure. These data lack sufficient detail of the smaller-scale structure of the cyclones, although satellite infrared temperature measurements do provide accurate gradients of sea surface temperature. Remote measurements can be supplemented, however, by a more rapid and extensive data collection system from merchant shipping.

Problems

10.1 Intense cyclones often display very large values of vorticity and temperature gradients, yet may have ceased development. Explain how this seeming paradox can occur.

10.2 Explain in terms of quasi-geostrophic theory why surface highs tend to move with an equatorward component and surface lows with a polar component.

10.3 Why do surface cyclones tend to follow low-level temperature gradients? (Explain in terms of quasi-geostrophic theory.) What about a high?

10.4 At the center of a 1000–500 mb thickness maximum, the 500 mb absolute vorticity advection is positive and the sea-level pressure tendencies are negative. In order for the surface pressure to decrease, however, it is necessary for the air column to experience a net warming between the surface and some upper level (say 10 mb). Where (at what level) and how would that warming take place and what type of chart would you present to illustrate your argument?

Further reading

Anthes, R. A., Y.-H Kuo and J. R. Gyakum 1983. Numerical simulations of a case of explosive marine cyclogenesis. *Mon. Wea. Rev.* **111**, 1174–88.

Bosart, L. F. 1981. The President's Day snowstorm of 18–19 February 1979: a subsynoptic scale event. *Mon. Wea. Rev.* **109**, 1542–66.

Bjerknes, J. 1921. On the structure of moving cyclones. *Geofys. Publ.* **1**, 1–8.

Bjerknes, J. and H. Solberg 1926. Life cycle of cyclones and the polar front theory of atmospheric circulation. *Geofys. Publ.* **3**, 1–18.

Danard, M. B. and G. E. Ellenton 1980. Physical influences of east coast cyclogenesis. *Atmosphere-Ocean* **18**, 354–63.

Forbes, G. S. and W. D. Lottes 1985. Classification of mesoscale vortices in polar airstreams and the influence of the large-scale environment on their evolutions. *Tellus* **37A**, 132–55.

Gall, R., R. Blakeslee and R. C. Somerville 1979. Baroclinic instability and the selection of the zonal scale of the transient eddies of middle latitudes. *J. Atmos. Sci.* **36**, 767–84.

Gyakum, J. R. 1983a. On the evolution of the QE II storm. I: Synoptic aspects. *Mon. Wea. Rev.* **111**, 1137–55.

Gyakum, J. R. 1983b. On the evolution of the QE II storm. II: Dynamic and thermodynamic structure. *Mon. Wea. Rev.* **111**, 1156–73.

Hanson, H. P. and B. Long 1985. Climatology of cyclogenesis over the East China Sea. *Mon. Wea. Rev.* **113**, 697–707.

Harrold, T. W. and K. A. Browning 1969. The polar low as a baroclinic disturbance. *Q. J. R. Met. Soc.* **95**, 719–30.

Hirschberg, P. A. and J. M. Fritsch 1991a. Tropopause undulations and the development of extratropical cyclones. Part I: Overview and observations from a cyclone event. *Mon. Wea. Rev.* **119**, 496–517.

Hirschberg, P. A. and J. M. Fritsch 1991b. Tropopause undulations and the development of extratropical cyclones. Part II: Diagnostic analysis and conceptual model. *Mon. Wea. Rev.* **119**, 518–50.

Kocin, P. J. and L. W. Uccellini 1985. A survey of major East Coast snowstorms, 1960–1983. Part I: Summary of surface and upper-level characteristics, NASA TM 8615. 101 pp. NTIS N85 N85 27471.

Kutzbach, G. 1979. *The thermal theory of cyclones* (Hist. Monogr. Ser.). Boston, MA: Am. Meteor. Soc.

Mansfield, D. A. 1974. Polar lows: the development of baroclinic disturbances in cold air outbreaks. *Q. J. R. Met. Soc.* **100**, 541–9.

Miller, J. E. 1946. Cyclogenesis in the Atlantic coastal region of the United States. *J. Meteor.* **3**, 31–44.

Mullen, S. L. 1979. An investigation of small synoptic-scale cyclones in polar airstreams. *Mon. Wea. Rev.* **107**, 1636–47.

Nuss, W. A. and R. A. Anthes 1987. A numerical investigation of low-level processes in rapid cyclogenesis. *Mon. Wea. Rev.* **115**, 2728–43.

Pagnotti, V. and L. F. Bosart 1984. Comparative diagnostic case study of east coast secondary cyclogenesis under weak versus strong synoptic-scale forcing. *Mon. Wea. Rev.* **112**, 5–30.

Petersen, R. A., L. W. Uccellini, A. Moslek and D. A. Keyser 1984. Delineating mid- and low-level water vapor patterns in preconvective environments using VAS moisture channels. *Mon. Wea. Rev.* **112**, 2178–98.

Reed, R. J. 1979. Cyclogenesis in polar air streams. *Mon. Wea. Rev.* **107**, 38–52.

Reed, R. J. and M. D. Albright 1986. A case study of explosive cyclogenesis in the eastern Pacific. *Mon. Wea. Rev.* **114**, 2297–319.

Roebber, P. J. 1984. Statistical analysis and updated climatology of explosive cyclones. *Mon. Wea. Rev.* **112**, 1577–89.

Rogers, E. and L. F. Bosart 1986. An investigation of explosively deepening oceanic cyclones. *Mon. Wea. Rev.* **114**, 702–18.

Sanders, F. 1986a. Explosive cyclogenesis in the west-central North Atlantic Ocean 1981–1984. Part I: Composite structure and mean behavior. *Mon. Wea. Rev.* **114**, 1781–94.

Sanders, F. 1986b. Explosive cyclogenesis over the west-central North Atlantic

Ocean 1981–1984. Part II: Evaluation of LFM model performance. *Mon. Wea. Rev.* **114**, 2207–18.

Sanders, F. and J. R. Gyakum 1980. Synoptic–dynamic climatology of the "bomb". *Mon. Wea. Rev.* **108**, 1589–606.

Sardie, J. M. 1984. On development mechanisms for polar lows. PhD thesis, Dept of Meteorology, Pennsylvania State University. 220 pp.

Sardie, J. M. and T. T. Warner 1983. On the mechanism for the development of polar lows. *J. Atmos. Sci.* **40**, 869–81.

Young, M. V., G. A. Monk and K. A. Browning 1987. Interpretation of satellite imagery of a rapidly deepening cyclone. *Q. J. R. Met. Soc.* **113**, 1089–116.

11

Optimum wavelength and growth rate of baroclinic waves

In looking at weather maps, one is struck by the fact that there is approximately one disturbance every 3000–4000 km along a latitude circle. The number of clearly identifiable trough axes in Figure 11.1 is about seven waves around the Northern Hemisphere. Thus, we can say that the global *wavenumber* is 7, which corresponds to a wavelength of about 3500 km. However, in many cases of explosive cyclone disturbances, the distance between trough and ridge or between surface and low and high pressure centers (one half-wavelength) is between about 1000 and 2000 km. Chapters 2 and 4 introduced the idea that the wavelength of disturbances affects the vertical motion and the surface pressure tendency. Since there is a variation in vertical motion and surface pressure tendency with wavelength, it follows that some wavelengths must grow faster than others. Indeed, there is an optimum wavelength for growth, which one supposes must be about 3500 km from a casual inspection of global 500 mb charts.

In fact, the favored wavelength seems to vary with time and space. Inspection of such global charts also suggests that there are fluctuations in the number of waves with time, reflecting the index cycle referred to in Chapter 5. Over a period of some weeks the number of identifiable waves may vary from four to 15.

One impression gained from inspection of rapidly developing cyclones, such as those found along coastal zones or over the ocean, is that they possess wavelengths less than 3500 km. Moreover, cyclogenesis seems to involve a narrowing of the distance between trough and downstream ridge, corresponding to the change in tilt of the wave; conversely, the later stages of decay (e.g. merging of an occluded cyclone with a semipermanent vortex) involve a broadening of the system. Thus, there is evidence of a variation in wavelength during the lifetime of a weather system.

Since the wavelength dependence appears explicitly in our scaled version of the linearized equations in Chapter 4, it seems reasonable to suppose that

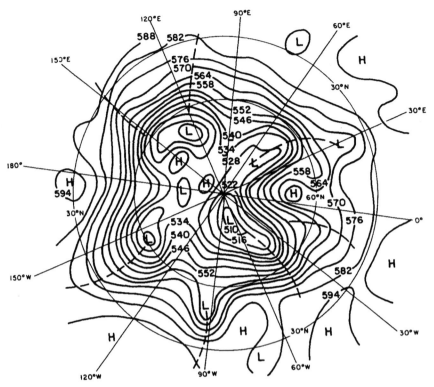

Figure 11.1 The 500 mb height (full curves in dam). The global pattern shows seven waves (denoted by broken curves drawn along the trough axes) for 0000 GMT 23 October 1979.

the quasi-geostrophic omega and pressure tendency equations can also explain the atmosphere's preference for selecting certain wavenumbers for development. In this chapter we will expand further on the simple sinusoidal model presented in Chapter 4 in order to demonstrate, without recourse to a formal and rigorous stability analysis, that the omega and surface pressure tendency equations imply a selectivity by the atmosphere for a preferred scale of cyclone growth.

Let us begin by reviewing the results of classic linear baroclinic instability theory, as illustrated by Figure 11.2. Theoretical analyses of the growth rate of waves in a baroclinic atmosphere have been made since the 1940s. The hatched region (marked unstable in this figure) shows where the amplitude of the baroclinic wave amplifies with time. Outside the region, the amplitude of the wave decreases with time. The rate of wave amplification increases inside the unstable region with increasing distance from the edge.

Although the various models differ slightly, they all tend to exhibit the properties illustrated in Figure 11.2. These models show that:

(a) the growth rate of waves is sensitive to vertical wind shear and therefore to the horizontal temperature gradient;

(b) larger vertical wind shears yield larger growth rates (and, by implication, more intense disturbances) and longer wavelengths;

(c) below some critical vertical wind shear no growth occurs;

(d) growth fails to occur below a minimum wavelength, about 2000 km in Figure 11.2;

(e) the optimum growth rate occurs at about 3000–4000 km, corresponding to a wavenumber of about 7 or 8 around the Earth at 45° latitude; and

(f) except under conditions of extraordinary vertical wind shear, there is an effective long-wave cutoff at about 8000–10 000 km.

Figure 11.2 depicts the vertical wind shear as a difference between the 500 and 1000 mb mean zonal wind speeds (U_5 and U_0). Maximum growth tends to occur at wavelengths of 3000–4500 km (broken curve labeled L_{max}). Some analyses of the stability problem show both a lower and upper critical wavelength respectively (L_0 and L_c), outside of which no exponential growth occurs (stable). In Figure 11.2, the long-wavelength cutoff, L_c, increases with increasing vertical wind shear, as does the rate of wave growth, which is larger and takes place over a wider range of wavelengths as L_{max} shifts to higher wavelengths.

The results of classic linear baroclinic instability theory are supported by

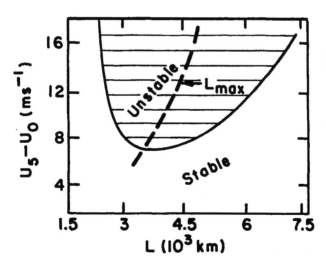

Figure 11.2 Baroclinic stability regimes as a function of wavelength (L) and thermal wind ($U_5 - U_0$) for a two-level model. L_{max} is the wavelength of maximum growth rate. (Based on a figure by Haltiner, 1971.)

observations. Results from non-linear (numerical) investigations show additional aspects of wave growth that are not evident in Figure 11.2:

(a) The lower critical wavelength below which no growth can occur (L_0) is not related to friction, but to the effects of vertical stratification of static stability. In some linear models, L_0 occurs at zero wavelength.

(b) Measures of the growth rate in the meridional and vertical fluxes of sensible heat are $(\overline{v' T'})$ and $(\overline{\omega' T'})$, which are shown in Chapter 5 to be related to the conversion of zonal to eddy potential energy and eddy available potential to eddy kinetic energy. The former is manifested on weather charts as the advection of warm air northward and cold air southward.

(c) The maximum growth rate, which occurs at L_{max}, increases with decreasing static stability ($\bar{s}$).

(d) Shorter waves seem to respond more sensitively to changes in stability than do long waves. L_{max}, the wavelength of most rapid wave growth, decreases with decreasing static stability.

(e) L_c, the long-wave cutoff, is related to the beta effect.

(f) Wave growth is more sensitive to stability in the lower troposphere than in the upper troposphere.

(g) Condensation and latent heating increase the growth rate of waves.

(h) In non-linear theory, shorter waves of 3000–4500 km grow faster initially, but in time they impart their energy to longer waves, which eventually dominate. Apparently the non-linear growth of longer waves also tends to suppress the growth of shorter waves by destroying the north–south temperature gradient and increasing the static stability by bringing in warmer air aloft and colder air at the surface. This cascade of wave energy, with time, is related to the fluctuations in the index cycle.

In the rest of this chapter we will try to explain some of the above aspects of baroclinic instability using the omega and pressure tendency equations as discussed in Chapter 4.

11.1 A simple two-level model of cyclone growth in a baroclinic atmosphere

It can be shown with the aid of some simple schematic relationships that the essentials of linear baroclinic instability theory are inherent in the quasi-geostrophic omega and pressure tendency equations presented in Chapter 4. Classical baroclinic instability theory expresses the growth rates of disturbances in terms of the doubling (or "e-folding") time, which is the time for a disturbance to increase its amplitude by a factor of 2 (or e = 2.72). We will adopt as a measure of disturbance growth the *maximum surface pressure tendency* in the disturbance. The mathematical treatment of Chapter 4 and the one-layer model derived therefrom will serve as a basis for examining the preferred wavelength and wavelength dependence of cyclone development.

In deriving (4.7) and (4.11), a simple assumption is made, which is that fields of vertical motion and advections are sinusoidal. The linearized omega and surface pressure tendency equations contain terms a and b (and a' and b'), which are dependent on wavelength. For given values of advections one can analyze the wavelength dependence of the surface pressure tendency, for example, as a function of wavelength and stability. Cursory inspection of (4.11) shows that pressure tendency vanishes at zero wavelength because the factors a' and b' vanish as L approaches zero; however, it is also evident from examination of the form of a' and b' that pressure tendency increases without bound with increasing wavelength (because a' increases without bound). This is clearly incorrect, but the problem can be rectified by considering the dependence of the advections themselves on wavelength.

Instead of simply specifying the advections as sinusoidal functions as we did in Chapter 4, a more basic model is proposed in which the 500 mb geopotential height and thickness patterns in (4.7) and (4.11) are assumed to be sinusoidal, rather than the advections. On this basis we derive sinusoidal functions for these two advections.

To begin this derivation, the linearized omega and pressure tendency equations of Chapter 4 are rederived with reference to the amplitude and wavelength of the 500 mb geopotential height and 1000–500 mb thickness fields. Let us consider the basic atmospheric state for Z and h. The periodic expression for Z is that of (2.14), which is

$$Z = Z_r + a_z y + \hat{Z} \sin \left(\frac{2\pi x}{L_x} \right) \sin \left(\frac{2\pi y}{L_y} \right) \tag{11.1a}$$

where a_z is the zonally averaged meridional geopotential height gradient at 500 mb and Z_r is a reference value of geopotential height at the equatorward border of the domain ($y = 0$). Recall from Chapter 2 that the advection of relative vorticity depends on there being a zonal wind speed component, which is expressed as a_z.

Similarly, 1000–500 mb thickness is also specified as a sine function, but with a phase lag of distance ϕ and angle $2\pi\phi/L_x$ between the troughs (or ridges) in the 500 mb geopotential height and thickness patterns. This is written as

$$h = h_r + a_h y + \hat{h} \sin \left(\frac{2\pi x}{L_x} + \frac{2\pi\phi}{L_x} \right) \sin \left(\frac{2\pi y}{L_y} \right) \tag{11.1b}$$

where a_h is the zonally averaged 1000–500 mb meridional thickness gradient and h_r is a reference thickness analogous to Z_r. Gradient a_h is equated to the mean zonal wind shear by the thermal wind relationship as follows:

$$a_h \equiv \frac{\overset{\cdots}{\partial h}}{\partial y} = -\frac{f_0}{g}(U_5 - U_0) \tag{11.2}$$

where U_5 and U_0 are defined as in (2.15). (Capital letter U signifies a zonal average.) Note that y is the distance along the north–south axis (measured positively toward the north from some arbitrary reference latitude), and h is the amplitude of the sinusoidal thickness perturbation. The parameter a_z is related to the mean zonal geostrophic wind by the relationship (2.15),

$$a_z \equiv \frac{\overset{\cdots}{\partial Z}}{\partial y} = -\frac{f_0 U_5}{g} \tag{11.3}$$

where the dotted overbars for Z (and h) denote a longitudinal average of 500 mb geopotential height and thickness. The phase-lag distance between the thickness and 500 mb height waves ϕ is positive when the thickness trough lies to the west of the 500 mb trough. All other symbols have the same meaning as in Chapter 4.

These sinusoidal patterns obeying (11.1) are shown in Figure 11.3 for the case of a wave of 120 m amplitude represented in the 500 mb height and thickness fields, displaced from each other by a phase lag of one-quarter of a wavelength. The 500 mb zonal wind speed (U_5) is taken as 25 m s^{-1} at 500 mb and zero at the surface. On the right-hand side of the figure, the 1000 mb geopotential height pattern is the difference between the 500 mb geopotential height and 1000–500 mb thickness fields shown on the left-hand side; it is displaced one-eighth of a wavelength east of the 500 mb trough axis.

Now let us consider first the advection of vorticity at 500 mb in this pattern. From the definition of the relative vorticity (2.8) one obtains the following relationship:

$$-V_{g5} \cdot \nabla(\zeta_{g5} + f) = -\left(\frac{16\pi^3 g^2 a_z \hat{Z}}{f_0^2 L^3} + \frac{2\pi g \hat{Z} \beta}{f_0 L}\right) C_x S_y \tag{11.4}$$

where the sine and cosine functions are henceforth abbreviated as S and C, respectively, with their arguments presented by subscripts. Thus, $C_x = \cos(2\pi x/L_x)$ and $S_y = \sin(2\pi y/L_y)$.

For a sinusoidal pattern and in the absence of the beta effect, (11.4) indicates that absolute vorticity advection is inversely proportional to the cube of the wavelength, as previously shown in Chapter 2. Long waves,

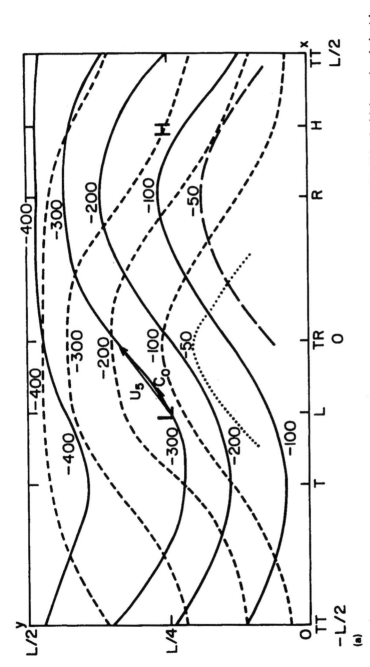

Figure 11.3 (a) Geopotential height contours (full and long broken curves, labeled in m) and 1000–500 mb thickness isopleths (short broken and dotted curves, labeled in m) in x and y for sinusoidal patterns expressed in equations (11.1a) and (11.1b). The x locations of the thermal trough (TT), the 500 mb trough (T), the surface low pressure center (L), the thermal ridge (TR), the 500 mb ridge (R) and the surface high pressure center (H) are indicated along the abscissa. Long and short arrows near low show direction and relative speed of the 500 mb geostrophic wind and surface low, respectively, labelled U_5 and C_0.

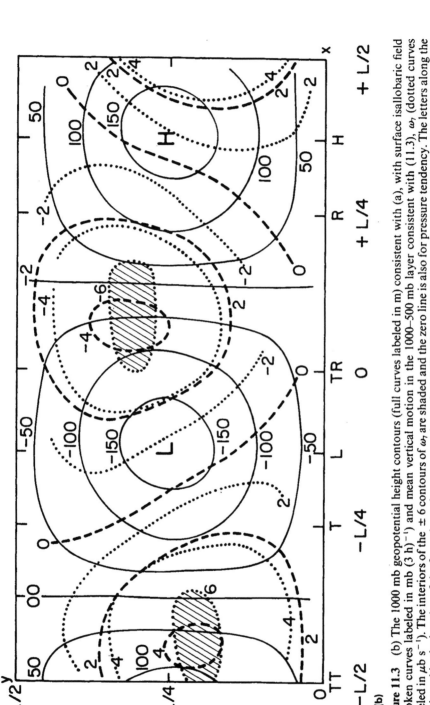

Figure 11.3 (b) The 1000 mb geopotential height contours (full curves labeled in m) consistent with (a), with surface isallobaric field (broken curves labeled in mb (3 h)$^{-1}$) and mean vertical motion in the 1000–500 mb layer consistent with (11.3), ω, (dotted curves labeled in μb s^{-1}). The interiors of the ± 6 contours of ω, are shaded and the zero line is also for pressure tendency. The letters along the x axis and in the interior of the figure have the same meaning as in (a). For these calculations, the atmospheric conditions are $\tilde{s} = 20\,^{\circ}C$ (500 mb)$^{-1}$, latitude $= 40°$, $h = \tilde{Z} = 120$ m, $p_* = 1000$ mb, $U_s = 25$ m s^{-1} and $\phi = 0.25L$. The wavelength (L) is 3500 km.

therefore, should possess weaker vorticity advection than short ones. This conclusion is generally supported by observation. Note that the two terms in parentheses are opposite in sign. The beta effect, represented by the second term in the parentheses, is always positive whereas the first term, which represents the relative vorticity advection, is negative because a_z is negative in the Northern Hemisphere; in the Southern Hemisphere a_z is positive (heights decreasing poleward in a westerly current) and f_0 is negative.

With increasing wavelength, the beta term becomes as large as the relative vorticity advection term in (11.4). This occurs at some large value of L, beyond which the absolute vorticity advection would change sign with increasing wavelength. The result that long waves are slow to develop and their rate is offset (and may be cancelled) by the beta effect is analogous to the opposing effect of the beta wind on motion of troughs and ridges discussed in Chapter 6. The relationship (11.4) also implies that maximum vorticity advection occurs exactly half-way between the trough and ridge, the consequence being that surface cyclones intensify most rapidly when they are centered one-quarter of a wavelength ahead of the 500 mb trough axis.

Similarly, the thickness advection can be determined by substituting the sinusoidal functions in the equation defining the thickness advection. Here it should be recalled from Chapter 3 that any geostrophic wind between 1000 and 500 mb can be used to compute the 1000–500 mb thickness advection provided that the isotherms at any level in this layer lie parallel to those at any other level (see (3.17)). Thus, the 500 mb geostrophic wind velocity (V_{gs}) is here considered as the advecting wind for 1000–500 mb thickness advection, which is equal to

$$- V_{gs} \cdot \nabla_p h = \left[\left(\frac{2\pi g \hat{Z} S_x C_y}{f_0 L} + \frac{g}{f_0} a_z \right) \left(\frac{2\pi h}{L} C_{x,\phi} S_y \right) \right.$$

$$\left. - \left(\frac{2\pi g \hat{Z}}{f_0 L} \right) \left(a_h + \frac{2\pi \hat{h}}{L} S_{x,\phi} C_y \right) \right] \qquad (11.5a)$$

where the subscripts (x, ϕ) refer to the argument $(2\pi x/L_x + 2\pi\phi/L_x)$. The expression for thickness advection is somewhat more complex than the one for vorticity advection because products of height and temperature amplitude are formed, and because it contains the phase lag between temperature and height fields. One inconsistency in this derivation is that the thickness advection contains products of sine and cosine functions in x and y, implying that there are higher-order sine or cosine terms (argument $4\pi y/L_y$) than prescribed in the imposed solution for omega (2.28). This can be seen by manipulation of the terms in (11.5a) using trigonometric identities to obtain

273

$$-V_{gs} \cdot \nabla_p h = -\frac{2\pi^2 g \hat{Z} \hat{h}}{f_0 L^2} S_\phi S_{2y} + \frac{2\pi g \hat{h} a_z}{f_0 L} C_{x,\phi} S_y - \frac{2\pi g \hat{Z} a_h}{f_0 L} C_x S_y \quad (11.5b)$$

which contains the term S_{2y}, defined as $\sin(4\pi y/L_y)$, a wave component with half the wavelength of the prescribed wavelength in height, thickness and omega.

The linearized omega equation (4.7) assumes that the thickness advection pattern is sinusoidal, thereby allowing a simple expression for the Laplacian of the thickness advection to be computed to determine the parameters b and b' in the linearized omega and surface pressure tendency equations. Taking the Laplacian of the thickness advection, as expressed by (11.5b), one obtains

$$\nabla_p^2 (-V_{gs} \cdot \nabla_p h) = -\frac{8\pi^2}{L^2} \left(-V_{gs} \cdot \nabla_p h - \frac{2\pi^2 g \hat{Z} \hat{h}}{f_0 L^2} S_\phi S_{2y} \right). \quad (11.6)$$

This relationship for the Laplacian of the thickness advection governing a sinusoidal pattern of height and temperature is not exactly proportional to minus the advection (as with vorticity advection) because of the presence of the second term in the parentheses of (11.6), which has a higher-wavenumber component S_{2y}. The S_{2y} term vanishes, however, if the perturbation thickness amplitude $\hat{h}$ is small compared to $\hat{Z}$ or is evaluated at $y = L/4$, which is at the latitude of the center of the disturbance (Fig. 11.3). Therefore, if one assumes that the temperature perturbation is small compared to that of geopotential height, or solutions are sought near the central latitude ($y = L/4$) of the disturbance, the higher-order components can be neglected in the expansion for vorticity and thickness advections. A higher-order term still resides implicitly in the definition of $-V_{gs} \cdot \nabla_p h$, however. (The effect of this higher-order term in Figure 11.3 is very small, however, if the perturbation amplitudes are small.)

Equations (11.4) and (11.5) are now substituted for their respective advections in the linearized omega equation (4.7) and pressure tendency equations (4.11) of Chapter 4. These solutions are shown in Figure 11.3b for the wave patterns of Figure 11.3a. Thus, the cyclone deepening rates determined from (11.4) and (11.5) in combination with (4.11) do not constitute a rigorous solution to the quasi-geostrophic equations, although they are nevertheless complex.

The resulting substitutions yield some lengthy equations. Let us first simplify these equations. First, since the deepening rate of the cyclone is expressed as the maximum surface pressure tendency, let us evaluate cyclone growth where the sine and cosine functions are equal to their maximum value of 1.0, which occurs at $x = 0$ and $y = L/4$. Second, we will assume that the mean zonal flow at 500 mb (U_5) is much greater than at 1000 mb (and therefore $|U_5| \gg |U_0|$. The result is a simplified equation for $(\partial p/\partial t)_s$, which is

$$-\left(\frac{\partial p}{\partial t}\right)_s = \left[\frac{f_0 p_*^2 \hat{Z} L^2}{8\pi^4 g \Delta p_5 \Delta p_3 K_0}\left(\frac{16\pi^3 g U_5}{f_0 L^3} - \frac{2\pi g}{f_0 L}\beta\right)\right.$$

$$\left. + \left(\frac{2p_*^2 \hat{Z} U_5}{\Delta p_5 \Delta p_3 \pi L K_0}\right)\right]\left(1 + \frac{L_R^2}{L^2}\right)^{-1}. \tag{11.7}$$

The two groupings of terms within the square brackets in (11.7) correspond to the absolute geostrophic vorticity advection (first term) and 1000–500 mb thickness advection. These two terms can be combined to form the expression

$$-\left(\frac{\partial p}{\partial t}\right)_s = \frac{4}{\Delta p_5 \Delta p_3 \pi K_0}\left(\frac{p_*^2 \hat{Z} U_5}{L(1 + L_R^2/L^2)}\right)$$

$$- \frac{1}{4\pi^3 \Delta p_5 \Delta p_3 K_0}\left(\frac{p_*^2 \hat{Z} L \beta}{(1 + L_R^2/L^2)}\right). \tag{11.8}$$

The two terms on the right-hand side of (11.8) constitute a version of the quasi-geostrophic surface pressure tendency equation scaled to show the relevant physical variables, wind speed (or vertical shear), Coriolis parameter, etc. Recall that the important length scale in the omega equation is the Rossby radius of deformation (defined previously in equation (4.7)), $L_R = [(8R_d/\bar{p})(\bar{s}p_*^2/f_0^2)]^{1/2}$. The Rossby radius of deformation appears as a governing parameter in many theoretical investigations of baroclinic instability. It is a function of three variables: static stability, Coriolis parameter (latitude) and depth of the disturbance.

11.2 Wavelength dependence for cyclone growth

The function (11.8) is graphed in Figure 11.4. Maximum surface pressure tendency is at wavelength L_{max}, the minimum (lowest possible) wavelength is at L_0 and the maximum wavelength for cyclone deepening is at L_c. In the remainder of this chapter we will examine the behavior of the function (in terms of the curve in Figure 11.4 and equation (11.8)) in order to show how static stability, latitude, depth of disturbance, wind speed and amplitude affect the cyclone deepening rate, wavelength of maximum growth rate and the maximum possible wavelength of the wave/cyclone. (The minimum wavelength for cyclone deepening (L_0) is always zero in this function.)

Consider the first term on the right-hand side of (11.8), the shear term, which contains the expression $[L(1 + L_R^2/L^2)]^{-1}$. The maximum surface pressure tendency exhibits a rapid rise with increasing wavelength to a maximum at L_{max}, above which the pressure tendency decreases more

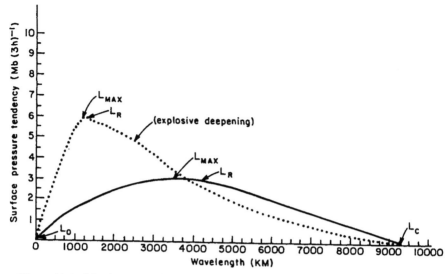

Figure 11.4 Maximum surface pressure tendency versus wavelength (full curve) from solution of (11.8). Relevant parameters are the wavelength of maximum growth rate (pressure tendency) labeled L_{max}, the Rossby radius of deformation (L_R), the lower cutoff wavelength (L_0) and the upper cutoff wavelength (L_c). The dotted curve illustrates the case of small Rossby radius of curvature, such as would occur with low static stability, high latitude and significant diabatic heating ($\mu = 0.3$).

gradually to zero at L_c; beyond L_c the pressure tendency is negative. This function explains the shape of Figure 11.4. Its mathematical properties can be appreciated by differentiating the first term on the right-hand side of (11.8) with respect to L and setting it equal to zero to solve for the maximum in the function in order to obtain the wavelength of maximum cyclone deepening. In so doing, one obtains the result that the maximum in the function (L_{max}) is at $L = L_R$. Thus, *in the absence of the beta effect, the maximum growth rate of cyclones, as expressed by the magnitude of the surface pressure tendency, occurs at the Rossby radius of deformation.* With the beta effect included, L_{max} occurs at a slightly smaller wavelength than L_R. At $L < L_R$, increasing L causes an increase in $(\partial p/\partial t)_s$ because $(1 + L_R^2/L^2)$ decreases faster than L increases in the denominator of (11.8). At wavelengths above L_R, increasing the wavelength involves a decrease in $(1 + L_R^2/L^2)$, but, because L is in the numerator, there is a decrease in the magnitude of $(\partial p/\partial t)_s$. The beta effect tends to work against the cyclone deepening, increasing the surface pressure tendency, as indicated by the vorticity advection relationship in (11.4). The beta effect is negligible at short wavelengths but increases with increasing wavelength. Consequently, the beta effect not only exerts a slowing effect on wave growth, but also shifts the wavelength of maximum cyclone deepening (L_{max}) to shorter wavelengths. Importantly, the beta effect in this equation accounts for a long-wave cutoff for cyclone development (L_c), beyond which the sign of $(\partial p/\partial t)_s$ reverses. By setting $(\partial p/\partial t)_s = 0$ and solving for $L\ (= L_c)$ one obtains the result that

$$L_c = (16\pi^2 U_5/\beta)^{1/2}$$

which is also analogous to setting the vorticity advection equal to zero in (11.4). Although the short-wave cutoff wavelength (L_0) is exactly zero in this model, the long-wave cutoff (L_c) is proportional to the square root of the zonal wind speed and inversely proportional to $\beta^{1/2}$. Therefore, weak westerly winds tend to inhibit the growth of longer waves and shift the wavelength of maximum cyclone deepening to somewhat smaller sizes, in accordance with Figure 11.2. Moreover, examination of (11.8) shows that the magnitude of the surface pressure tendency is smaller with weaker wind speeds and smaller amplitude of the 500 mb wave.

Let us see how surface pressure tendency increases at the optimum wavelength $L = L_R$. To do this we set $L = L_R$ in (11.8) and ignore the β term. The result is that at this optimum wavelength $(\partial p/\partial t)_s$ is proportional to $p_*^2 \hat{Z} U_5/L_R$. Thus, the pressure tendency increases inversely with the magnitude of L_R. Since the disturbance depth p_* appears in both the numerator and in L_R in this expression, the maximum surface pressure tendency is roughly proportional to $p_*^2 \hat{Z} U_5 f_0$ and inversely proportional to $\bar{s}^{1/2}$. Thus, the surface cyclone deepening rate increases with increasing amplitude of the disturbance, with increasing 500 mb wind speed (vertical wind shear or meridional temperature gradient), with increasing latitude and with decreasing static stability. Although p_* appears both inside and outside of the expression for L_R, a decrease in L_R solely due to the decrease in p_* causes both L_R and $(\partial p/\partial t)_s$ to decrease. Generally, however, the maximum surface cyclone growth rate is almost inversely proportional to the magnitude of L_R.

Since neither latitude nor disturbance depth vary by more than a factor of 2 over mid-latitudes, the dominant influence on surface pressure tendency, for a given zonal wind speed, is the static stability. Static stability affects the vertical motion and therefore the surface divergence and the surface pressure tendency because it determines how hard the atmsophere must work to move air vertically against the effects of buoyancy. The more stable the atmosphere, the more difficult it is to move air vertically. Stated alternatively, the smaller the static stability, the more steeply inclined are the isentropic surfaces for a given horizontal temperature gradient (equation (12.10)) and the faster air can ascend or descend. A given vertical displacement therefore corresponds to a smaller horizontal scale over which the ascent takes place. This result that cyclone growth rate is very much dependent on the static stability lies at the heart of explosive cyclogenesis discussed in Chapter 10.

The linearized equation for ω_d with the advections (11.4) and (11.5) substituted in (4.9) behaves differently from that for surface pressure tendency. In the vertical motion equation, the magnitude of ω_d increases continuously with decreasing wavelength, becoming infinite at zero wavelength, but decreases to zero at a finite upper wavelength, as in the case of pressure tendency. Paradoxically, infinite vertical motion does not imply infinite pressure tendency, although the vorticity tendency is unbounded for decreasing wavelength.

For small values of L_R, as would occur in the presence of small static stability, the preferred wavelength of maximum cyclone deepening rate is small but the cyclone deepening rate is large. The result of low static stability on the deepening of a cyclone and on the wavelength of maximum deepening rate is illustrated by the difference between the two curves in Figure 11.4; the dotted curve is meant to simulate the case of a polar low. The distribution of L_R for a case of coastal cyclogenesis is presented in Figure 11.5. Note that the values of L_R tend to be largest over the interior of the continent and over the northeast, and smallest off the southeastern coast.

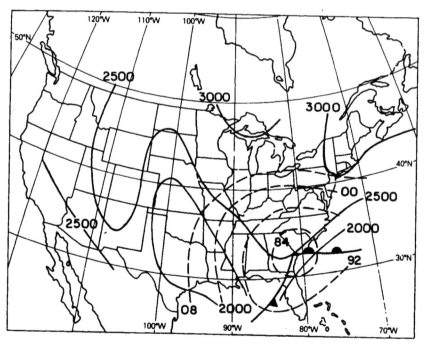

Figure 11.5 Distribution of sea-level isobars (broken curves in mb with 1000s digit omitted) and isopleths of Rossby radius of deformation (L_R) (full curves labeled in km) for 0000 GMT 29 March 1984. (Figure provided by Timothy Dye.)

The relationship between latitude and cyclone deepening rate is also contained in L_R. As latitude decreases, the pressure tendency becomes smaller, although there is little sensitivity to latitude poleward of about 55°. The wavelength of maximum cyclone deepening rate is slightly dependent on wind speed (more specifically, vertical wind shear), L_{max} increasing with increasing U_S; surface pressure tendency, however, increases linearly with increasing U_S.

Changing the vertical phase lag also affects the cyclone deepening rate but not the wavelength dependence. Pressure tendency increases with increasing phase lag for phase angles less than 90°; this is somewhat misleading because the relative location of the surface low center changes with changing phase angle. (Evaluating (11.8) at the surface low center, the surface pressure tendency is largest when the vertical phase lag between the surface and 500 mb height fields is one-quarter of a wavelength.)

Decreasing the depth of the disturbance (p_*) shifts the wavelength of maximum deepening rate to shorter wavelengths but also decreases the cyclone deepening rate because p_* is directly proportional to sea-level pressure tendency in the numerator of (11.8). An effective top of cyclones probably lies close to the tropopause. In the case of polar lows, which occur in the cold air near upper troughs, the depth of the tropopause is less than that for the frontal cyclone. The dotted curve in Figure 11.4 suggests conditions appropriate to explosive maritime cyclones, such as the polar low, which form at high latitudes and in conditions of low static stability.

As discussed in Chapter 9, a simple assumption can be made that diabatic heating is proportional to the dynamically forced vertical motion (9.4), which leads to a solution for surface pressure tendency analogous to (11.8), but with the scaled static stability $\bar{s}' = (1 - \mu)\bar{s}$ in L_R. Increasing μ causes the effective stability ($\bar{s}'$) to diminish, the wavelength of maximum cyclone deepening rate to decrease and the surface pressure tendency to increase. A value of $\mu = 0.5$, for example, will produce a value of L_{max} less than 1000 km for typical values of static stability ($\bar{s}$), latitude and disturbance depth. As the heating factor (μ) approaches 1.0, diabatic heating begins totally to dominate the dual baroclinic–diabatic system depicted in Figure 9.15, the wavelength approaches zero and the surface pressure tendency increases boundlessly. The troublesome result is that the disturbance deepens at an infinite rate on the smallest resolvable scale. Lack of resolution of this problem constituted a temporary block in modeling convectively driven storms until the advent of the theory of "conditional instability of the second kind" (CISK) in the 1960s. For tropical cyclones, however, quasi-geostrophic theory is no longer valid because the effective Rossby number of the system (U/fL) approaches or exceeds 1 and the effective static stability becomes negative. It is, therefore, remarkable that the simple theory discussed in this chapter remains credible even at small wavelengths (e.g. $L < 1000$ km).

The vanishing of cyclone deepening rate at zero wavelength is a result peculiar to other simple models and may be due to an inability to resolve vertical variations in static stability. It is easy to see, physically, why there is a decrease in the magnitude of $(\partial p/\partial t)_s$ and a reversal of sign at large wavelengths. Centers of maximum and minimum vorticity and temperature are relatively far apart at long wavelengths and therefore the advections are likely to be feeble and offset by the beta effect. At shorter wavelengths it is more difficult to illustrate the physical basis for the lower-wavelength cutoff. Although with increasingly shorter wavelengths the vorticity tendency increases boundlessly, geopotential heights do not behave in such fashion because of the $L_R{}^2/L^2$ term in the denominator of the Laplacian of geopotential height. Thus small perturbations can spin up their vorticity faster without a corresponding increase in the magnitude of $(\partial p/\partial t)_s$ because of the small size of the system. A more physical explanation is that the vorticity tendency of small systems may be large and the horizontal pressure gradients are large but the net difference in pressure between the high and the low is nevertheless small because the scale is small.

When $L = L_R$, the two terms on the left-hand side of the omega equation (4.3) are of equal importance and the advections are the most efficient in driving the local pressure tendency. When $L < L_R$, the horizontal Laplacian of ω increasingly dominates the left-hand side of the omega equation with decreasing wavelength. Consequently, at short wavelengths ($L < L_R$), the advections force the vertical motion mainly through the term $\nabla_p^2 \omega$ on the left-hand side of the omega equation, and so the wind field exerts a greater influence on the Laplacian of the temperature tendency and therefore on the distribution of temperature. Conversely, at long wavelengths, when $L > L_R$, the vertical stretching term dominates the left-hand side of the omega equation, and the advections exert a greater effect on the local vorticity tendency, and therefore on the distribution of the wind field, than on the temperature field. A balance occurs between the two terms on the left-hand side of the omega equation at $L = L_R$. Stated alternatively, at long wavelengths ($L \gg L_R$) the vorticity (wind) field responds more sensitively to the advections than does the mass (temperature) field, and at short wavelengths the mass (temperature) field responds more sensitively to the advections than does the wind field. At the Rossby radius of deformation (L_R), both terms on the left-hand side of the omega equation possess equal importance. *Accordingly, an alternative definition of what constitutes a long or short wave is the size of L with regard to L_R.* The ratio (L/L_R) govern the importance of *geostrophic adjustment* to either vorticity or temperature advection.

11.3 Summary: fundamental influences on cyclogenesis

Let us now summarize what these equations are saying. The surface pressure tendency in (11.8) increases linearly with increasing value of the 500 mb wind speed (U_5) and amplitude of the perturbation (Z). Surface pressure tendency also increases with the latitude (with f), with decreasing stability ($\bar{s}$), with increasing depth of the disturbance and with increasing diabatic heating. This analysis explains why cyclones develop much more rapidly and possibly possess shorter wavelengths over the ocean (where stability is relatively small and diabatic heating large), and why baroclinic development is more rapid at middle and upper latitudes (where the wind speeds are greater and the Coriolis parameter larger), than in the Tropics.

Fundamentally, the necessary conditions for rapid baroclinic storm development are:

(a)   a strong meridional temperature gradient;
(b)   a vertical phase lag between geopotential height and temperature fields;
(c)   a location at middle or high latitudes; and
(d)   a relatively low static stability.

That cyclogenesis tends to favor certain geographical locations is due to the tremendous inhomogeneity of lower-tropospheric baroclinicity and static stability. Rapid intensification of a surface cyclone along the coast or over warm ocean or land surfaces (e.g. the Great Plains of the United States in springtime (see Fig. 9.12)) occurs in response to a migrating wave and a pre-existing pattern of forcing. Cyclogenesis does not occur continuously following the movement of a seemingly potent 500 mb trough but occurs preferentially in certain locations. This suggests that there are regions where the atmosphere is much more efficient in allowing 500 mb absolute vorticity and 1000–500 mb thickness advections to be realized as stronger vertical motions and surface pressure tendencies.

The dominance of the Rossby radius of deformation on the scale of cyclone deepening underscores the difference between the formation of a typical trough–ridge system and an explosively deepening cyclone. When the static stability possesses a typical globally averaged value, growth rate is slow and occurs at wavelengths of 3000–4500 km; growth rate may be negligible because of internal friction. On the other hand, if quasi-geostrophic forcing occurs in a region of low static stability, rapid cyclone deepening at relatively small wavelengths may occur, even in the absence of large vorticity or temperature advections.

Although standard treatments of baroclinic instability theory refer to an optimum wavelength for wave growth between 3000 and 4500 km, one does not often observe explosive growth of waves at these longer wavelengths, but at scales of 1000–2000 km. The reason for this preferred cyclone deepening at smaller wavelengths is that static stability (and, therefore, L_R) is highly variable in space and depends on geographic factors that influence surface and convective warming. Rapid cyclone growth can occur only at small L_R and, therefore, at small wavelengths.

Thus, explosive growth occurs only in those regions where the Rossby radius of deformation is small. Large variations in the Rossby radius of deformation are imposed indirectly as the result of terrain influences on static stability, e.g. a warm ocean surface beneath cold air aloft on the poleward side of the polar front jet. Maps showing the distribution of L_R (e.g. Fig. 11.5) would constitute a useful tool for forecasting cyclone development, but they are currently not in operational use.

Problems

11.1 For a value of wavelength $L = L_R$, calculate the maximum negative surface pressure tendency for the case of 17 January 1979 (shown in Fig. 14.1). Estimate the size of the advection solenoids using Table 4.1. Confine your analysis to the earlier time period.

11.2 Suggest a stability argument why surface highs do not grow explosively.

11.3 (a) Using equation (11.8) and ignoring the beta effect, find the ratio of surface pressure tendencies for the following three cases: (i) wind speeds (U_s) of 20 and 40 m s^{-1}; (ii) amplitudes ($\hat{Z}$) of 120 and 240 m; and (iii) $L_R = L$ at 4000 and 2000 km.

    (b) What will be the upper cutoff wavelength (L_c) in the model for wind speeds of 10 and 20 m s^{-1} at 40° latitude? What happens to L_c as the latitude approaches 90°?

Further reading

Carlson, T. N. 1982. A simple model illustrating baroclinic development. *Bull. Am. Met. Soc.* **63**, 1302–8.

Gall, R. 1976. The effects of released latent heat in growing baroclinic waves. *J. Atmos. Sci.* **33**, 1686–701.

Gall, R., R. Blakeslee and R. C. Somerville 1979. Baroclinic instability and the selection of the zonal scale of the transient eddies of middle latitudes. *J. Atmos. Sci.* **36**, 767–84.

Green, J. S. A. 1960. A problem in baroclinic stability. *Q. J. R. Met. Soc.* **86**, 237–51.

Haltiner, G. J. 1971. *Numerical weather prediction.* New York: Wiley.

Lilly, D. 1960. On the theory of disturbances in a conditionally unstable atmosphere. *Mon. Wea. Rev.* **88**, 1–17.

Moorthi, S. and A. Arakawa 1985. Baroclinic instability with cumulus heating. *J. Atmos. Sci.* **42**, 2007–31.

Rasmussen, E. 1979. The polar low as an extratropical CISK disturbance. *Q. J. R. Met. Soc.* **105**, 531–49.

Simmons, A. J. and B. J. Hoskins 1978. The life cycles of some nonlinear baroclinic waves. *J. Atmos. Sci.* **35**, 414–32.

Staley, D. O. and R. Gall 1977. On the wavelength of maximum baroclinic instability. *J. Atmos. Sci.* **34**, 1679–88.

12

Airflow through mid-latitude synoptic-scale disturbances

A classic problem in synoptic meteorology is to relate mesoscale patterns of vertical air motions, cloudiness and precipitation to routinely observable synoptic-scale, three-dimensional fields of horizontal wind (momentum), temperature and pressure (mass). Historically, numerous models were proposed during the nineteenth and early twentieth centuries, culminating in the Norwegian polar-front cyclone model of the Bergen school (see Ch. 10), in which precipitation patterns are related to vertical motions resulting from the relative movement of air along inclined frontal surfaces. The strength of the Norwegian cyclone model, coupled with modern theoretical innovations, has led to an understanding of how weather systems evolve.

Nevertheless there are situations where the classic cyclone models do not seem to explain the presence of cloud or precipitation patterns. There are many instances in which precipitation is widespread in the absence of a well-defined front or surface cyclone. Although the vertical motions are consistent with the quasi-geostrophic forcing, the distribution of cloudiness and precipitation does not conform exactly with regions of ascent, since clear air may ascend without producing cloud. Thus the quasi-geostrophic dynamics are not always able to explain the shape of cloud and precipitation patterns or the distribution of humidity about the cyclone. Moreover, the sloping surfaces of ascent depicted in the Norwegian cyclone model are not readily appreciated in prescribing the quasi-geostrophic forcing.

Movement of air along sloping surfaces in synoptic-scale weather systems is termed "slantwise ascent or descent" or quasi-horizontal slantwise convection, to distinguish the motion from that in smaller-scale convective storms in which the air can ascend vertically from the bottom to the top of the troposphere in an hour or less, as compared to days in a wave/cyclone. This slantwise motion of air streams is not evident on conventional weather maps.

A great deal of research concerning isentropic vertical motions and trajectories has been carried out over the past half-century and, at times, the method of isentropic analysis has undergone periods of relative popularity followed by intervals of declining interest. Modern methods of isentropic analysis, however, were developed during the 1930s as the result of the pioneering work of J. Namais. Its application to the analysis of airflow through synoptic-scale weather systems was revitalized in the late 1950s and 1960s by a number of researchers, such as E. Danielsen, E. Reiter, R. Bleck, R. Petersen, K. Browning, F. H. Ludlam, T. N. Carlson and others. Presently, isentropic analysis is once more undergoing a revival.

Analyses of isentropic trajectories of airflow through mid-latitude systems, such as have been made since the middle 1960s, tend to show that slantwise motion deviates markedly from that on isobaric surfaces. This departure from horizontal isobaric flow is most evident when the winds are taken relative to the velocity of the system (front or cyclone). One of the most revealing aspects of this "relative wind" isentropic flow is the existence of sharply defined boundaries, which differentiate air streams of vastly differing moisture contents. Air streams tend to contain relatively narrow ranges of θ and θ_w peculiar to the air stream's origins. Furthermore, the shapes of cloud bands correspond to airflow in a coordinate system moving with the system. In this relative wind framework, sharply defined cloud edges mark the lateral boundaries of air streams of differing origins. Isentropic analysis offers an opportunity for revising the antiquated *air mass* (static) concept in favor of an *air stream* (Lagrangian) one.

In this chapter, air motion through the mature wave/cyclone is described in relative wind isentropic coordinates in order to illustrate various aspects of slantwise motion and to relate it to the three-dimensional structure of cyclones and their attendant cloud patterns. In so doing, we will present case studies to illustrate different aspects of the wave/cyclone and cloud patterns that are not apparent on conventional isobaric charts. Most of the discussion is concerned with the structure of the wave/cyclone in relative wind isentropic coordinates (with references to the quasi-geostrophic dynamic framework discussed in previous chapters). We begin by illustrating the relative wind motion through non-developing baroclinic cloud leaves and subsequently show how the airflow and cloud pattern evolve during cyclogenesis. We will make an analogy between airflow through synoptic-scale weather systems and the airflow through a cumulonimbus cloud. Finally, some aspects of cyclogenesis on downstream development of cyclones is presented within the context of the airflow concept.

12.1 Isentropic analysis

There are various advantages and disadvantages in the use of isentropic coordinates. Disadvantages are that:

(a) the atmosphere is not completely adiabatic, especially in the boundary layer and in the vicinity of strong vertical mixing or convection;
(b) isentropic surfaces may intersect the ground;
(c) isentropic surfaces may extend from low to high levels in the atmosphere and thereby represent no single quasi-horizontal surface; and
(d) meteorologists are unaccustomed to interpreting isentropic weather maps.

Advantages are that:

(a) as a first approximation the motion is adiabatic (as much as it is quasi-geostrophic) and this motion is related to the configuration and origin of air streams;
(b) vertical motion can be shown explicitly on a quasi-horizontal isentropic chart; and
(c) isentropic flow presents a truer picture of the three-dimensional air motion than isobaric surfaces and preserves the quasi-horizontal behavior of the three-dimensional flow.

Moreover, some of the fundamental relationships or parameters, such as the vorticity equation or the vertical motion, can be expressed more simply on isentropic charts than in p coordinates.

For example, a version of the vorticity equation can be written in potential temperature (θ) coordinates for frictionless, adiabatic flow by combining the vorticity equation (3.6) (with ζ retained in the development term) with the continuity equation (1.8e). In isentropic coordinates and for the adiabatic motion, this equation is written

$$d(\zeta_\theta + f)/dt = -(\zeta_\theta + f)\, \nabla_p \cdot V_\theta \qquad (12.1)$$

or

$$\frac{d}{dt}\left((\zeta_\theta + f)\, \frac{\partial \theta}{\partial p}\right) = 0$$

which is to say the Lagrangian change of absolute vorticity on the potential temperature surface is equal to the development term; the tilting terms disappear in the transformation from pressure to isentropic coordinates. The above equation is analogous to (9.1). The significance of $(\zeta_\theta + f)\, \partial\theta/\partial p$, the potential vorticity, is discussed in more detail in Chapter 15.

In the Eulerian system (with respect to a fixed point on the ground), the change of a scalar variable with respect to time is expressed as a local derivative. In the *Lagrangian* system, however, the change of a scalar variable following a point in space is expressed by the total derivative. Thus, restating (1.29),

$$\frac{d\theta}{dt} = \frac{\partial\theta}{\partial t} + u\,\frac{\partial\theta}{\partial x} + v\,\frac{\partial\theta}{\partial y} + \omega\,\frac{\partial\theta}{\partial p} = \dot\theta$$

$$= \frac{\partial\theta}{\partial t} + V\cdot\nabla_p\theta + \omega\,\frac{\partial\theta}{\partial p} = \left(\frac{\dot Q_{nd}}{c_p}\right)\left(\frac{p_0}{p}\right)^\kappa. \tag{12.2a}$$

When there is no condensation, the principal contributors to diabatic heating are sensible heating at the surface, absorption or emission of radiant flux energy by the air and turbulent mixing. Turbulent mixing of the air can change the parcel's value of θ, although the process merely rearranges the vertical distribution of dry or moist static heat energy in a column without adding or subtracting any heat. While all of these effects can be quite important in the planetary boundary layer, the dominant contribution to diabatic heating above the boundary layer in clear air is long-wave (thermal infrared) radiation. Let us recall that radiant energy losses in clear air at middle levels are typically the equivalent of about 1–2°C per day cooling. This amount is equal to a temperature advection brought about by a wind speed of 10 m s^{-1} blowing across a gradient of temperature of 1°C per 500 km, a rather small value. Therefore, a good first approximation is to neglect the total diabatic heating in (12.2a) above the boundary layer and in regions of relatively tranquil, cloud-free air.

Equation (12.2a), written as a conservation equation, $d\theta/dt = 0$, suggests a method by which the motion of air through weather systems can be analyzed. Further, as a corollary to (12.2a), the mixing ratio (r) is also conserved following the flow in dry, adiabatic processes, such as will occur in clear areas above the boundary layer. Thus,

$$dr/dt = 0 \qquad d\theta/dt = 0 \tag{12.2b}$$

and

$$\omega = \frac{\partial\theta/\partial t + V\cdot\nabla_p\theta}{-\partial\theta/\partial p}. \tag{12.3}$$

From (12.3), one can see that vertical motion on a pressure surface is equal to the sum of the advection of potential temperature and the local (temporal) derivative of potential temperature divided by the static stability. The local derivative can be determined from a succession of isentropic charts made at differing times. This form of the potential temperature equation is very useful for computing isentropic vertical motion from isobaric temperature charts.

Alternatively, a transformation of the terms in (12.3) from derivatives of a constant-pressure surface to those of pressure on a constant-θ surface is accomplished by means of the following relationships:

$$\left(\frac{\partial\theta}{\partial t}\right)_p = -\frac{\partial\theta}{\partial p}\left(\frac{\partial p}{\partial t}\right)_\theta; \qquad \left(\frac{\partial\theta}{\partial x}\right)_p = -\frac{\partial\theta}{\partial p}\left(\frac{\partial p}{\partial x}\right)_\theta$$

$$\left(\frac{\partial\theta}{\partial y}\right)_p = -\frac{\partial\theta}{\partial p}\left(\frac{\partial p}{\partial y}\right)_\theta \qquad (12.4a)$$

from which the expression for the adiabatic vertical motion (ω_θ) is obtained as

$$\omega_\theta = (\partial p/\partial t)_\theta + (V_\theta \cdot \nabla_\theta p). \qquad (12.4b)$$

Equation (12.4b) states that vertical and horizontal motion on a θ surface is expressed in terms of the advection of pressure and the pressure tendency on that surface.

An approach to applying (12.4b) has been to calculate the three-dimensional motion of the air from a succession of maps and, with the aid of constraints, the trajectories of air motion through weather systems. One of these constraints is analogous to the geostrophic wind in which the air parcels obey a geostrophic balance on the θ surface but the height lines are replaced by a streamfunction M, called the *Montgomery streamfunction*, represented by the full curves in Figure 12.1. This streamfunction is written as

$$M = (c_p T + gz)_\theta \qquad (12.5)$$

which is equal to the dry static energy of a parcel. The analogous equations for geostrophic balance on the isentropic surface are

$$v_{g\theta} = (1/f_0)\, \partial M/\partial x \qquad (12.6a)$$

$$u_{g\theta} = -(1/f_0)\, \partial M/\partial y. \qquad (12.6b)$$

The horizontal gradient of potential temperature is proportional to that of the temperature

$$\nabla_p \theta = (p_0/p)^\kappa \nabla_p T \qquad (12.7)$$

and is virtually equal to the horizontal temperature gradient in the lower troposphere ($(p_0/p)^\kappa = 1$). The slope of the isentropic surface with respect to

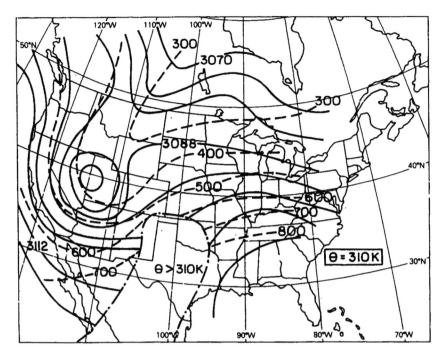

Figure 12.1 Isentropic flow on a 310 K surface at 0000 GMT 14 April 1972. Full curves are isopleths of the Montgomery streamfunction (M) at intervals of 60×10^2 m^2 s^{-2}; broken curves are isobars in mb. Chain curve is intersection of isentropic surface with ground; in region labeled $\theta > 310$ K, the isentropic surface is below the ground. (Based on figure by Danielsen, 1974.)

that of the pressure surface, which is nearly horizontal, is simply the isobaric temperature gradient divided by the static stability ($\bar{s}$).

An important corollary to the above argument, which can be verified by inspection of a typical vertical temperature sounding, is that, *for a given positive static stability, higher values of θ or θ_w are found at higher elevations (lower pressure)*. Alternatively stated, *in a statically stable atmosphere, isentropic surfaces slope upward toward colder air*. The pressures in Figure 12.1 generally decrease poleward in the direction of the trough, whose closed center in the Montgomery streamfunction isopleths corresponds to relatively cold air. Isentropes with values greater than 310 K intersect the ground in a salient along the lower part of the diagram.

In a cross section (Fig. 12.2), the isentropes on the right slope upward toward the left (southwest) in the direction of the trough center and toward a dome of cold air. At the trough axis the tropopause is relatively low (about 350 mb) compared to that on the eastern side of the figure (about 150 mb). Strongest winds occur close to the tropopause in the isentropes and in the

289

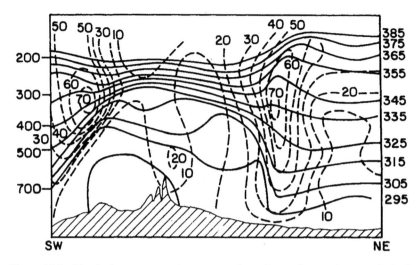

Figure 12.2 Vertical cross section approximately orthogonal to wind flow southwest–northeast at 0000 GMT 20 April 1973. Full curves are isentropes in K and broken curves are isotachs in m s^{-1}. (Based on figure by Danielsen, 1974.)

flare just below the tropopause. The flare marks the change in slope of the isentropes upward in the direction away from the cold dome above the jet and downward away from the cold dome below the jet. The atmosphere in the stratosphere above the cold dome is relatively warm compared to the air in the high troposphere to the west and east, as discussed with regard to the formation of tropopause undulations in Chapter 10. There is a discontinuity along the sloping part of the tropopause. At the level of the jet the horizontal potential temperature gradient is approximately zero, which is consistent with a maximum in the wind profile ($\partial u_g / \partial p = 0$) at that level.

Dry isentropic motion is not strictly valid in regions of condensation, although the vertical distribution of moisture is such that the adiabatic approximation is not greatly in error in the middle and upper troposphere. If one defines "active cloud" as that which is producing continuous precipitation, it can be seen from inspection of satellite photographs (e.g. Fig. 1.3) that areas of active cloud cover a relatively small fraction (perhaps 10–20%) of the Earth's surface. Except for convective clouds, which are usually confined to small areas, most of the precipitating cloud in mid-latitude weather systems consists of bands of layer cloud, with well-defined lower and upper boundaries typical of stably stratified flow. Most of this stable condensation is due to quasi-horizontal slantwise ascent and can be represented mathematically as a pseudo-adiabatic process.

For moist or dry adiabatic processes the wet-bulb potential temperature θ_w (or, alternatively, the equivalent potential temperature θ_e) is conserved, and either θ_w or θ_e can be substituted for θ in (12.2)–(12.4) for adiabatic flow, provided the static stability ($- \partial\theta/\partial p$ or $- \partial\theta_w/\partial p$) is positive. Accordingly, the vertical motion can be computed from moist isentropic surfaces using the moist static stability ($(T/\theta_e) \partial\theta_e/\partial p = \bar{s}_m$). A moist adiabatic streamfunction similar to that for the Montgomery streamfunction (12.5) can be derived using virtual equivalent potential temperature in place of temperature. The gradient of this moist Montgomery streamfunction is differentiated with respect to x or y on the equivalent potential temperature surface to obtain the moist adiabatic geostrophic wind components. In practice, however, use of the moist Montgomery streamfunction is impeded by its great sensitivity to uncertainty in the moisture and geopotential height data. Because of errors inherent in the data, objective procedures for determining either the

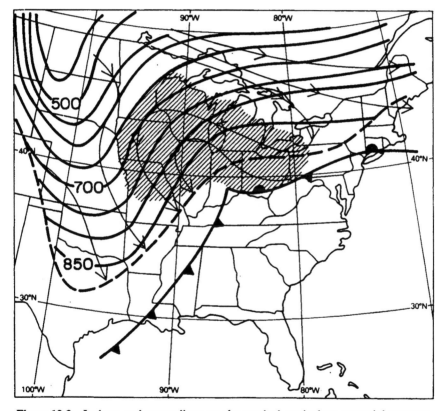

Figure 12.3 Isobars and streamlines on the vertical equivalent potential temperature surface ($\theta_e = 291$ K) for 0000 GMT 26 March 1964. Shading represents precipitation; the broken border defines the intersection of the potential temperature surface with the ground. Surface frontal boundaries are shown in conventional notation. (Based on a figure by Danielsen and Bleck, 1967.)

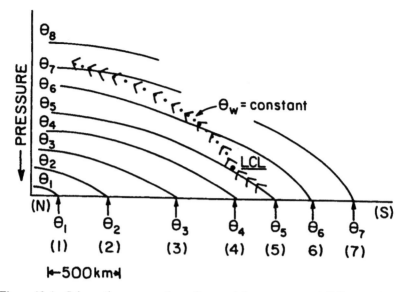

Figure 12.4 Schematic cross section of potential temperature θ (full curves) in the Northern Hemisphere. The vertical coordinate is pressure and the horizontal one is distance from north (N) to south (S) in the Northern Hemisphere. Subscripted isentropes refer to potential temperature surfaces, which intersect the ground at locations designated by numbers in parentheses. A parcel moving northward with potential temperature θ_5 (arrows) from a surface location at point 5 reaches its lifting condensation level at location 4 and thereafter ascends along the barbed streamline following a moist adiabat (θ_w = constant).

dry or moist isentropic streamfunctions may not afford any improvement in determining air trajectories over subjective methods.

Consider a moist isentropic flow on an equivalent potential temperature surface poleward and west of a developing cyclone, illustrated in Figure 12.3. Because the horizontal pressure gradients represent gradients of both moisture and temperature, the isobaric gradient is much stronger than for the dry case shown in Figure 12.1. By combining both temperature and humidity, θ_w becomes an excellent indicator of fronts. Generally, ascent is occurring in the region of precipitation (shaded), except for the southwestern portion of the precipitation shield where moist air appears to be descending on a trajectory from a relatively dry region to the west.

Now, imagine a schematic meridional cross section of potential temperature, as in Figure 12.4. If the isentropes are fixed, the streamlines and material trajectories are identical. The sloping potential temperature isotherms (θ_1, θ_2, θ_3, etc.) intersect the surface at locations designated as 1, 2, 3, ... Because it is colder toward the pole, the isentropes slope upward in that direction. Thus, the potential temperature isentrope (θ_5) that intersects the surface at location 5 is found aloft at location 2.

Air moving poleward dry adiabatically ascends along the isentropic surfaces and, if sufficient lifting takes place, achieves condensation at the lifting condensation level (LCL) of the air parcel. In Figure 12.4, the air that leaves point 5 and the surface achieves condensation above location 4. Subsequently, the air parcel rises along a θ_w surface, depicted by the arrows in the figure, and arrives at upper levels where the potential temperature of the air is no longer θ_5 but is higher due to the release of latent heat of condensation along the path of the parcel.

The maximum value of potential temperature achievable by the condensation process for an air parcel lifted to the upper troposphere is the equivalent potential temperature, θ_e. In this example, the equivalent potential temperature for the surface-level air parcel at location 5 is about θ_7 and the parcel at location 5 undergoes a gain of potential temperature during ascent by air of about ($\theta_7 - \theta_5$). For example, for moist air with an initial potential temperature (θ_0) of 30°C and a mixing ratio of 11 g kg^{-1}, characteristic of tropical air entering mid-latitudes at low levels, the condensation level occurs at 810 mb at $\theta_w = 20$°C. The equivalent potential temperature (θ_e) of this parcel is 63°C and ascent to the high troposphere results in a gain in θ of about 33°C, which is the value of ($\theta_e - \theta_0$) for the parcel.

Unsaturated descent on dry potential temperature surfaces occurs in Figure 12.4 when air moves southward. Since the figure represents a typical meridional cross section, it is reasonable to conclude that poleward-moving air generally is ascending and equatorward-moving air descending. It is evident in Figure 12.4 (and equation 12.3)) that the vertical displacement of an air parcel depends upon the baroclinicity and static stability of the atmosphere and on the configuration of streamlines on potential temperature surfaces.

Furthermore, the air parcel reaches a higher level in the atmosphere following θ_w than on the original θ surface. In a potentially stable atmosphere ($\partial\theta_w/\partial z > 0$), there are no ambiguities in identifying an isentropic surface. A serious problem in analyzing airflow on θ_w surfaces is that θ_w may decrease with height over portions of the atmosphere. Two or more levels may have the same value of θ_w. This situation is analogous to the use of a negative stability in the quasi-geostrophic equations; therefore, isentropic analysis in such a layer breaks down, being inconsistent with quasi-geostrophic theory and adiabatic motion. (Let us recall that a negative $\partial\theta_w/\partial z$ implies that convection can occur at saturation of the layer.) Diabatic influences also complicate the use of these equations and thus invalidate the adiabatic assumption. The planetary boundary layer is one region where the adiabatic assumption breaks down.

12.2 The frozen-wave approximation

We can account for the time derivative in (12.4b) by using the "frozen-wave" approximation (so-named by one of my students, John Takacs), which states that the system translates horizontally *without change of shape or intensity*, with any scalar quantity remaining unchanged at that point in the system. To show how this is done, let us imagine that the isentropes in Figure 12.4 move northward (poleward) at a speed greater than the poleward component of the wind. Individual air parcels would appear to be moving southward with respect to a frame of reference attached to any point on the isentropic surface. In this case, the air would be descending rather than ascending. Ascent occurs as long as the horizontal wind speed is greater than the poleward velocity of the isentropic surfaces. Clearly, the relative motion of the air with respect to that of the potential temperature surface must be considered by evaluating the local derivative in (12.4).

Let us now consider a moving feature on a weather map (a trough axis, front, cyclone center, etc.), as illustrated in Figure 12.5a. Imagine that we are riding atop a wave axis and are able to survey the surrounding air streams. The equation governing the change of a scalar variable (S) following the translatory motion of this wave is

$$\delta S/\delta t = \partial S/\partial t + C \cdot \nabla S \tag{12.8}$$

where the partial derivative on the left-hand side of the equation (the delta operator δ) denotes the local change in a scalar quantity (S) following the wave and the phase velocity (C) is the horizontal velocity of the system, i.e. $C = C_x i + C_y j$. This frozen-wave approximation is equivalent to saying that no value of the wave changes following its motion. This means that $\delta S/\delta t = 0$ and that (for a θ surface)

$$\partial S/\partial t = - C_\theta \cdot \nabla_\theta S. \tag{12.9}$$

Of course, the wave will really be changing with respect to the viewer riding on the wave, but (12.9) is a reasonable assumption over periods that are small compared to the lifetime of the wave.

The frozen-wave approximation is supported by casual inspection of weather charts in which one finds that, indeed, highs and lows, troughs and ridges, migrate from west to east and at upper levels change their shape and intensity only slowly over a period of several days. If a synoptic-scale trough–ridge pattern is considered as the feature of interest, it is sufficiently accurate to equate the local derivative in (12.4b) with the eastward velocity component of phase velocity $(C = C_x i)$, so that

$$(\partial p/\partial t)_\theta = - C_x(\partial p/\partial x)_\theta.$$

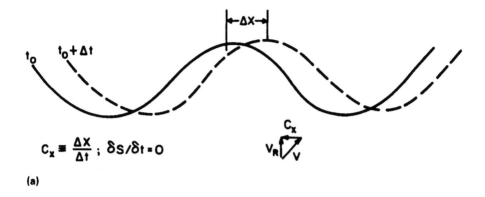

$$C_x \equiv \frac{\Delta X}{\Delta t} \; ; \; \delta s/\delta t = 0$$

(a)

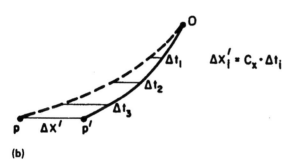

$$\Delta x_i' = C_x \cdot \Delta t_i$$

(b)

Figure 12.5 (a) Schematic illustration of eastward translation by the 'frozen' wave. During the time interval (dt) between t_0 and $t_0 + \Delta t$, the wave moves with an eastward phase speed C_x from its initial location (full curve) to that of the broken curve, covering a distance Δx over the time interval Δt. A graphical illustration of the transformation from real wind (V) to relative wind (V_R) is shown with regard to the small vector triangle. (b) Illustration of transformation from relative wind (full curve) to absolute (ground reference; broken curve) positions for a trajectory reaching point O. The trajectories (O–p; O–p') are divided into segments of equal time increments from the origin; the total elapsed time from the origin is Δt_i. The trajectory O–p represents that with respect to the Earth's surface and is displaced a distance $\Delta x_i' = C_x \Delta t_i$ to the west of the relative trajectory (for an eastward moving feature of phase speed C_x).

Substitution of this result in (12.4b) leads to a time-dependent equation for the vertical motion on moist or dry isentropic surfaces along the direction of the wind (s). It is written

$$\omega_\theta = V_{R\theta} \cdot \nabla p = V_{R\theta}\left(\frac{\partial p}{\partial s}\right)_\theta = V_{R\theta}\left(\frac{\partial \theta}{\partial s}\right)_z \bigg/ \left[-\left(\frac{\partial \theta}{\partial p}\right)\right] \quad (12.10)$$

where

$$V_{R\theta} = V_\theta - C_x i - C_y j$$

and subscript R refers to the relative wind. Equation (12.10) is appropriate for dry or moist processes. For moist adiabatic ascent, θ_w or θ_e would be the appropriate isentrope, rather than the dry potential temperature, and the stability (in (12.3)) is the vertical gradient of θ_w (or θ_e). The method of relative wind isentropic analysis lends itself to the study of a variety of phenomena on differing scales such as waves, cyclones, fronts and convective storms. The vertical motions are expressed by the magnitude of isobaric advection on an isentropic surface.

By removing the local derivative and using the adiabatic temperature equation (12.4b), an instantaneous picture of the three-dimensional (x, y, p) streamlines can be obtained. In the frozen-wave system, the relative wind streamlines are identical to trajectories relative to the moving system. Relative trajectories can be converted to absolute trajectories on the Earth's surface. This is illustrated in Figure 12.5 for an eastward-moving wave of phase speed C_x. The displacement to the west is equal to the system's phase speed (C_x) times the time elapsed (Δt) between the initial and terminal positions, as illustrated in Figure 12.5b. Assessment of phase speed is imprecise, however, because the movement must be ascertained from a succession of previous charts, and that movement may not be steady or uniform with regard to latitude or time.

Despite these approximations, relative wind isentropic analyses retain the correct vertical motion and trajectory pattern for time periods that are small with respect to the lifetime of weather systems not undergoing explosive development, i.e. 1–2 days for mid-latitude cyclones. Indeed, experience shows that replacing the local time derivative by the phase speed in the frozen-wave approximation is superior to direct calculation of the local derivative from a succession of 12 h charts. More exact numerical methods than the relative wind method exist for calculating trajectories, but their construction requires extensive calculation. Moreover, it is not clear whether methods involving more exacting dynamic or energetic constraints yield more accurate results, in view of the inaccuracies in the data. To be sure, the relative wind method is visually appealing for depicting the three-dimensional motion of air through weather systems without great loss of accuracy.

The technique of relative wind isentropic analysis was first introduced in the late 1950s in order to study the air motion through cumulonimbus clouds. In the middle 1960s the method of relative wind analysis was applied to large-scale weather patterns by Ludlam, Browning and others in the U.K. The relative wind isentropic analyses for synoptic-scale weather systems tend to show several well-defined air streams, which have been identified by names (the warm conveyor belt and the cold conveyor belt). It is worth restating that these air streams differ from the outdated air mass concept of the 1930s and 1940s.

12.3 Relative wind isentropic flow through baroclinic waves

Unlike the flow on isobaric surfaces, the streamlines (which are identical to trajectories in the relative wind system) describe meridional exchanges of air between lower and upper levels. Let us define the phase speed of a wave as the eastward speed of a particular trough–ridge pattern. In the relative isentropic system, meridional flow associated with troughs and ridges tends

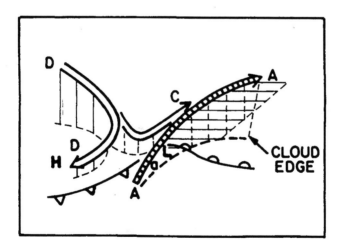

Figure 12.6 Schematic four-dimensional flow pattern for a frozen trough–ridge system, shown in perspective. The descending branch of the circulation (labeled D–D) splits to join the ascending branch of the circulation (A–A) at C. The ascending branch achieves condensation, reaching its lifting condensation level at point "a", and thereafter remains saturated, with the formation of layer cloud (horizontal hatching). The broken line labeled cloud edge is the projection of the westernmost boundary of the ascending branch on the surface and corresponds to a visible cloud edge on satellite photographs.

to resemble the flow pattern depicted in Figure 12.6. This pattern arises because:

(a)   isentropic surfaces slope upward toward the pole;
(b)   air on these surfaces is overtaken by troughs and ridges at low levels;
(c)   the air at high levels overtakes troughs and ridges from the west; and
(d)   air ascends ahead of troughs and descends behind troughs.

(Recall the argument of the vertical coherence of troughs and ridges presented in Chapter 6.)

The circulation shown in Figure 12.6 has two branches: one ascending and one descending. Branch A–A ascends along an anticyclonic trajectory, reaching saturation at point "a" (beyond which there is a layer cloud). Branch D–D descends (also anticyclonically) toward low levels west of the surface cold front. Part of this flow (D–C) splits near the trough axis and turns cyclonically to flow alongside the ascending flow.

The sharp edge of the western and poleward borders of the cloud corresponds to a confluence of the rising air stream with dry air. The dry air stream originates in the high troposphere or lower stratosphere near the upstream ridge. In Figure 12.6, air stream D–D undergoes a large net downward displacement; some air originates near the base of the polar tropopause. Air stream D–C descends through a large vertical displacement but undergoes ascent adjacent to air stream A–A while retaining its low relative humidity. The early stages of cyclone development, in which there is no identifiable surface cyclone, are characterized by a band of clouds east of the trough. These bands (the baroclinic cloud leaves) correspond to major confluence zones within large-amplitude synoptic-scale waves and may extend over considerable distances between low and high latitudes (Fig. 12.7a), or be situated on the anticyclonic side of a jet near the ridge (Fig. 12.7b).

A distinctive feature of baroclinic cloud leaves is that the western and poleward edges tend to be quite well defined. In some instances the baroclinic cloud leaf consists solely of high cirrus and middle cloud. These cloud bands tend to cross geopotential height lines, with lower heights on the poleward side of the cloud band. Typically, most of the cloud on the equatorward side of the leaf consists of middle and low cloud, usually accompanied by precipitation. Cloud leaves may persist in the absence of an identifiable surface cyclone or frontal boundary.

The sharp western boundary of the upper-tropospheric baroclinic cloud leaf often coincides with a wind speed maximum or with a zone of strong winds. In the examples shown in Figure 12.7, the strongest winds at 300 mb coincide with the western or poleward edges of the cloud. (Figure 12.7a is the typical case of a deep trough and strong polar front jet with no well-defined surface cyclone.) In the relative motion system, *sharp cloud boundaries*

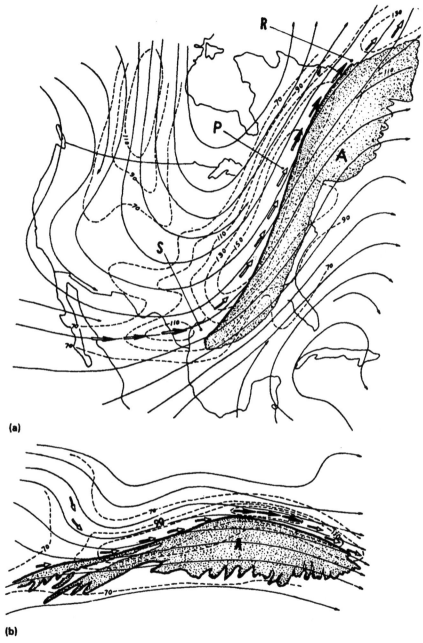

(a)

(b)

Figure 12.7 (a) Schematic 300 mb geostrophic flow (full curves), isotachs (broken curves labeled in kt), jet axes (arrows along locus of points S–P–R) and baroclinic cloud leaf (shading labelled A) exhibiting a sharp cloud edge. (b) Same as (a) but for a cloud leaf located in the ridge of a longer-wave system. (From Weldon, 1979.)

correspond to streamlines at cloud level. Alternatively stated, *air does not cross sharp cloud edges in the relative motion system. Cloud edges are, therefore, zones of confluence.* A cross-wind component may exist over or under a cloud edge, but not at the level of the cloud edge. Ragged cloud boundaries correspond to airflow across those boundaries in the relative wind system *at the level of the cloud.*

For the purpose of discussion (and because it is a readily identifiable feature of relative wind isentropic analysis) a confluent asymptotic streamline will be referred to as the *limiting streamline.* The limiting streamline in Figure 12.6 (A–A) constitutes a boundary between air streams of widely differing origins: between moist air, in relative easterly flow at lower latitudes ahead of the trough, and dry air, in the upper troposphere west of the trough. The contrast in the moisture contents of the two air streams is readily evident in the gradient of relative humidity across the limiting streamline.

The rising, moist air stream east of the trough axis is here referred to as the "warm conveyor belt". Although the warm conveyor belt constitutes a

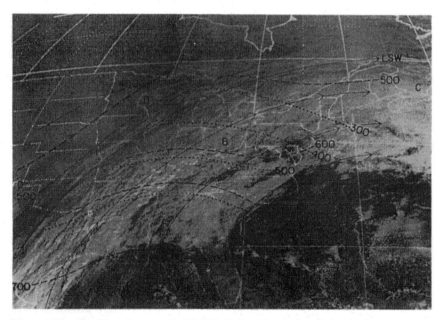

Figure 12.8 Satellite photograph (GOES; visible) for 1430 GMT 8 December 1978, for $\theta = 37\,°C$ potential temperature surface and $\theta_w = 20.5\,°C$ surface. The principal ascending air stream is labeled A–B–C. Relative wind streamlines are drawn using a full curve in the cloudy region and a broken curve in the clear air west of the streamline labeled D–D. Isobars, labeled in mb, are broken curves for the dry surface and dotted curves for the moist.

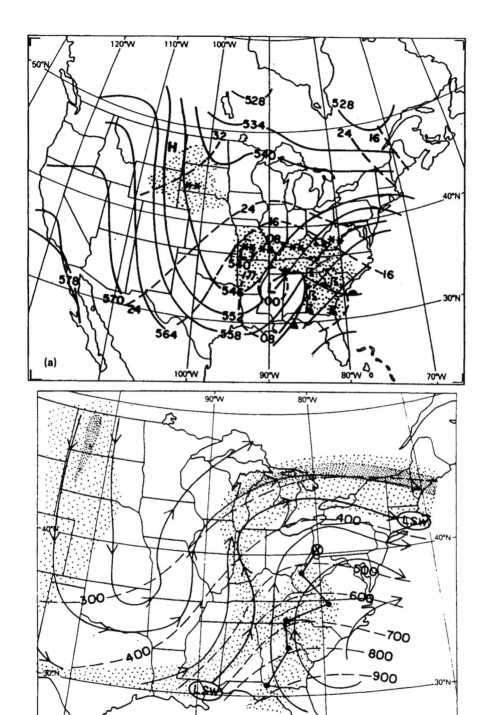

Figure 12.9 (see p. 303 for caption.)

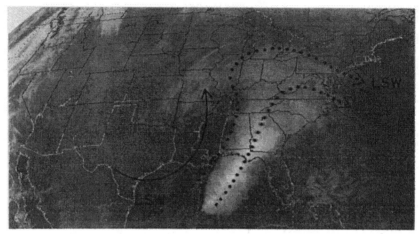

(c)

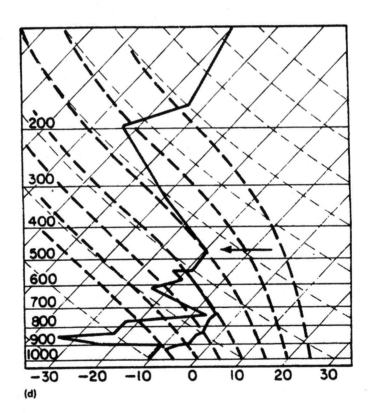

(d)

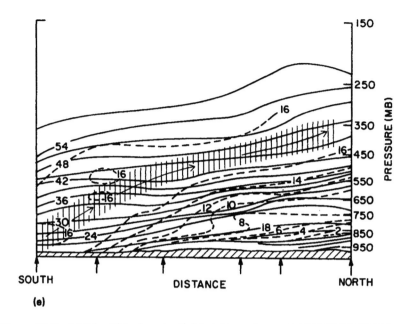

(e)

Figure 12.9 (a) The 500 mb potential height contours at 1200 GMT 27 February 1984 (full curves labeled in dam) and sea-level isobars (broken curves) in mb above 1000 mb). Surface features are denoted by conventional weather symbols. The letter L denotes the locations of the sea level and 500 mb cyclone centers; shading denotes continuous precipitation. Surface fronts are indicated in conventional symbols. (b) Relative wind isentropic analysis for the $\theta_w = 15°C$ isentropic surface (phase speed of 500 mb wave is 11 m s^{-1}). Thin barbed full curves are streamlines and broken curves are isobars in mb. The heavy full streamline labeled LSW is the limiting streamline for the warm conveyor belt, which is visible on the satellite photograph in (c). Shading denotes the region where the relative wind speeds exceed 25 m s^{-1}. The chain of filled circles connected by line segments denotes the location of the cross section in (e). The location of the Pittsburgh radiosonde (see (d)) is denoted by the circled cross. (c) Infrared satellite photo (GOES) for 1030 GMT 27 February 1984. Flow in warm conveyor belt is represented by dotted streamlines. The westernmost dotted streamline (labeled LSW) is the limiting streamline in (b) and the full stream- line is that of the dry air stream. (d) Temperature and dew-point sounding on a skew T–log p sounding diagram for 1200 GMT 27 February 1984 at Pittsburgh, PA (station symbol marked with a circled cross in (b)). Note that the temperature and dew-point lines merge above 550 mb and that the base of the warm conveyor belt ($\theta = 15°C$) air is located near 480 mb (horizontal arrow). (e) Cross section of potential temperature (full curves labeled in °C) and wet-bulb potential temperature (broken curves labeled in °C) along the chain of stations denoted by filled circles in (b) for 1200 GMT 27 February 1984. Hatching represents axis of maximum relative humidity (greater than 90%) in the cross section. Station locations are given by the arrows at the bottom; that of Pittsburgh (the circled cross in (b)) is on the extreme right-hand side. Arrow segments represent isentropic relative flow on $\theta_w = 16°C$ surface.

family of isentropic trajectories over some depth and breadth of the atmosphere, it is recognizable as a cloud mass with a discrete (on the synoptic scale, at least) western border. The warm conveyor belt is defined as a layer of air that (a) originates in relative easterly flow in the lower troposphere ahead of a trough, (b) ascends generally poleward, forming a cloud shield by condensation and (c) reaches the middle or upper troposphere in relative westerly flow near the downstream ridge. In the warm conveyor belt, air rises laterally in slantwise fashion from the lower to the upper troposphere; once it has achieved saturation along its path, the flow is marked by a layer cloud, which persists until after the air starts to descend. The western boundary of the warm conveyor belt coincides with the limiting streamline.

Figure 12.8 shows a warm conveyor belt (A–B–C), which is marked by a wide band of cloud with a sharp western edge; the cloud extends to the ridge along its northern border (the point labeled C). Further east (out of the picture) the cloud is marked by streamers of high cirrus, which become increasingly less dense and less visible downstream from the main cloud mass. The relative sharpness of the upwind cloud edge suggests a uniformity of the saturation (lifting condensation) level in the ascending air stream and in the vertical displacement required to saturate the air stream.

In Figure 12.8, air moves through the warm conveyor belt and cloud leaf and ascends along a moist adiabat through isentropic surfaces above that of $\theta = 37\,°C$ (the path indicated by the letters A–B–C), its ascent being more rapid and through a greater depth than on the dry surface. Accordingly, the cloud is increasingly more dense along the direction of flow as successive strata reach saturation and ascend on moist surfaces (B). In the middle part of the cloud leaf, cirrus masks much of the lower-level cloud; however, further downstream the baroclinic cloud leaf consists primarily of high and middle cloud (C).

The limiting streamline (LSW) separates the dry and moist air streams; it is located along the cloud edge between the moist air stream entering the cloud leaf along its southern part (A), and that following streamlines immediately to the west of the cloud edge (D–D). Dry air is also ascending, though somewhat less rapidly than that in the cloud leaf.

Consider the conventional analysis in Figure 12.9a in contrast with the relative wind isentropic analysis in Figure 12.9b. Here, layer cloud extends poleward for a distance of about 2600 km (Fig. 12.9c). (The narrow band of bright cloud within the main cloud mass is due to the existence of a squall line south of the warm front.) Relative wind isentropic flow on the moist surface is shown in Figure 12.9b for $\theta_w = 15\,°C$. This isentropic surface slopes upward toward the north and toward the center of the pressure trough. Air entering the cloud mass from the south reaches saturation at the southern end of the cloud and ascends. The western cloud edge corresponds to the limiting streamline, which is the equatorward (and westernmost) streamline in the relative easterly flow to move poleward ahead of the trough. A belt of strong winds coincides with the rising, saturated air stream (shaded region). Greatest vertical displacement of the air within the conveyor belt lies north of the warm front near the limiting streamline (just inside the cloudy air). Because of its southernmost origin, the limiting streamline tends to possess the highest moisture content of all trajectories in the warm conveyor belt. Note that *the relative wind flow closely conforms to the shape of the cloud.*

Soundings made along the northern end of the cloud mass (Fig. 12.9c), at the location of the arrow labeled "north" in Figure 12.9e, show that the wet-bulb potential temperature in the saturated air is generally between 15 and 17°C in the layer above approximately 480 mb. Cloud cover at this location consisted of mid- and high-level layers of altocumulus or altostratus and cirrus. The warm conveyor belt is bounded below in Figure 12.9d by a stable layer within which there can be seen a large vertical gradient of wet-bulb potential temperature. The altitude of the warm conveyor belt decreases toward the south (Fig. 12.9e). South of the surface warm front the saturated air ($\theta_w = 15°C$) lies in the boundary layer, above which the wet-bulb potential temperature decreases slowly with height (potential instability).

Moist isentropic streamlines cross the dry isentropes (Fig. 12.9e) at a steeper angle with the horizontal, while following an air stream that coincides approximately with the envelope of nearly saturated air, $\theta_w = 15$–$17°C$. This air originates south of the warm sector at low levels and arrives above 450 mb in the north, cutting across dry adiabats toward higher values of θ. Except for the southern part of the cross section, the air is potentially stable.

Figure 12.9 and 12.10 both illustrate cases of deep troughs and strong upper-tropospheric winds without notable surface cyclogenesis. The limiting streamline coincides with a well-defined western edge of the conveyor belt and cloud leaf. Maximum winds at 300 mb lie close to the cloud edge, as suggested by Figures 12.7a and b. Sharp differentiation between dry and moist air streams, existing at altitudes above about 650 mb, is shown clearly in Figure 12.10c. As pointed out in regard to Figure 10.8, *the tropopause is higher over the moist ascending air stream than over the dry air.*

The cloud edge in Figure 12.10a coincides with a limiting streamline and a surface cold front. In back-bent cold fronts, the western cloud edge and limiting streamline may lie a considerable distance to the west of the surface front. The cloud edge may also lie ahead of the cold front inside the warm sector, as is often the case for systems over the British Isles. The limiting streamline is evidently identical to the "upper cold front", which has been well documented for cases of cyclone passages over the British Isles. An example of such an upper cold front is shown in Figure 12.11. The upper

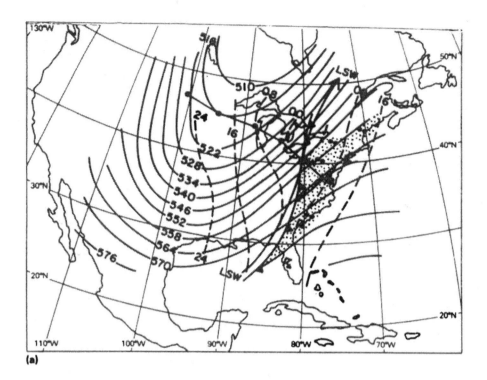

(a)

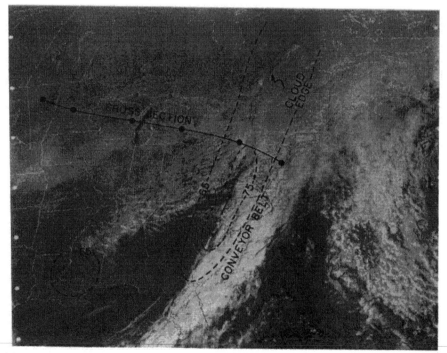

(b)

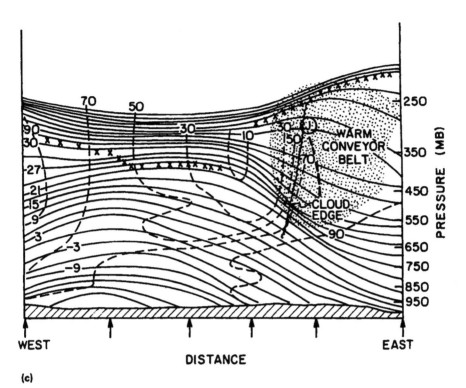

WEST   EAST

DISTANCE

(c)

Figure 12.10 (a) The 500 mb analysis (full curves are geopotential heights labeled in dam) and surface isobaric chart (broken curves labeled in mb above 1000 mb) for 1200 GMT 2 February 1981. Shading denotes region of sustained precipitation (with conventional weather symbols), terminated along poleward border. The heavy full curve labeled LSW is the limiting streamline for warm conveyor belt (the cloud edge). The chain of connected filled circles denotes the location of the cross section in (c). (b) Visible (GOES) satellite picture for 1530 GMT 2 February 1981. The line of filled circles labeled "cross section" refers to the location of (c). The conveyor belt and cloud edge are labeled. Thin broken curves are isotachs for 65 and 75 m s^{-1} at 300 mb; the arrow denotes the direction of wind at that level. (c) Transverse cross section through the conveyor belt along an east–west line segment drawn in (a) and (b) for 1200 GMT 2 February 1981. Full curves are isentropes in °C and broken curves are isopleths of relative humidity in percent. Other features are: (1) shading for the region where the wind speed exceeds 50 m s^{-1}; (2) the letter J at the location of the jet core (70 m s^{-1}); (3) a scalloped border indicating the cloud edge and limiting streamline; (4) a heavy broken curve extending downward from the jet core showing the axis of maximum wind speed; and (5) a broken chain of crosses denoting the tropopause.

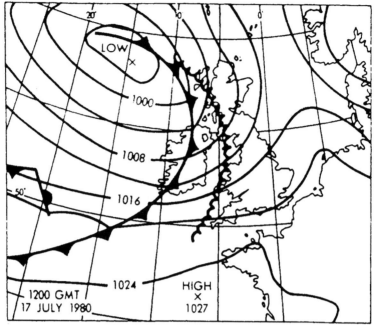

Figure 12.11 Modified frontal analysis for 1200 GMT 17 July 1980, with the surface cold front drawn conventionally and the upper cold front represented by a scalloped line. (From Browning and Monk, 1982.)

front coincides with a gradient of temperature and a very sharp gradient in humidity. Showers and continuous precipitation tend to occur just east of the upper cold front. In upper cold fronts that lie ahead of the surface cold front, a shallow layer of moist air is located in a zone between the two frontal boundaries. Showery bands tend to lie parallel to the upper front. In forward-sloping fronts (limiting streamline east of the cold front) cessation of precipitation precedes the passage of the upper cold front, although further precipitation of a showery or transient nature can still occur with the passage of the surface cold front.

Conveyor belt cloud bands are not always associated with fronts or major troughs. This point is illustrated with regard to Figure 12.12, which is a case where there was no distinct surface feature or well-developed trough at 500 mb. Heaviest precipitation (snowfall denoted by double stars) was taking place equatorward of the highest cloud tops (cross-hatched region corresponding to a radiometric temperature of $-50°C$) over a region where the cloud tops were relatively warm (-20 to $-30°C$; light shaded to heavy shaded region). (A separate system can be seen equatorward along the Gulf of Mexico coast.) The 500 mb chart (Fig. 12.12b) shows a strong wind speed and vorticity maxima.

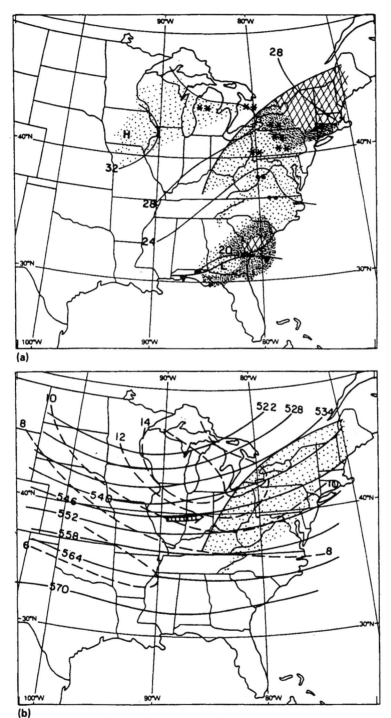

(a)

(b)

Figure 12.12 (see p. 310 for caption.)

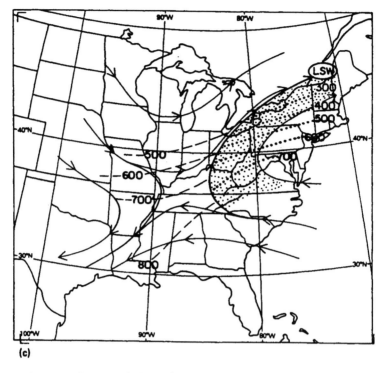

(c)

Figure 12.12 (a) Sea-level isobars (full curves labeled in mb above 1000 mb) and cloud analysis for 1200 GMT 9 January 1980: light shading corresponds to mid-level cloud tops, heavy shading to mid- to upper-level cloud tops and hatching to high-troposphere cloud tops. Conventional symbols denote locations of steady precipitation (rain or snow). (b) The 500 mb geopotential height contours (full curves labeled in dam) and absolute vorticity isopleths (broken curves labeled in units of 10^{-5} s^{-1}) for 1200 GMT 9 January 1980. Shading denotes region of warm conveyor belt cloud shown in (a) and the arrow denotes the location of the maximum winds at 300 mb. (c) Relative wind isentropic flow on $\theta = 32°C$ and $\theta_w = 9°C$ isentropic surfaces for 1200 GMT 9 January 1980. Full curves are streamlines on both isentropic surfaces; the limiting streamline is labeled LSW; they are identical for the dry and moist surfaces in region of cloud (shading). Broken and dotted curves, respectively, are the isobars on the dry and the moist isentropic surfaces (labeled in mb). (The system's phase speed was 20 m s^{-1}.)

The two circulation branches for $\theta = 32°C$ (Fig. 12.12c) resemble the flow pattern in Figure 12.6. Air approaches the trough from the east along the mid-Atlantic coast, moves northward, saturates and moves along the $\theta_w = 9°C$ surface altitudes above 300 mb (dotted isobars), as compared to 500 mb for the dry $\theta = 32°C$ surface. Precipitation occurs where ascent is rapid from the lower to middle troposphere.

The limiting streamline (LSW in Fig. 12.12c) coincides with the sharp western edge of the cloud mass and with a large horizontal gradient of moisture in the middle and upper troposphere (Fig. 12.10c). Note that strong ascent occurs in conjunction with strong positive vorticity advection at 500 mb (Fig. 12.12b). *The dipole pattern of ascending and descending anticyclonic trajectories on relative isentropic surfaces is a manifestation of a positive and negative vorticity advection couplet associated with a traveling wave disturbance.*

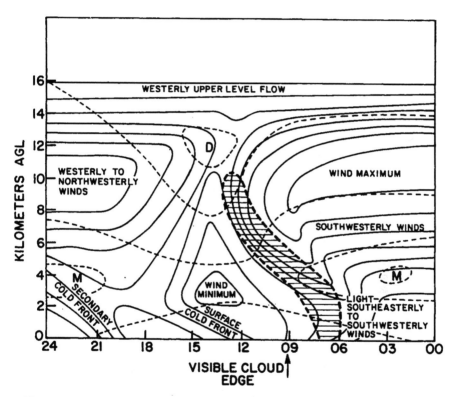

Figure 12.13 Partially schematic analysis of the wind speed determined from Doppler wind (profiler) measurements made for 30 conveyor belt cases over central Pennsylvania. Vertical axis is height (km) above local terrain and horizontal axis is time (h). The heavy broken curve and shading represent a zone of confluence, which coincides with a southerly wind maximum. Each 3 h segment corresponds to approximately 150 km horizontal distance. Thin broken curves are isopleths of profiler power return, which correlates with specific humidity. (Thus, dry and moist regions, whose centers are indicated by the symbols D and M, are associated with low and high power returns, respectively). The visible cloud edge as viewed by satellite is indicated by an arrow in the lower margin. (Courtesy of M. Hemler, 1988.)

311

The structure of cloud edges

To the west and poleward of the limiting streamline is a zone of very dry air. This dry zone tends to lie parallel with an axis of maximum wind speeds in the upper troposphere between the high tropopause over the warm conveyor belt and low tropopause over the trough. This wind speed maximum tends to lie inside the visible cloud edge (viewed from above) at lower levels and in the clear air in the high troposphere. A composite wind analysis, based on profiler measurements through conveyor belts over Pennsylvania (Fig. 12.13), shows that the cloud edge lies in a southerly wind confluence, which slopes inward beneath the warm conveyor belt. This confluence zone coincides closely with the visible cloud edge.

(a)

Figure 12.14 (a) Visible (GOES) satellite picture for 2101 GMT 24 November 1980, showing a cloud edge illuminated by the Sun from the west.

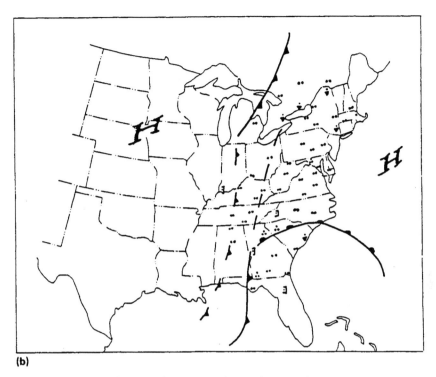

(b)

Figure 12.14   (b) Elements of the sea-level chart for 1200 GMT 24 November 1980. (Courtesy of G. Forbes.)

Cloud edges possess considerable vertical depth and extent, as shown in Figures 12.14a and 12.15. In one case, the western face of the cloud (Fig. 12.14a) is illuminated by the Sun; in another (Fig. 12.15), it projects a shadow, an indication of the great depth of the western side of the cloud near the limiting streamline. The ruler-sharp boundary in Figure 12.14a extends from the Gulf of Mexico into Canada, where it is lost in the sub-Arctic twilight. Shallower cloud in the boundary layer is present west of the cloud edge and there is an intermediate-level cloud mass along the northern part of the system. The surface chart (Fig. 12.14b) shows an ill-defined cyclone, with steady rain on both sides of the front.

The dry air stream

The cloud edge corresponds to a large-scale confluence between dry and moist air streams of widely differing origins:

(a)   the warm conveyor belt originates in relative easterly flow in the lower

313

Figure 12.15 Visible (GOES) satellite picture for 1430 GMT 28 January 1981, showing cloud edge shadowed by the morning Sun. The 110 kt (55 m s^{-1}) isotach at 1200 GMT on the 300 mb surface is shown by the broken curve; the arrow denotes the location and direction of the 300 mb jet maximum along the cloud edge.

troposphere ahead of the trough axis and sometimes as far south as the Tropics; and

(b)   the dry air stream, which originates in the high troposphere and lower stratosphere west of the trough, is identifiable on soundings as a deep layer of low relative humidity in the middle and upper troposphere (e.g. Fig. 12.10c).

Cloud cover and precipitation

Cases such as that shown in Figure 12.16 illustrate the limitations in the classical cyclone/front model. In the relative wind isentropic model, rainfall occurs where ascent and water vapor content are maximized, which occurs just inside the limiting streamline in the low to middle troposphere in well-developed baroclinic waves (horizontal hatching). Note that the winds are strongest on isentropic surfaces just inside the visible cloud edge at mid-levels. Thus, the most rapid ascent occurs a little to the right of the limiting streamline (C in Fig. 12.16). Where a surface warm front is present, the air ascends most rapidly just to the right of the limiting streamline, poleward of the warm front. The broad area of ascent in the vicinity of the warm front is consistent with the anabatic nature of the warm front. (Some investigators have observed a jet-like feature at the top of the planetary boundary layer in the warm sector. The warm conveyor belt flows almost parallel to the cold front, except near the surface where the warm moist air in the boundary layer is overtaken by the cold front.) Highest clouds are to be found along the poleward end of the cloud leaf, where the conveyor belt is

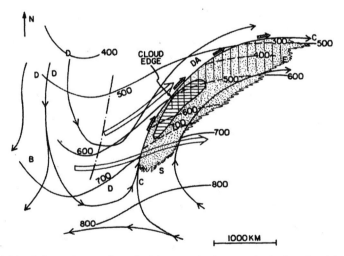

Figure 12.16 Schematic relative wind isentropic flow analysis for cloud leaf and conveyor belt. Streamlines are represented by the thin barbed full curves. Isobars on the dry potential temperature surface are thin full curves labelled in mb, and on the moist isentropic surface are broken curves labeled in mb. Shading represents layer cloud cover, vertical hatching the region of highest (coldest) cloud as seen on a satellite picture, horizontal hatching the region of heaviest precipitation, hatched arrows the axis of maximum wind speed on the relative isentropic surface, double-shafted arrows the relative wind direction at 300 mb, and the chain curve the axis of the 300 mb trough. The letters represent features discussed in the text, in particular the confluent axis (C), called the limiting streamline for the warm conveyor belt.

315

reaching the high troposphere (the vertically hatched lines in Figure 12.16). An observer situated along the northern end of the conveyor belt (at E) might find the sky covered with cirrus and altocumulus clouds, the so-called "mackerel sky", which is known to be a harbinger of stormy weather.

An axiom is that relative winds blow parallel to the cloud edges at the levels of the conveyor belt. Alternatively stated, relative flow does not cross sharp cloud boundaries, although it can cross the cloud edge above or below the cloud. Figure 12.16 shows that the 300 mb relative winds (double-shafted arrows) lie parallel to the cloud edge on the poleward (downstream) end (DA), but where the conveyor belt is not reaching that level, the 300 mb winds cross above the warm conveyor belt along its equatorward side. Since θ_w tends to be low in the dry air crossing the trough, the decrease of wet-bulb potential temperature with height between moist and dry air constitutes a potentially unstable lapse rate. This results in the formation of convective cells, which locally enhance the rainfall and also mix the conveyor belt air with that at higher levels. Such interleaving of dry and moist air streams creates multiple cloud layers.

Flow across cloud boundaries at the level of the cloud occurs near the locations of outflow or inflow along the sides of the conveyor belt cloud, respectively marked with an E and an S in Figure 12.16. This is consistent with the observation that cirrus clouds are usually found on the anticyclonic side of jets (Fig. 12.7). *Note that the highest-layer cloud tops on the infrared satellite photograph do not necessarily coincide with the heaviest precipitation.*

The eastern side of the warm conveyor belt cloud, however, is ragged and not as easily identifiable on satellite photos because there is a flow across the cloud boundaries, into or out of the cloud.

12.4 Cyclogenesis: the cold conveyor belt

With the formation of a closed circulation in the lower troposphere, the relative wind streamlines (for the simple cloud leaf model shown in Figure 12.16) become modified. An S-shaped cloud pattern forms with maximum cyclonic curvature close to the location of the surface low. This progression is shown in Figure 12.17a. Development of a cyclonic bend in the S-pattern indicates the formation of an easterly component at low levels poleward of the surface low and warm front, which constitutes air that descends in the downstream surface high-pressure cell. This air is moistened by evaporation at the surface and by precipitation falling from the warm conveyor belt. The flow emerges from under the warm conveyor belt beneath the dry air in the middle and upper troposphere, as shown in Figure 12.17b. Note that in Figure 12.10c the isentropic surfaces ahead of the trough slope upward toward the west. Thus, air that moves westward beneath the warm conveyor

(a)

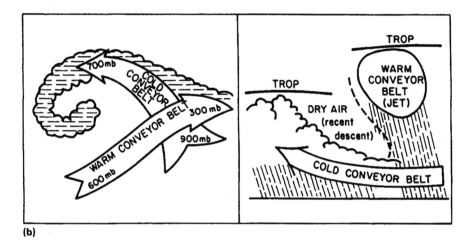

(b)

Figure 12.17 (a) Configuration of the warm conveyor belt undergoing modification from a baroclinic cloud leaf to a comma (left to right) and from a rearward-sloping to warm front type conveyor belt. (b) Schematic model of a mature cold conveyor belt or instant occlusion in plan view (left) and cross section (right). (Based on figures from Browning, 1989.)

belt and beneath the upper wind speed maximum first ascends dry adiabatically and then, following saturation, ascends moist adiabatically. Except for the air's origins this "cold conveyor belt" is very similar to the warm conveyor belt. By contrast, however, the cold conveyor belt moves directly across the region of the polar front jet, as shown in Figure 12.17b. For this reason, the cold conveyor belt is sometimes called the "polar trough conveyor belt".

In the analysis of a cold conveyor belt shown in Figure 12.18, the western edge of the warm conveyor belt cloud is highlighted in the satellite photograph (Fig. 12.18b) by a shadow made by the sun in the early morning; a shadow is also projected on the tongue of protruding cloud. The S-shaped left edge of the smooth-textured high clouds extending along the limiting

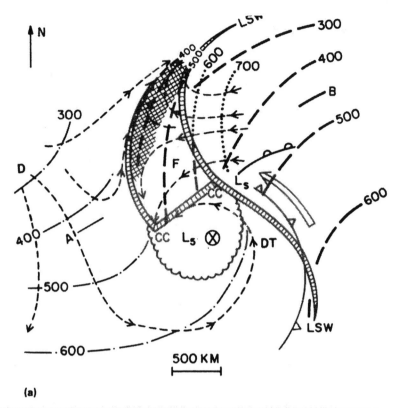

(a)

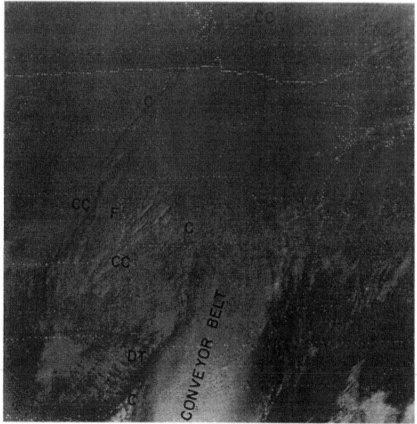

(b)

streamline (LSW in Fig. 12.18a) marks the boundary between the warm conveyor belt and the dry air stream (see also Fig. 12.10b).

Cloud formed by the rising cool air stream whose origin lies north of the warm front (F in Fig. 12.18a) protrudes from under the warm conveyor belt. Viewed from above, this cold conveyor belt cloud is a rough-textured middle and high cloud, possessing a relatively flat top, which is clearly differentiated from (and at a lower height than) the warm conveyor belt cloud.

The equatorward and western edges of the cold conveyor belt cloud cover that is exposed to view from above are sharply defined; they are, respectively, shaded and illuminated by the sun in Figure 12.18b. These cloud edges, like their counterpart in the warm conveyor belt, constitute relative wind deformation lines, material barriers along which air streams of differing origins merge but do not mix. Cold and warm conveyor belts both have limiting streamlines (CC and LSW in Fig. 12.18a), although that for the cold conveyor belts splits, allowing some air to turn equatorward in response to a developing closed circulation at middle levels. The cyclonic flow around the middle-level vortex is suggested in Figure 12.21a. Thus, the formation of a cyclonic cusp in the cold conveyor belt cloud is an indication of middle-level cyclogenesis, as described in Chapter 10; the cold conveyor belt air stream reaches upper levels poleward and west of the upper-level cyclone center. While some of the cold conveyor belt moves cyclonically around the upper vortex and descends behind the cold front, a part of this flow continues to ascend along an anticyclonic path. Further movement of this flow is probably along the direction of the warm conveyor belt cloud at its poleward end.

Figure 12.18 (a) Partially schematic isentropic analysis over the central part of the United States (modified for 1200 GMT 22 October 1979), corresponding to a cold conveyor belt, whose visible cloud mass, labeled F, is formed during cyclogenesis. (The model closely corresponds to the satellite picture in (b).) The warm conveyor belt cloud is raised, and its direction of flow in the relative wind system is shown by the double-shafted arrow. Isobars on a moist isentropic surface for the warm conveyor belt are shown by heavy broken curves (mb). The cloud edge and limiting streamline for the warm conveyor belt is labeled LSW. Relative streamlines on the moist isentropic surface within the cold conveyor belt are drawn with thin broken curves and the isobars as broken curves (except dotted in the region of the warm conveyor belt). Hatching denotes the region of high cloud in the cold conveyor belt and scalloping encloses boundary-layer stratus. Pecked streamlines depict relative wind flow on a dry isentropic surface that originates near the tropopause (chain isobars). The symbols L_s and L_5 denote the location of the surface and 500 mb low pressure centers and the circled cross is the 500 mb absolute vorticity maximum. Letters DT, CC and F are referred to in the text. C denotes the edge of the warm conveyor belt. Segments labeled A and B refer to the end points of the cross section shown in Figure 12.19b. Surface fronts are shown by conventional symbols. (b) GOES satellite photograph for 1430 GMT 22 October 1979 (see (a)).

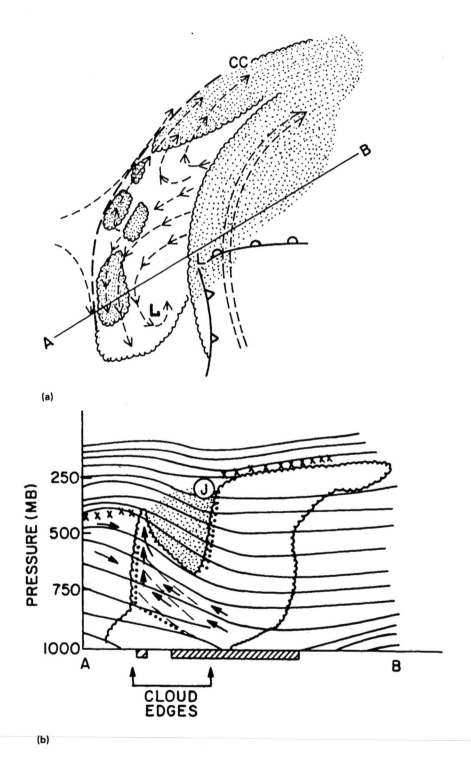

(a)

(b)

CLOUD
EDGES

Movement of the dry air stream also becomes more complex during cyclogenesis. This airstream splits: a part of it moves over the top of the cold conveyor belt cloud (high troposphere) or (at lower levels) is confluent with the cold conveyor belt and moves around it. Another branch of the dry air stream descends behind the cold front (near the surface). Large-scale ascent occurs almost everywhere within the region of cold conveyor belt cloud, except equatorward of the 500 mb vortex. In Figure 12.18a clear dry air, labeled DT and bounded by low stratus or convective, showery cloud near the cold front, is the visible portion of the "dry tongue". Note that ascent is implied along the poleward end of the dry tongue.

Figure 12.19b shows that along line A–B the cold conveyor belt slopes upward toward the west following moist adiabats below the dry air stream; the latter forms a canyon between the warm and cold conveyor belt cloud masses. (Enhanced infrared satellite photographs may even show that cloud above a given level develops first west of the warm conveyor belt and then expands eastward, ultimately joining with the warm conveyor belt cloud over an increasingly deep layer). A composite view of the relative wind model is shown in Figure 12.20.

Wind, temperature and vorticity changes during cyclogenesis

Evolution of a comma-shaped cloud pattern from that of a cloud leaf is accompanied by changes in the wind and temperature patterns in the conveyor belt patterns. During cyclogenesis, wind maxima form at two locations along the western and poleward edges of the cloud in accordance with changes in the geopotential height pattern at upper levels, as previously discussed in earlier chapters.

The 500 mb jet is situated in the warm conveyor belt just east of the limiting streamline and oriented parallel to the cloud edge in the relative system. A second wind speed maximum develops at high levels along the

Figure 12.19 (a) Schematic cloud distribution around a surface cyclone and frontal system in the case of a cold conveyor belt (e.g. Fig. 12.18) and along the cross section (A–B) depicted in (b). Shading denotes cloud tops higher than a certain level (approximately 500 mb). Thin broken curves denote relative wind flow in the cold conveyor belt, and the bold-faced L denote the location of the 500 mb low center. Surface fronts and the location of the surface low are shown in conventional notation. Scalloping denotes the outline of dense cloud as seen on a visible channel satellite photograph. (b) Schematic cross section of potential temperature (full curves) and selected moist adiabats (thin broken curves) along the segment A–B in (a) and Figure 12.18a. Scalloping represents cloud, crosses the tropopause level, J the location of the polar front jet and shading the region of very dry air. Dots signify boundaries (lateral or bottom) of the warm and cold conveyor belt cloud. Arrows suggest motion of air, both vertically and horizontally, in the plane of the cross section. Hatching below denotes the region of precipitation at the ground.

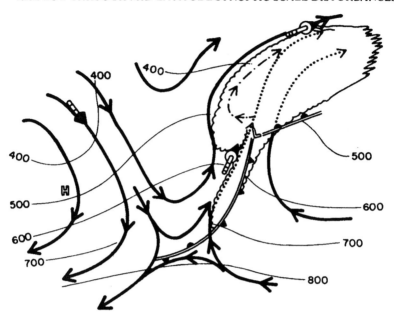

Figure 12.20 (a) Schematic illustration of relative wind flow on a dry isentropic surface. Thin full curves are isobars; these are discontinued in regions of saturation, where the flow is drawn as a dotted curve for the warm conveyor belt, and as a chain curve for the cold conveyor belt. The dry air stream originates at high levels in the northwest and descends toward the trough axis. There it splits into two branches (see discussion in text). One branch descends into low levels west of the trough and the other flows around the trough to ascend in a narrow stream over the western extension of the cloud shield (scalloped border). The symbols L and H refer to the locations of the surface low and high pressure centers. Fronts are depicted in conventional symbols. Jet maxima are represented by solid arrow segments.

cloud edge in the region where the latter is curved anticyclonically (Fig. 12.20). The poleward wind maximum forms in the high troposphere along the cloud edge near the downstream ridge after the onset of cyclogenesis. Note that these changes correspond to the evolution of the vorticity pattern discussed in Chapter 10.

Recall that rapid cyclogenesis is accompanied by the migration of the 500 mb wind speed maximum into the region equatorward of the cyclone in association with 500 mb trough development. Similarly, ridge development corresponds to an increasing temperature gradient on the poleward side of the cloud and a weakening of the high-level baroclinicity and the 300 mb winds equatorward of the poleward wind speed maximum (in the region of the bold-faced arrows in Figure 12.16). This is consistent with an increase in warm air advection in the lower and middle troposphere along the warm front, in the high troposphere along the poleward edge of the warm conveyor belt cloud and at mid levels over the cold conveyor belt cloud. Simultaneously, there is an increase in positive vorticity advection at 500 mb over the visible portion of the cold conveyor belt cloud and, consequently, an increase in ascent west of the surface cyclone.

A reasonable question is why such discrete air stream boundaries should exist in a continuous medium. Fronts constitute an obvious example of an air stream discontinuity on the synoptic scale. Like surface fronts, separate conveyor belt cloud belts occur because of the widely differing origins of the air streams and large-scale confluence and shearing (see Ch. 13). In the case of the cold conveyor belt, the cloud boundaries reflect not only the large-scale confluence but the sharp contrast in temperature and moisture between the warm sector and the cold side of the warm front.

Airflow through occluded cyclones

Airflow through occluded cyclones differs from that through developing cyclones in that no separate synoptic-scale air stream boundaries are clearly evident in this later stage. Figure 10.7 clearly shows that the western cloud

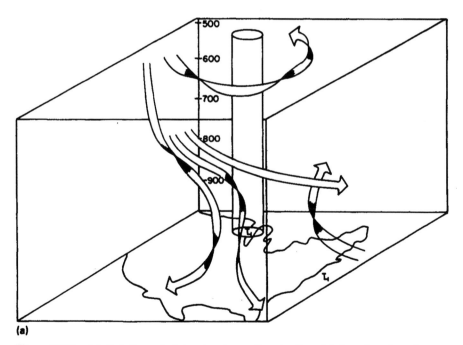

(a)

Figure 12.21 (a) Relative wind isentropic motion in the vicinity of a deep low between the surface and the 500 mb level (vertical scale); the vertical cylinder is the location of the low center (L).

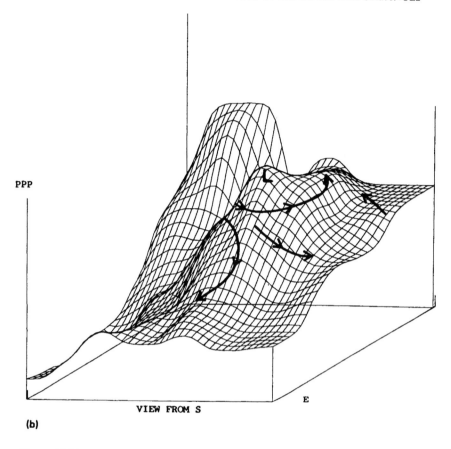

PPP

VIEW FROM S

E

(b)

Figure 12.21 (b) Representation of an isentropic surface in three-dimensional perspective for the occluded cyclone of (a). The arrows show the relative wind isentropic motion on that surface with respect to the cyclone center, which is labeled by an L. (Figures courtesy of J. Belles.)

edge of the occluded cyclone is bent almost completely around the low. Because of their size and complexity, occluded lows are poor subjects for relative wind isentropic analysis. Such analyses show that the air streams near the center of the occluded low tend to move parallel to vorticity and thickness contours without much vertical motion. Patterns of ascent and descent are nevertheless recognizable south and east of the cyclone in Figure 12.21. A cold dome, represented by an upward bulge in the isentropic surfaces (Fig. 12.21b), is evident near the low center. Consequently, air sinks as it moves away from the low center and ascends as it moves towards it; but there is no systematic exchange of air between the center of the low and its surroundings. In effect, the inner region of the occlusion has ceased to play a role in the conversion of potential to kinetic energy.

Some types of occlusions are called "instant occlusions" because a cold conveyor belt forms separately from the warm conveyor belt or develops rapidly away from the warm conveyor belt. A cloud pattern resembling that shown in Figure 12.17b may occur at any time when air near the polar front moves directly across the direction of the upper winds toward colder air.

12.5 Parcel theory analog for conveyor belts

A point made earlier in this chapter is that satellite pictures show that cloud cover normally occupies a small fraction of the total image. At middle latitudes, most of the important precipitation-producing cloud is confined to relatively narrow bands, which are associated with baroclinic zones (or fronts) in the lower and middle troposphere. These baroclinic cloud leaves result from ascent due to large-scale baroclinicity as expressed by gradients of potential temperature between low and high latitudes. The two branches of the circulation shown in Figure 12.6 illustrate the meridional exchange that occurs in response to this baroclinicity. Superficially, these two branches resemble the rising and sinking air streams inside of a cumulonimbus cloud, as suggested in Figure 12.22. In a cumulonimbus cloud, air rises from the lower to the upper troposphere. In baroclinic clouds this ascent occurs over a period of two days (a mean ascent of about $3 \ \mu b \ s^{-1}$), a vertical displacement of 350 mb per day. The point made with regard to Figure 12.4 is that the potential temperature of the parcel increases due to release of latent heat (by about 30°C between the lower and upper troposphere). If that same parcel were to return immediately to the planetary boundary layer, it would do so at its equivalent potential temperature (assuming that all the moisture is precipitated (see Fig. 1.4)). Clearly, this surface temperature implied by adiabatic decent would be exceedingly warm. Therefore, in order for a complete cycling to take place in which air initially in the planetary boundary layer rises to the upper troposphere in the conveyor belt and subsequently returns to the planetary boundary layer at its *initial* temperature, the air must undergo a diabatic cooling during descent.

Unlike the cumulonimbus downdraft, which is cooled by evaporation of rain water by entraining dry air, the large-scale cooling is accomplished by radiative cooling. For a typical rate of cooling of 1–2°C per day in the upper and middle troposphere, an air parcel undergoing a complete cycle requires 15–30 days in order to cool by 30°C; during this time the parcel passes through many disturbances and circulates the globe several times.

Although the original identity of an air parcel is, of course, obliterated

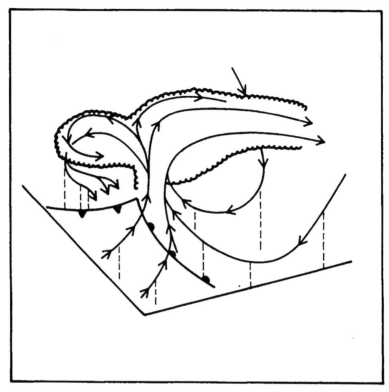

Figure 12.22 Three-dimensional streamflow in a mature wave/cyclone showing warm and cold conveyor belt air streams. Perspective suggests analog with cumulonimbus cloud circulation. (Based on figure by Danielsen and Bleck, 1967.)

within a few days, we can press this argument in a statistical sense. If ascent is, say, 15 times more rapid than descent, continuity requires that the area of ascent be equal to one-fifteenth the area of descent at a rate of about $-0.4 \, \mu b \, s^{-1}$. (Small-scale descent in upper-tropospheric fronts may be quite large, however.) The fact that active cloud occupies a relatively small fraction of a satellite image supports the conclusion that the average rate of ascent is generally much more intense than descent and takes place over smaller areas than descent.

The similarity of the relative isentropic flow pattern, illustrated in Figure 12.20, to the cumulonimbus cloud is suggested in Figure 12.22. Warm moist air ascends from the planetary boundary layer to the upper troposphere, achieves condensation and rises along a moist adiabat. Unlike convective updrafts, which are driven by buoyant forces, the atmosphere at middle latitudes is normally potentially stable but baroclinically unstable, at least selectively so according to the ambient values of static stability, wavelength

326

and other factors discussed in Chapter 11. An air parcel in the statically stable mid-latitude atmosphere is therefore obliged to rise slantwise rather than vertically.

The difference in size between areas of rising and sinking air is due to the condensation process. In a perfectly dry atmosphere, areas of rising and sinking motion and the rates of ascent and descent would be nearly equal. As in a cumulonimbus cloud, the vertical displacement of the conveyor belt air and the instantaneous vertical motion depend on the initial wet-bulb potential temperature at the source and the temperature at mid-levels at the trajectory terminus. In large, highly baroclinic systems, where the warm conveyor belt air is of tropical origin and possesses a high value of θ_w, the conveyor belt is latitudinally and vertically extensive. Unlike the cumulonimbus, the air parcels in a statically stable atmosphere must move sideways in a slantwise fashion to reach their equilibrium level, which is generally higher in the troposphere the higher is the wet bulb potential temperature (θ_w) near the surface in the source region at lower latitudes.

Total precipitation that falls from a conveyor belt is a function of the moisture content in the planetary boundary layer at the source of the conveyor belt. Let us assume that, in the source of the warm conveyor belt, the depth of the planetary boundary layer is Δp_f. If this planetary boundary layer contains a specific humidity (q), relative wind speed (V_R) and width (l), the horizontal flux of moisture per unit surface area into the back end of the conveyor belt (F_C) is $V_R \Delta p_f q l$. This flux F_C is related to the average rate of precipitation over an unspecified surface area downstream. Estimates of the precipitation rate based on this type of simple reasoning have proved successful in studies of conveyor belts over the U.K. In general the simple flux method of assessing rainfall rates over large areas yields an overestimate by up to a factor of 2 because much of the moisture is re-evaporated or remains as cloud droplets. In regions of mid-latitudes where there is potential instability, some of the moisture may be precipitated in thunderstorms, particularly within the warm sector; this constitutes a "short-circuit" to the large-scale baroclinic flow pattern depicted in Figure 12.20.

Let us now consider the mathematical similarity between the cumulonimbus and the large-scale flow. The equation that governs the total specific energy of an air parcel moving in the relative wind system can be written as

$$\frac{d}{dt}(\tfrac{1}{2}V_R^2 + gZ + c_p T + L_e r) + 2V_R \cdot (\boldsymbol{\Omega} \times C) = \frac{1}{\rho}\left(\frac{\partial p}{\partial t}\right)_R \quad (12.11)$$

where V_R is the relative velocity, r is the mixing ratio of condensed water, $\boldsymbol{\Omega}$ is the Earth's angular rotation rate, and the remaining terms have their usual meaning. For a steady state, i.e. translation (the frozen-wave approximation), the local derivative vanishes. The last term on the left-hand side of

(12.11) accounts for the translation of the system with the phase velocity C. If the phase velocity of the system is small compared to that of the wind, or if the meridional displacement of the air parcel following the trajectory is relatively small, the kinetic energy of the relative system can be neglected and the following approximation (subject to these restrictions) for the change in kinetic energy of a parcel following the relative flow streamlines is obtained:

$$\tfrac{1}{2}V_{Rf}^2 - \tfrac{1}{2}V_{Ri}^2 = g(\Delta z' - \Delta z). \qquad (12.12)$$

Here, V_{Ri} is the initial (relative) wind speed at the start of the trajectory, V_{Rf} is the final wind speed at some location along the streamline, Δz is the thickness between the initial and final pressures for an ambient vertical temperature distribution at the location of the parcel, and $\Delta z'$ is the thickness between the same pressure surfaces for a sounding having a vertical temperature distribution identical to the adiabatic path taken by the ascending air parcel. The above equation represents an "extended parcel theory".

To illustrate the extended parcel theory, consider Figure 12.23a, in which an air parcel originates in the Tropics at level p_i (800 mb) at height z_i and relative wind speed V_{Ri} and reaches level p_f (300 mb) at height z_f with a relative speed of V_{Rf}. The thermodynamic path of the parcel is dry adiabatic until reaching saturation, and, subsequently, it moves along a moist adiabat. On the thermodynamic diagram, in Figure 12.23b, this path forms the positive area A between the dry and moist process lines and the actual sounding made at the trajectory terminus marked with a filled circle in Figure 12.23a. The gain in kinetic energy of the parcel between its start at 800 mb and its terminus in the downstream ridge, where the relative wind speed is 80 m s^{-1}, is given approximately by equation (12.12). This kinetic-energy gain is proportional to the shaded area labeled A (equal to $g(\Delta z' - \Delta z)$ in Figure 12.23b).

Area A on the sounding in Figure 12.23b represents the kinetic energy generated by the use of the air parcel. Note that this kinetic energy is expressed by the left-hand side of (12.12). The right-hand side represents the potential energy converted to kinetic energy. Thus, (12.12) expresses another aspect of available potential energy. In a rough sense, the meridional difference in θ_w is related to area A in Figure 12.23b and is thus a measure of available potential energy. The kinetic energy realized by an ascending air parcel in the conveyor belt is independent of the path taken to arrive at a given location; but it depends on the difference in temperature between that of a moist adiabat of θ_w at the source and the temperature of the ambient sounding along the trajectory. Not surprisingly, therefore, large kinetic energy (strong winds) is achieved in the high troposphere in regions of greatest baroclinicity and where the conveyor belt source region possesses a

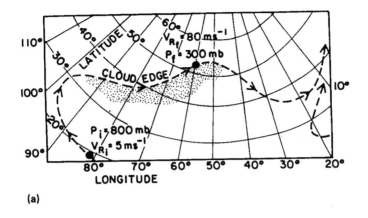

(a)

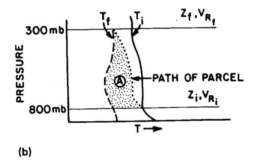

(b)

Figure 12.23 (a) Relative isentropic path taken by an air parcel (broken streamline) originating at a location in the Tropics (filled circle with initial relative wind speed (V_{Ri}) of 5 m s^{-1} at $p_i = 800$ mb). The parcel reaches the ridge, at $p_f = 300$ mb, and has a relative wind speed (V_{Rf}) of 80 m s^{-1} (filled circle). The trajectory constitutes the limiting streamline in the relative wind system, and the conveyor belt cloud is shaded. (b) Schematic vertical soundings of temperature at two locations along the trajectory in (a) (filled circles). The soundings at the two points are labelled T_i and T_f to denote initial and final soundings. The dotted curve represents the path of the parcel ascending from 800 mb (at height z_i), where its relative wind speed is V_{Ri}, to 300 mb (at height z_f), where its relative wind speed is V_{Rf}. The parcel gains kinetic energy by an amount proportional to the positive area A formed between the adiabatic path of the parcel (dotted curve) and temperature profile at the trajectory terminus. Mathematically, this gain in kinetic energy is expressed by equation (12.12).

relatively high wet-bulb potential temperature.

The most favorable regions and times for achieving high wind speeds in the conveyor belt are along the eastern sides of the great land masses, North America and Asia, during wintertime. Indeed, extraordinarily fast-moving jets, exceeding 100 m s^{-1}, have been reported in wintertime for the high troposphere over Japan, which lies between the cold Siberian land mass and the warm waters of the southwestern North Pacific Ocean.

Each wave or vorticity couplet is associated with a vertical and meridional exchange, in which "warm" air is lifted to higher latitudes and replaced by "cool" air sinking from higher latitudes (Figs 12.6 & 12.20). Each successive warm conveyor belt travels into the downstream waves and descends, losing its identifying cirrus cover. Although it is not clear exactly where the conveyor belt goes after reaching the downstream ridge, we can envision successive exchanges of air in the manner of Figure 12.23a in successive downstream troughs. Each intrusion of lower-latitude air moves through the successive downstream disturbances as it undergoes gradual radiational cooling.

12.6 Downstream development

It is possible that cyclogenesis can affect the behavior of weather systems both downstream and upstream. This conjecture is supported both by observation and by the results of numerical modeling experiments. Explosive cyclogenesis along the east coast of North America is sometimes followed by the breakdown of zonal flow over Europe. Changes from zonal to azonal flow and the establishment of a low-index blocking pattern in the upper flow are often preceded by the amplification of an upstream thermal trough and the formation of a northwest–southeast (negative) tilt of the trough in the height pattern and the intensification or generation of wind speed maxima, such as the wind speed maximum in the high troposphere near the downstream ridge (Fig. 12.20).

Observations further show that the development of a downstream trough is preceded by the amplification of the upstream ridge, events that immediately follow cyclogenesis. Moreover, the period between development in the trough and that in the downstream ridge, or between formation of the upper-tropospheric jet in the ridge and development of the downstream trough, is about a day or two. Awareness of downstream development has led some forecasters in the U.K. to establish rules for the southward displacement of thermal troughs over the Atlantic Ocean according to the magnitude of the southerly wind component in the upstream 500 mb trough, which would be located over the western North Atlantic Ocean or eastern North America.

Enhancement of the thermal trough west of the downstream trough occurs about 60–100° of longitude east of the upstream trough between one and three days following cyclogenesis in the upstream trough and the development of a strong southerly wind core in the warm conveyor belt. The propagation speed of the developmental impulse is about 30–40° of longitude per day, implying that cyclogenesis in one location also affects cyclogenesis downstream in the next wave in about as much time as it takes for the air at upper levels to flow from one trough to another.

Downstream development can be studied using Hovmoller diagrams (see Fig. 5.3). Figure 12.24 shows that trough B formed about 60° of longitude farther east than trough A. As trough B moved slowly eastward, trough C formed about 100° of longitude farther east, three days after the inception of trough B. By the last day of the record B and C are in evidence, and the geopotential heights are considerably more perturbed than on the first day.

Figure 12.25 illustrates downstream development on a series of 500 mb charts. In Figure 12.25a, a migrating 500 mb trough (labeled A) and attendant surface cyclone were located over North America. During the first 48 h

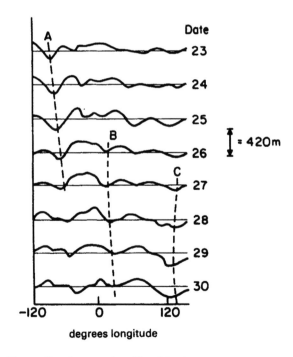

Figure 12.24 Hovmoller diagram (see Fig. 5.3) showing downstream development of troughs B and C following the appearance of trough A. The geopotential height of the 500 mb surface between 40 and 55 °N is in the vertical scale and the horizontal axis is longitude. The period covers several days in November 1945. (From Cressman, 1948.)

331

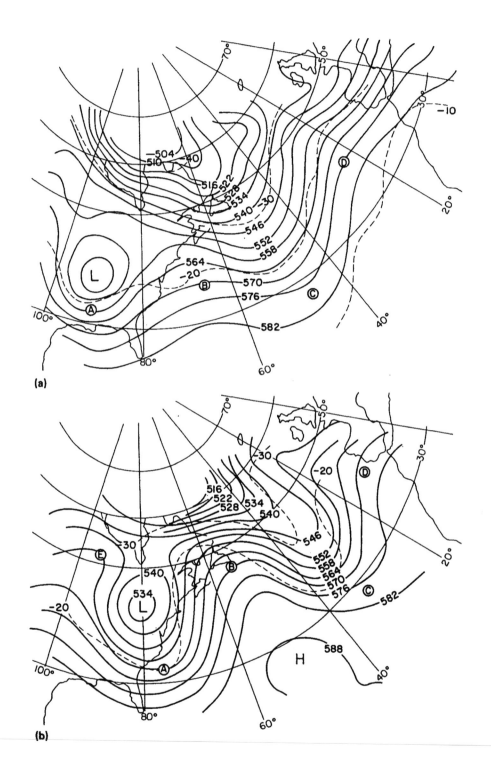

(a)

(b)

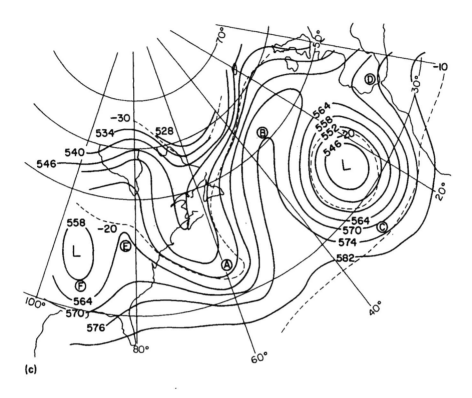

(c)

Figure 12.25 (a) The 500 mb geopotential height (full curves labeled in dam) and temperature (broken contours in °C) for 1200 GMT 4 April 1984. Troughs and ridges are labeled with letters discussed in text. (b) Same as (a) but for 0000 GMT 6 April 1984. (c) Same as (a) but for 0000 GMT 9 April 1984.

(Figs 12.25a & b), the North American trough deepened and acquired a negative tilt south of the closed center in the height field. Southerly winds ahead of the trough had intensified markedly and the isotherms had been advected northward in association with rapid ridge formation downstream. The downstream trough near 30°W (C) and its downstream ridge near 15°W (D) also intensified. Over the following 72 h (Figs 12.25b & c) the original trough A weakened, the first downstream ridge B amplified noticeably and the downstream trough C had become a rather intense eddy near 25°W. After $4\frac{1}{2}$ days, the initial zonal pattern of Figure 12.25a is highly azonal, and there is also evidence of some upstream development in the ridge (E) and trough (F) over North America (Fig. 12.25c).

Events accompanying downstream development can be summarized as follows:

(a) Following cyclogenesis, downstream development occurs in the high troposphere near the downstream ridge. An indication of ridge development is thickening cirrus cloud cover and increased warm air advection at high levels along the poleward end of the warm conveyor belt. This is accompanied by the increase of negative vorticity in the ridge and the formation of a strong wind speed maximum along the poleward end of the conveyor belt cloud. This jet is entirely confined to the upper troposphere, although its speed may come to exceed that of the upstream jet, which exists at any level along the western edge of the conveyor belt south of the surface cyclone center. Just as the jet west and equatorward of the surface cyclone is formed by the processes of cold air advection and strong confluence in the lower troposphere, coupled by height falls aloft in the region where the 500 mb trough is developing a negative tilt, so also does the jet form in the high troposphere near the ridge in accordance with warm air advection and height rises in the upper troposphere along the poleward end of the conveyor belt cloud.

(b) There is confluence in the ridge at high levels. Numerical modeling experiments show that the formation of the downstream trough follows from changes occurring at high levels near the ridge. These observations bear strongly on subsequent discussions concerning tropopause folding and upper-tropospheric frontogenesis (Ch. 15). They are also compatible with the early theory of Namias and Clapp, who postulated that upper-tropospheric fronts and jets form in the confluence between the ridge and the straight or cyclonically curved flow on the poleward side of the ridge.

(c) Cold advection occurs downstream from the ridge at low levels, bringing cold thicknesses into the downstream trough.

(d) Cold advection into the downstream trough results in the development of the trough and an increase in positive vorticity advection ahead of the downstream trough.

(e) Subsequent cyclone development occurs ahead of the next downstream trough. In general, downstream development seems to be related to the propagation of a relatively small (short-wave) upper vorticity maximum from the upstream cyclogenesis region. Initial movement of the 500 mb vorticity maximum may be associated with the concentration of that vorticity on the eastern side of the developing trough, as discussed in Chapter 10.

12.7 Blocking

Fluctuations in the amplitude of waves, shown in Figure 12.24, are part of the index cycle of the atmosphere referred to in Chapter 5. In view of the steering concepts discussed in Chapter 6, it is easy to see why such patterns are called blocking, since height patterns associated with large eddies correspond to a relatively weak steering flow. The transition from zonal to azonal flow is one in which strong steering is replaced by weak steering. Chapter 6 suggests that blocking patterns are subject to retardation or retrogression of their eastward motion because of the greater influence of the beta wind. Blocking is accompanied by the growth of very long, large-amplitude waves, which derive their energy from the non-linear interaction of smaller waves with the larger ones. These non-linear interactions are responsible for shifting the wavelength of maximum growth from shorter to longer waves, the latter feeding on the energy of the cyclone-scale disturbances. The mechanism appears to be a combination of baroclinic and barotropic processes.

Although the index cycle and blocking possess a predominant random aspect, one can distinguish certain behavioral traits. Blocking activity seems to be most frequent in the Northern Hemisphere over (a) the Pacific Ocean just west of North America, (b) the Atlantic Ocean just west of Europe and (c) over Europe just west of the Ural mountains. In the Southern Hemisphere, blocking seems to be most favored when the 500 mb ridges lie close to 55°E, 175°E and 65°W. Blocking seems to be related to topography (particularly in the Northern Hemisphere where the blocking duration is longer than in the Southern Hemisphere).

Blocking has been simulated in numerical models, even in the absence of terrain. As in the actual atmosphere, models show that large positive geopotential height anomalies seem to be more prevalent than large negative anomalies, the latter being more closely associated with an equatorward displacement of the jet stream rather than a breakdown of the zonal flow. During blocking, a split in the jet stream often forms in which the poleward branch is associated with the establishment at low levels of a blocking anticyclone, typically near 50°N along the western sides of the continents. At the same time cyclone tracks may be further south than usual and there may be unusual cold advection at low latitudes, in the case where the southerly branch of the jet is in evidence. Such episodes often bring unseasonably cold and sometimes snowy weather at lower latitudes.

Blocking is more frequent during the winter and spring months, and reaches a maximum frequency in springtime. In many instances blocking follows the formation of an intense cyclone and 500 mb trough development upstream. Blocking cycles tend to last from one to four weeks before zonal flow is re-established. A preferred period of between three and four weeks has been identified by some researchers. Prediction or recognition of blocking patterns can greatly help forecast weather conditions over extended periods. A particular blocking regime occurring in the middle of a winter may determine the character of that season.

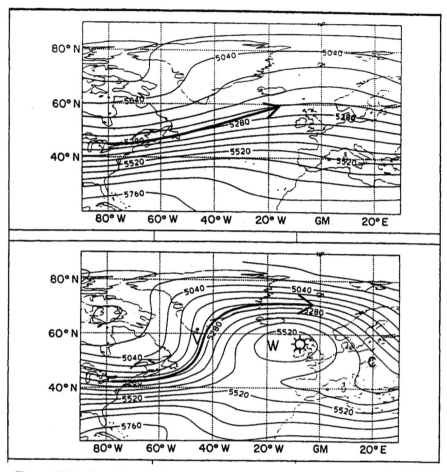

Figure 12.26 Simulations of blocking ridges. The average 500 mb geopotential height pattern (labeled in dam) has been simulated using the NCAR community climate model for a perpetual month of January, and two-week averages for different periods are shown here in the two maps. The heavy full streamline denotes the direction of the jet stream and the sun symbol represents fair weather conditions. (From Mullen, 1985.)

Once blocking is established, cyclone growth is retarded by the inevitable decrease in the strong zonal current and meridional gradient of temperature and increase in static stability, which favors slower growth of cyclones and a larger wavelength. Diminution of cyclone meridional mass exchange inevitably allows the meridional temperature gradient to increase with time, once more establishing a strong zonal flow. Numerical simulations of blocking ridges and the index cycle have been achieved in long-range general circulation models, which place the blocking ridge at the correct longitudes when realistic surface boundary conditions and topography (e.g. the land–sea contrast) are specified. An example of such a simulation is provided in Figure 12.26. In some instances, blocking cycles involve only a hemispheric redistribution (rather than a creation or destruction) of energy and energy conversions. Chapter 5 discusses a possible relationship between blocking and index cycles and components in the potential- and kinetic-energy equation.

Problems

12.1 Figure 12.27 is a form of nephanalysis (cloud analysis abstracted from a satellite photo). Data are plotted on the map at a number of station locations. At each station the *relative* wind, pressure (p_θ) and mixing ratio (r_θ) are plotted (see station key) for the dry potential temperature surface $\theta = 28°C$. Winds are plotted in conventional Beaufort notation (1 flag = 5 m s^{-1}); pressure is plotted as a three-digit number

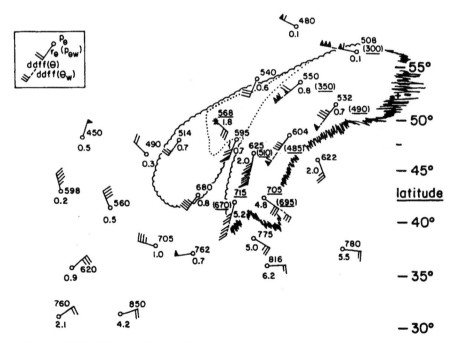

Figure 12.27 Distribution of winds, pressure and mixing ratios on dry and moist isentropic surfaces and a nephanalysis with station locations; see problem 12.1 and key (top left) for details.

in mb; mixing ratio is in g kg^{-1}. Where the saturation exists on the $\theta_w = 16°C$ surface, the winds are plotted as dashed barbs and the pressure is in parentheses. Underlined pressures signify saturation at that level. Areas within the scalloping represent overcast cloud. Dotted curves suggest sharp divisions between higher and lower cloud decks.

Perform the following analysis:

(a)   Construct an isobaric analysis at some reasonable contour interval (say 100 mb) on the dry isentropic surface.

(b)   Draw in some streamlines on the $\theta = 28°C$ surface and label the limiting streamline (LSW) of the warm conveyor belt.

(c)   Construct an isobaric analysis of the $\theta_w = 16°C$ flow in the saturated region(s) using a different symbol or color for the isobars on the moist surface.

(d)   Dash in some streamlines on the moist isentropic surface.

Answer the following questions:

(e)   Point out (by labeling) on the map where you would expect to find the heaviest precipitation in the warm conveyor belt and explain how you arrived at the answer. Equating isobars on the isentropic surface with isotherms on isobaric surfaces, indicate where you would expect to find the greatest likelihood of frozen precipitation and explain why.

(f)   Point out where on the map you would expect to find saturation first occurring following an air parcel along a streamline moving through the warm conveyor belt; explain your reasoning.

(g)   Locate the region of strongest upward vertical motion (on a moist or dry surface) and downward vertical motion, and explain how you arrived at your answer.

(h)   Find and label a streamline where relatively dry air of high-tropospheric origin is ascending.

(i)   Label a trajectory (streamline) that is undergoing the greatest vertical displacement from start to terminus: upward and downward (one each).

(j)   Indicate where large-scale descent occurs on the $\theta = 28°C$ surface in the presence of overcast cloud. Why is that cloud present?

(k)   One station (designated by a star; station pressure 568 mb) is saturated on the $\theta = 28°C$ surface but the wet-bulb potential temperature on that surface is 11°C. There is no value of $\theta_w = 16°C$ below the level of the tropopause at that station. What is the origin of the $\theta_w = 11°C$ air reaching this station? (A rough streamline or two and a couple of sentences will suffice.)

(l)   Some stations in the warm conveyor belt cloud region show relatively dry air on the $\theta = 28°C$ surface but saturated air on the $\theta = 16°C$ surface. Where does that unsaturated air originate?

(m)   In general, why is it that the troughs and ridges on θ surfaces overtake the air (relative wind easterly) along the equatorward side of relative wind isentropic surfaces and are overtaken by the air (relative wind westerly) on the poleward side?

(n)   Why is it that the western border of the middle and high cloud is so very sharp whereas the borders of most cloud masses are rather ragged?

(o)   What evidence do you see that indicates that cyclogenesis has been taking place?

12.2   Imagine a rather poor-quality visible-wavelength satellite image such as shown schematically in Figure 12.28. All that you can see is the general cloud outline (denoted as either sharp or ragged borders) and some internal boundaries (broken curves denoting abrupt changes in general cloud top elevation). Locate the following features on the diagram:

(a)   the surface low center,

(b)   the 500 mb low center,

(c)   the 500 mb absolute vorticity maximum,

(d)   the wind speed maxima at 300 mb,

(e)   the dry tongue,

(f)   the general direction of the warm and cold conveyor belt flow through the cloud system,

(g)   the highest (coldest) cloud tops and

(h)   a layer of low cloud in the presence of large-scale descent.

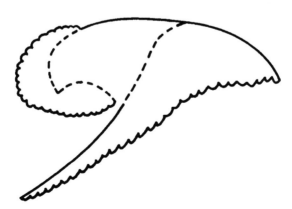

Figure 12.28   Schematic visible-wavelength satellite image showing internal and external cloud boundaries; see problem 12.2.

Further reading

Atkinson, B. W. and P. A. Smithson 1978. Mesoscale precipitation areas in a warm frontal wave. *Mon. Wea. Rev.* **106**, 211–22.

Austin, J. F. 1980. The blocking of mid-latitude westerly wind by planetary waves. *Q. J. R. Met. Soc.* **106**, 327–50.

Betts, A. H. and J. F. R. McIlveen 1969. The energy formula in a moving reference frame. *Q. J. R. Met. Soc.* **95**, 639–42.

Browning, K. A. 1974. Mesoscale structure of rain systems in the British Isles. *J. Met. Soc. Japan* **50**, 314–27.

Browning, K. A. 1985. *Conceptual models of precipitation systems.* Met. Off. RRL Res. Rep. no. 43.

Browning, K. A. 1990. Organization of clouds and precipitation in extratropical cyclones: Eric Palmen Memorial Volume. American Meteorological Society, Boston MA, pp. 129–53.

Browning, K. A., M. E. Hardman, T. W. Harrold and C. W. Pardoe 1973. The structure of rainbands within a midlatitude depression. *Q. J. R. Met. Soc.* **99**, 215–31.

Browning, K. A. and T. W. Harrold 1969. Air motion and precipitation growth in a wave depression. *Q. J. R. Met. Soc.* **95**, 288–309.

Browning, K. A., F. F. Hill and C. W. Pardoe 1974. Structure and mechanism of precipitation and the effect of orography in a wintertime warm sector. *Q. J. R. Met. Soc.* **100**, 309–30.

Browning, K. A. and G. A. Monk 1982. A simple model for the synoptic analysis of cold fronts. *Q. J. R. Met. Soc.* **100**, 435–52.

Browning, K. A. and C. W. Pardoe 1973. Structure of low-level jet stream ahead of midlatitude cold fronts. *Q. J. R. Met. Soc.* **99**, 619–38.

Carlson, T. N. 1980. Airflow through midlatitude cyclones and the comma cloud pattern. *Mon. Wea. Rev.* **108**, 1498–509.

Carlson, T. N. 1981. Speculations on the movement of polluted air to the Arctic. *Atmos. Env.* **15**, 1473–7.

Carr, F. H. and J. P. Millard 1985. A composite study of comma clouds and their association with severe weather over the Great Plains. *Mon. Wea. Rev.* **113**, 349–61.

Cressman, G. P. 1948. On the forecasting of long waves in the upper westerlies. *J. Meteor.* **5**, 44–57.

Danielsen, E. F. 1966. *Research in four-dimensional diagnosis of cyclonic storm cloud systems.* Penn. State Univ., Sci. Rep. no. 2, Project Rep. no. 6698 to Air Force Cambridge Research Laboratories, AF19(628)-4762.

Danielsen, E. F. 1968. Stratospheric–tropospheric exchange based on radioactivity, ozone and potential vorticity. *J. Atmos. Sci.* **25**, 502–18.

Danielsen, E. F. 1974. The relationship between severe weather, major dust storms and rapid large-scale cyclogenesis (II). Subsynoptic extratropical weather systems: observations, analysis and prediction. Notes from a colloquium. Summer 1974, Vol II. NCAR Rep. no. ASP-CO-3-V-2 226–241.

Danielsen, E. F. and R. Bleck 1967. *Research in four-dimensional diagnosis of cyclonic storm cloud systems.* Final Sci. Rep. to Air Force Cambridge Research Laboratories, AF19(628)-4762.

Dole, R. M. 1983. Persistent anomalies of the extra-tropical Northern Hemisphere winter-time circulation. In *Large-scale dynamical processes in the atmosphere,* 95–109. New York: Academic Press.

Durran, D. R. and D. Weber 1988. An investigation of the poleward edges of cirrus clouds associated with mid-latitude jet streams. *Mon. Wea. Rev.* **116**, 702–14.

Fredricksen, J. S. 1984. The onset of blocking and cyclogenesis in Southern Hemisphere flows: linear theory. *J. Atmos. Sci.* **41**, 1116–31.

Green, J. S. A., F. H. Ludlam and J. F. R. McIlveen 1966. Isentropic relative-flow analysis and the parcel theory. *Q. J. R. Met. Soc.* **92**, 210–19.

Grotjahn, R. 1979. Cyclone development along weak thermal fronts. *J. Atmos. Sci.* **36**, 2049–74.

Grotjahn, R. and C.-H. Wang 1989. On the source of air modified by ocean surface fluxes to enhance frontal cyclone development. *Ocean–Air Interact.* 257–88.

Harrold, T. W. 1973. Mechanisms influencing the distribution of precipitation within baroclinic disturbances. *Q. J. R. Met. Soc.* **99**, 232–51.

Harrold, T. W. and P. M. Anston 1974. The structure of precipitation systems – a review. *J. Rech. Atmos.* **8**, 41–57.

Hirschberg, P. A. and J. M. Fritsch 1991. Tropopause undulations and the development of extratropical cyclones. Part III: Diagnostic analysis and conceptual model. *Mon. Wea. Rev.* **118** (in press).

Jackson, M. L., D. A. Gillette, E. F. Danielsen, I. H. Blifford, R. A. Bryson and J. K. Syers 1973. Global dustfall during the Quaternary as related to environments. *Soil Sci.* **116**, 135–45.

Legras, B. and M. Ghil 1985. Persistent anomalies, blocking and variations in atmospheric predictability. *J. Atmos. Sci.* **42**, 433–71.

Ludlam, F. H. 1980. *Clouds and storms*. Pennsylvania: Pennsylvania State University Press.

Miles, M. K. 1959. Factors leading to the meridional extension of thermal troughs and some forecasting criteria derived from them. *Met. Mag.* **88**, 193–205.

Mullen, S. 1985. On the maintenance of a blocking anticyclone in a general circulation model. Ph.D. Thesis, University of Washington. NCAR Cooperative Thesis no. 86.

Namias, J. and P. F. Clapp 1949. Confluence theory of the high tropospheric jet stream. *J. Meteor.* **6**, 330–6.

Palmen, E. and C. W. Newton 1969. *Atmospheric circulation systems*. New York: Academic Press.

Reiter, E. and J. D. Mahlman 1965. Heavy radioactive fallout in the southern United States, November 1962. *J. Geophys. Res.* **70**, 4501–20.

Rex, D. F. 1950a. Blocking action in the middle troposphere and its effect on regional climate. Part I: An aerological study of blocking action. *Tellus* **2**, 196–211.

Rex, D. F. 1950b. Blocking action in the middle troposphere and its effect on regional climate. Part II: The climatology of blocking action. *Tellus* **2**, 275–301.

Saltzman, B. and C.-M. Tang 1985. The effect of finite-amplitude baroclinic waves on passive low-level, atmospheric constituents, with applications to comma cloud evolution. *Tellus* **37A**, 41–55.

Simmons, A. J. and B. J. Hoskins 1979. The downstream and upstream development of unstable baroclinic waves. *J. Atmos. Sci.* **36**, 1239–54.

Trenberth, K. E. 1986. The signature of a blocking episode on the general circulation in the southern hemisphere. *J. Atmos. Sci.* **43**, 2061–9.

Trenberth, K. E. and K. C. Mo 1985. Blocking in the Southern Hemisphere. *Mon. Wea. Rev.* **113**, 3–21.

Weldon, R. 1979. *Cloud patterns and the upper air wind field*. Satellite Training Service Course Notes, Part IV, Air Weather Service Document AWS/TR-79/003.

13

Kinematics of surface fronts

In the Norwegian polar front cyclone model, precipitation patterns are related to sloping frontal surfaces in an area of low pressure. Well before the classification of fronts by the Bergen school and their acceptance by forecasters in the 1920s, Margules was able to show that the slope of a front must be upward toward colder air and that there must be cyclonic shear and a local minimum of pressure along the front. The Bergen school classified four types of fronts: cold, warm, stationary and occluded. Inclined frontal surfaces were conceived as interfaces between wedges of cold and warm air in which the warm air may overrun the advancing or retreating cold air. Later, ana- and kata-fronts (see Ch. 12) were classified according to the direction of motion, respectively up or down sloping isentropes. Fronts were once thought to furnish the energy needed for cyclogenesis. This early concept of fronts provided a reasonably consistent model for the distribution of weather around cyclones and thereby led to great improvements in the ability to forecast mid-latitude weather.

With the advent of routine upper-air soundings in the 1930s and 1940s, meteorologists discovered that fronts are not infinitely thin interfaces between wedges of air but consist of sloping zones of finite width, within which gradients of temperature, vorticity (horizontal and vertical wind shear), convergence and static stability are very large compared to background synoptic-scale values. These "frontal zones" are relatively small in scale (50–200 km), although the lateral extent of the fronts is comparable to the scale of cyclones. Thus, fronts are associated with two scales of motion: one parallel and one perpendicular to the frontal zone.

Because of the disparity in scales along and perpendicular to a front, gradients of kinematic properties in the cross-front direction are very much greater than those along the front. The idea that fronts are kinematically controlled features dominated the research on frontal systems during the 1950s. This research led to the discovery that there are a variety of frontal structures in both the lower and upper troposphere. In some instances the surface front extends throughout the troposphere up to the lower stratosphere. More often, surface fronts are relatively shallow (less than 3 km thick) and may appear independent of upper-tropospheric fronts. Surface fronts, however, are intimately associated with cloudiness and precipitation and the migration of colder air toward warmer air. Upper-tropospheric fronts, which reflect deep tropospheric circulations, are less directly associated with surface weather conditions and with migrating cold and warm air masses. Rather, they are more closely related to upper-tropospheric jet streaks and the structure of the tropopause.

During the 1950s researchers concentrated on describing the kinematic nature of both the surface and upper-tropospheric fronts. By the 1960s it was realized that frontal development (frontogenesis) is not explainable solely on the basis of kinematic processes but requires the inclusion of dynamic arguments, specifically the effects of ageostrophic motion on the intensification or weakening with time of the temperature gradient. By the 1960s it had become evident that the Norwegian cyclone model could not explain frontogenesis nor could it account totally for the observed structure of fronts.

In this chapter we discuss lower-tropospheric fronts and their kinematics. We begin with the methodology of the 1950s, and illustrate how fronts were treated by investigators at that time. This methodology is based largely on a development of the kinematic frontogenesis equations. Later, we will relate these kinematic concepts to quasi-geostrophic forcing.

13.1 Synoptic aspects of surface fronts

Surface fronts are commonly thought of as a zone of strong horizontal temperature gradient on a sea-level pressure chart. The definition of a front is not always apparent from an examination of the isotherms. Fronts are recognized by the following kinematic properties:

(a) a zone of stronger temperature, moisture, vertical motion and vorticity gradients normal to the frontal boundary on the cold side of the front (as compared with the warm air side);

(b) frontal gradients that appear discontinuous from those of the synoptic-scale background;

(c) a relative minimum in pressure (giving the impression of a trough along the front);

(d) a relative maximum of vorticity along the front;

(e) a zone of confluence along the front;

(f) strong vertical and lateral (cyclonic) wind shear along the front; and

(g) rapid changes in cloud cover and precipitation (surface fronts).

Many of these characteristics are present in the frontal zone analysis in Figure 13.1, which is a dramatic example of frontogenesis. Over a period of

343

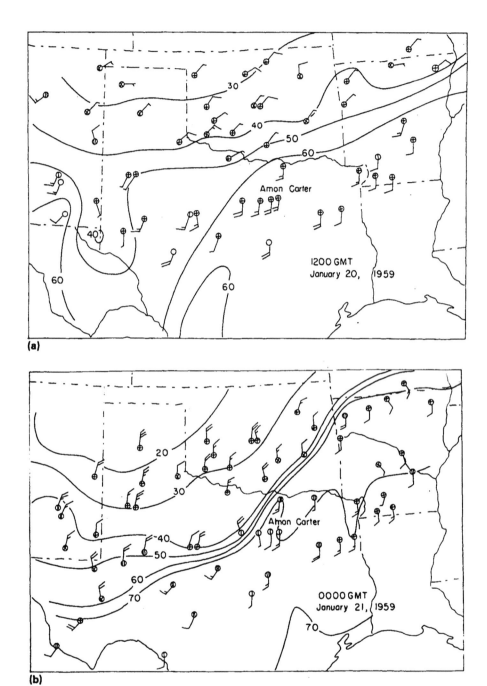

Figure 13.1 (a) Surface isotherms (°F) and wind barbs (one full barb = 5 m s^{-1}) for 1200 GMT 20 January 1959 for region over the southern United States. (b) Same as (a) but for 0000 GMT 21 January 1959. (Courtesy of Fred Sanders.)

12 h the initially weak front intensified rapidly and became very sharply defined; at Amon Carter, a town in north Texas, the temperature fell by about 24°F (13°C) in less than 1 h. Note the increasing confluence and cyclonic wind shear along the front with time. In Figure 13.1a, the weak temperature gradient is more readily diagnosed by the presence of a wind shift, a zone of confluence and lateral cyclonic wind shear. Note that the frontal zone exhibits a discontinuity in the temperature *gradient* rather than the temperature itself. Twelve hours later (Fig. 13.1b), a large gradient of temperature was concentrated within a narrow frontal zone, which had moved rapidly southward. Considerable cloudiness occurred on the cold side of the front but there was little precipitation.

Fronts can also be identified on vertical cross sections. Figure 13.2 shows the cold front in the temperature and wind patterns. The horizontal temperature gradient (diffuse at the earlier time) slopes backward at a steep angle (with the vertical) toward the cold air. At a relatively short distance on the cold side of the front, the horizontal temperature gradient has vanished although there is a significant vertical temperature gradient, which comprises a shallow layer of high static stability (less than 100 mb deep). In accordance with the thermal wind relationship (1.33), the westerly winds increase rapidly with height, most noticeably near the location of the strongest horizontal temperature gradient.

Intense cold fronts, such as those shown in Figures 13.1 and 13.2, tend to be very shallow. Although there are instances where the cold front is associated with a front extending up to the tropopause (as suggested in Figure 15.10a, for example), the vast majority of fronts that are well defined at the surface weaken rapidly with height. Surface cold or warm fronts rarely are identifiable at 500 mb, although the 1000–500 mb thickness and the 500 mb absolute vorticity fields are useful in locating the position of the surface front when only the output from numerical forecast models is available (see discussion below). In the case where the surface front does extend throughout the troposphere, the entire feature is sometimes referred to as the "polar front", although such definitions are relatively arbitrary. Indeed, some analysts take account of an "arctic front" poleward of the polar front. Such labels are more statistical than real. It is more likely that what appears to be the continuation of the front at middle and upper levels is a separate feature, dynamically linked with but of different origin from the surface front. In the general sense, latitudinal gradients in surface temperature, imposed by differential radiative heating of the surface, produce outbreaks of cold air equatorward and warm air poleward. The edges of these cold air masses are marked by shallow cold and warm fronts that have the largest horizontal temperature gradients near the lower boundary.

There have been fewer investigations of the kinematics of warm fronts than of cold fronts. The former have been studied mostly with regard to the

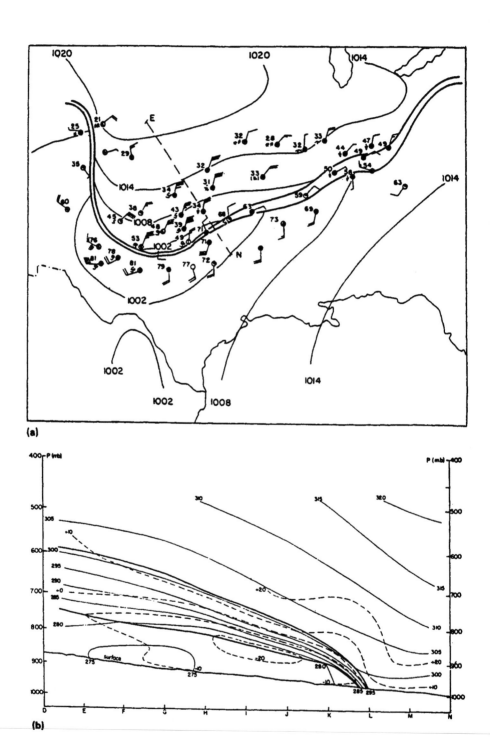

(a)

(b)

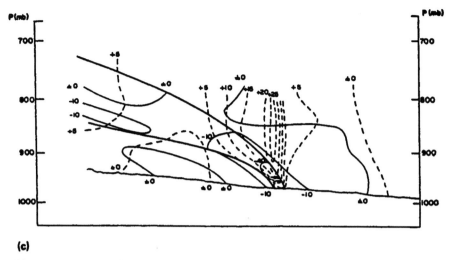

(c)

Figure 13.2 (a) Surface analysis for 0330 GMT 18 April 1953. The broken line (E–N) indicates the position of the vertical cross section in (b); heavy full curves denote boundaries of frontal zone; light full curves are isobars of sea-level pressure (contour interval 6 mb). Plotted reports follow conventional station model. (b) Distribution of potential temperature (light full curves, contour intervals 5 K) and wind component (broken curves, contour interval 10 m s^{-1}) normal to cross section (line labeled E–N in (a)) for 0300 GMT 18 April 1953. Heavier full curves indicate boundaries of frontal zone. Distance between adjacent letters on horizontal axis is 100 km. (c) Distribution of horizontal divergence (light full curves, selected contours in units of 10^{-5} s^{-1}) and vertical motion (broken lines, contour intervals 5 cm s^{-1} (positive rising)) for part of cross section E–N in (a) and (b) at 0300 GMT 18 April 1953. Heavy full curves indicate frontal boundaries. (From Sanders, 1955.)

distribution of precipitation. Warm fronts are represented typically as inclined isentropic surfaces along which air ascends in slantwise fashion. Warm fronts are considered to be ana-fronts, although there are examples of cold fronts that are also ana-fronts. Kinematically (and dynamically) there is no fundamental difference between cold and warm fronts, although the Norwegian school of meteorology made a sharp distinction between the role of cold and warm fronts in the precipitation process. By definition, cold fronts propagate toward warmer air and at a rate faster than warm fronts, which move toward colder air at the surface. This observation has led some meteorologists to conclude that there is an overrunning of one front by the other to form an occlusion. Studies of occluded fronts are rare, however, except with regard to the generation of small-scale precipitation patterns.

A misconception is that cold fronts tend to have steeper slopes than warm fronts. This is true, however, only over the lowest several hundred meters. Warm fronts exhibit extensive cloud cover and sustained precipitation on the cold side of the front, and cold fronts are more likely to be accompanied by

347

brief showery precipitation. Both types of surface fronts have similar slopes above the lowest several hundred meters; cold fronts, however, tend to bulge forward near the surface in the manner of a density bore. The difference in the large-scale precipitation patterns between warm and cold fronts is due to the synoptic-scale forcing, which governs whether air moves up or down isentropic surfaces relative to the motion of the front. The definition of ana- and kata-fronts reduces to a perception of whether the relative wind motion is up or down the isentropes, respectively. Thus, cold fronts displaying katabatic (descending) air motion can become ana-fronts, depending on the direction of the relative wind.

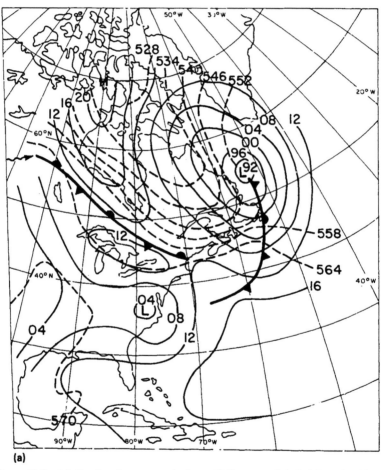

(a)

Figure 13.3 (a) Sea-level pressure isobars (full curves labeled in mb; 1000 digit omitted) and 1000–500 mb thickness contours (labeled in dam) from a 48 h numerical forecast (the United States Weather Service's Nested Grid Model (NGM)) for 0000 GMT 21 June 1987. Surface front (indicated by wiggly arrow on left of diagram) is placed along the edge of the large thickness gradient.

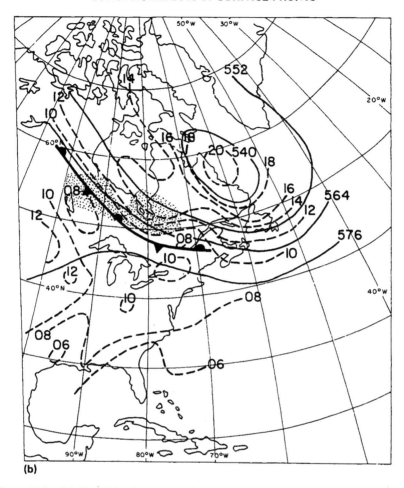

(b)

Figure 13.3 (b) The 500 mb geopotential height (dam) and absolute vorticity isopleths (labelled at intervals of 1×10^{-5} s^{-1}) for 0000 GMT 21 June 1987. Shading denotes predicted precipitation associated with front, which has been placed along the edge of the vorticity gradient.

Analysis of surface fronts from computer-derived maps

Fronts may be difficult to recognize in computer-generated products, particularly those from operational forecast models, because of a lack of vertical and horizontal resolution in both the numerical grid values and of observations. Many of the classic frontal indicators lose their definition during the process of objective analysis and numerical integration of the equations. The three most reliable parameters for locating surface fronts in conventional numerical output are time continuity and the fields of (a) surface pressure,

(b) 1000–500 mb thickness and (c) 500 mb absolute (or relative) vorticity.

Some simple rules for finding fronts can be made in using machine-generated patterns of vorticity and thickness to determine the location of surface fronts. Predicted 48 h fields for the case of a surface cold front are shown in Figure 13.3. This cold front has been moving across Canada prior to the forecast time, which was 0000 GMT 19 June 1987. The surface pressure trough had all but completely disappeared (Fig. 13.3a), except for a portion of the front south of a large cyclone. Because lower-tropospheric temperature gradients affect the 1000–500 mb thicknesses, it is possible to locate the front along the warm edge of a discontinuous gradient in the thickness pattern. In Figure 13.3a, the front is drawn along the surface pressure trough such that it hugs the warm side of the large thickness gradient over the interior of Canada.

Surface fronts also are related to the edges of 500 mb vorticity gradients. Figure 13.3b shows the location of the surface front along the edge of the large 500 mb vorticity gradient over central Canada. The location of this front closely coincides in both Figures 13.3a and b. Note that there are no advection solenoids along the front, although there is, nevertheless, a band of precipitation along the front. The precipitation is associated with frontal forcing, which is a subject addressed in the next chapter.

13.2 Frontogenesis and the kinematics of fronts

During the 1950s, J. Miller, F. Sanders, R. Reed and others investigated a number of lower-tropospheric frontal cases such as the one shown in Figure 13.2. They found that the isentropic structure of these fronts resembles that of Figure 13.2b. Figure 13.2b is taken across a weak low-pressure system (Fig. 13.2a), not far from the warm front, which lies farther to the east. The wind component normal to the cross section increases rapidly with height, especially near the location of the surface front where the isentropes are more vertically oriented.

In this classic study, Sanders analyzed the kinematic fields surrounding the shallow-surface cold front in Figure 13.2. Figure 13.2c shows that strong upward vertical motion occurs just above the front, most notably in a narrow updraft with speeds up to 20 cm s^{-1} (~ 20 μb s^{-1}). Sanders, following Miller's and Petterssen's earlier derivations of the kinematic frontogenesis equations, included vertical motion and evaluated components of the frontogenesis equation using analyzed fields of motion and temperature. Like Miller, he defined frontal strength as the total gradient vector of temperature (or potential temperature). His mathematical development made it possible to evaluate the front strength in terms of changes in both the horizontal and vertical potential temperature gradients. Here we will confine our attention to an analysis of the magnitude of the horizontal temperature gradient ($|\nabla_p \theta|$), which we will choose to represent the horizontal front strength. Further, we define frontogenesis (frontolysis) as the rate of increase (decrease) of the magnitude of the temperature gradient with time following a parcel.

Taking the total differential of the horizontal potential temperature gradient (1.13c) and applying the Eulerian expansion of the material derivative as in (1.9), one obtains an expression for the rate of frontogenesis, which is

$$
\frac{d}{dt}|\nabla_p \theta| = \nabla_p \theta^{-1} \left\{ -\left[\left(\frac{\partial \theta}{\partial x} \right)^2 \frac{\partial u}{\partial x} + \left(\frac{\partial \theta}{\partial y} \right)^2 \frac{\partial v}{\partial y} \right] - \left[\frac{\partial \theta}{\partial x} \frac{\partial \theta}{\partial y} \left(\frac{\partial v}{\partial x} + \frac{\partial u}{\partial y} \right) \right] \right.
$$

$$
- \left[\left(\frac{\partial \theta}{\partial x} \right) \left(\frac{\partial \theta}{\partial p} \right) \frac{\partial \omega}{\partial x} + \left(\frac{\partial \theta}{\partial y} \right) \left(\frac{\partial \theta}{\partial p} \right) \frac{\partial \omega}{\partial y} \right]
$$

$$
+ \left. \left[\frac{\partial \theta}{\partial x} \frac{\partial}{\partial x} \left(\frac{d\theta}{dt} \right) + \frac{\partial \theta}{\partial y} \frac{\partial}{\partial y} \left(\frac{d\theta}{dt} \right) \right] \right\} \tag{13.1}
$$

$$
= \nabla_p \theta^{-1} \{ + C + S + T + DB \}.
$$

In the above expression the coordinates are x, y, p and t instead of x, y, z and t as originally used by Sanders for evaluating the case study shown in Figure 13.2. The first three terms on the right-hand side of (13.1) represent the isentropic redistribution of potential temperature by the motion field. The first term (labeled C) represents the effect of confluence within a field of "stretching deformation", which differentially advects cold and warm air toward each other along an "axis of dilatation". (Deformation and dilatation are defined and briefly discussed later in this chapter.) The second term on the right (labeled S) describes the effect of horizontally sheared advection within a field of shear deformation. The third term on the right (labeled T) represents the effect of differential isentropic warming resulting from a cross-frontal gradient of vertical velocity. Because it describes the effect of rotation of vertical gradients of potential temperature into the horizontal plane, this term is often referred to as the "tilting term". (As is shown later in this chapter, the tilting term in the frontogenesis equation is analogous to the tilting terms in the vorticity equation.) The last term on the right of this equation (labeled DB) describes the effect of cross-frontal gradients of diabatic warming associated with heating at the lower boundary, small-scale turbulent mixing, phase changes of water and radiation.

Another such expression has been derived for the vertical potential temperature gradient (i.e. the static stability), which can also be considered a measure of frontal strength. This equation is discussed in Chapter 16. Most early frontal analysis concentrated on an evaluation of the horizontal fron-togenesis function.

As it stands equation (13.1) is rather cumbersome to evaluate by inspec-tion. In order to simplify the explanation of these terms, we rederive the equation by taking the total time derivative of the one-dimensional tempera-

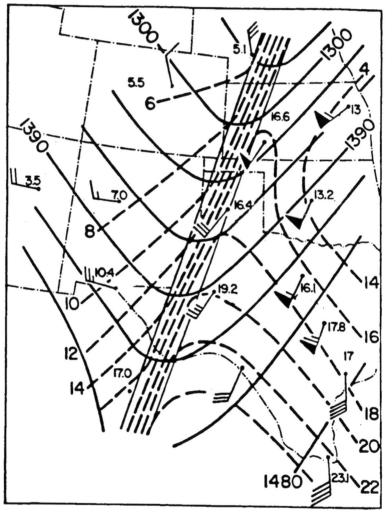

Figure 13.4 Geopotential height contours (m, full curves) and temperature (°C, broken curves) at 850 mb for 1200 GMT 3 April 1981 through a geostrophic shearing deformation cold front. Frontal boundaries are shown as thin full lines. Wind flags and barbs (one full flag = 5 m s⁻¹) are also shown. (From a figure by Shapiro, 1982.)

ture gradient, $-\partial\theta/\partial y$, which is considered a measure of front strength. Let us adopt the convention that (unless otherwise stated) the positive y axis is taken *normal to the frontal zone in the direction of cold air.* Figure 13.4 shows that the frontal zone need not be parallel to the isotherms. In order to avoid confusion with the meteorological x and y axes, the rotated coordinates are designated by primes. Note that the rotated u component lies parallel to the x' axis and the v component parallel to the y' axis in the same sense as in the meteorological coordinate system of x and y. Adoption of the negative sign before $\partial\theta/\partial y'$ is made to keep the front strength positive.

It is important to realize, however, that an axis normal to the frontal zone does not necessarily mean that the axis is also normal to the isotherms, because the latter may lie at an angle with the frontal zone, as in Figure 13.4. The time derivative of the one-dimensional gradient of temperature along y' is derived by expanding the term $d(-\partial\theta/\partial y')/dt$, which is

$$\frac{d}{dt}\left(-\frac{\partial\theta}{\partial y'}\right) = \left(\frac{\partial u}{\partial y'}\right)\left(\frac{\partial\theta}{\partial x'}\right) + \left(\frac{\partial v}{\partial y'}\right)\left(\frac{\partial\theta}{\partial y'}\right)$$

$$+ \left(\frac{\partial\omega}{\partial y'}\right)\left(\frac{\partial\theta}{\partial p}\right) - \frac{\partial}{\partial y'}\left(\frac{d\theta}{dt}\right) \qquad (13.2)$$

$$= S + C + T + DB$$

where the terms are labeled as in (13.1). Equations (13.1) and (13.2) apply to the horizontal gradient of any scalar quantity (e.g. specific humidity), but

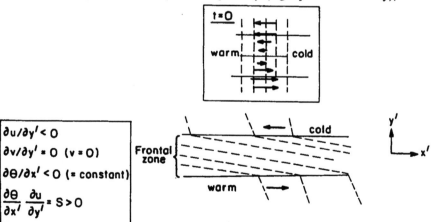

Figure 13.5 Schematic temperature deformation pattern for pure shear. Broken lines represent isotherms (cold toward positive y') and full lines the frontal boundaries. Arrows represent the direction and magnitude of the initial wind deformation ($t = 0$). Mathematical symbols enclosed in box at left indicate the signs of the derivatives and shearing term in equation (13.2).

can be applied in various ways to examine the time rate of change of vertical and horizontal gradients in a front.

Consider the terms separately in (13.2), making use of schematic diagrams. Figure 13.5 illustrates shear deformation, henceforth refered to as shear. The wind is parallel to the frontal zone, $\partial v/\partial y'$ is zero and, therefore, the confluence term vanishes. Now, the u component of the wind is increasingly negative (it blows toward negative x') in the y' direction and, therefore, $\partial u/\partial y'$ is negative. Since $\partial\theta/\partial x'$ is negative (isotherms crossing the frontal zone), the product of $\partial u/\partial y'$ and $\partial\theta/\partial x'$ is positive and the effect of horizontal shear is frontogenetic. It is not difficult to see from Figure 13.5 that the continued effect of the shear wind field is to push the isotherms together with time.

Shear is suggested as being important in forming and maintaining cold and occluded fronts, which tend to possess temperature and wind distributions with the essential properties illustrated in Figure 13.5. Shear is further illustrated in regard to the schematic representation of a typical isotherm distribution around an open wave cyclone (Fig. 13.6). The shear term is positive (frontogenetic) just behind the cold front in the region marked $S+$. In a real example of a cold front at 850 mb, shown in Figure 13.4, shear deformation strongly contributes to frontogenesis.

Confluence, sometimes referred to as stretching deformation, is illustrated in Figure 13.7. In this figure, isotherms lie parallel to the x' axis and the shear

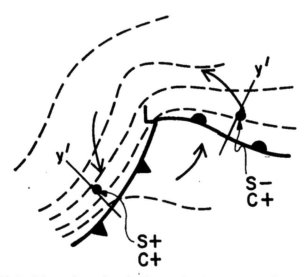

Figure 13.6 Schematic surface isotherms (broken curves), fronts and wind flow (arrows) around an open wave cyclone. The symbols S and C (with plus or minus signs) are at the location of the small filled circles and refer, respectively, to the sign of the shear and confluence terms in the kinematic frontogenesis equation (13.2).

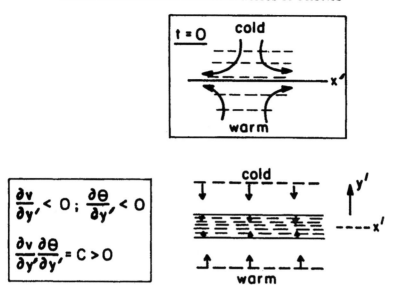

Figure 13.7 Schematic deformation pattern for pure confluence. As shown in Figure 13.5, the broken lines represent isotherms (cold toward positive y') and arrows the magnitude and direction of the wind. At the top is an initial pattern of confluence ($t = 0$) along the axis of dilatation. Mathematical symbols enclosed in the box refer to sign of derivatives and confluence term in equation (13.2).

term vanishes because $\partial\theta/\partial x'$ is zero. This figure shows that the flow pattern is such as to compress the isotherms with time, since $\partial v/\partial y'$ and $\partial\theta/\partial y'$ are both negative, and the product of the two, representing the confluence term, is positive.

Confluence is thought to be important in maintaining warm fronts, which often are frontolytic with regard to the shear. Thus, for a warm front with the cold air to the west and north (the typical case over continental North America), the presence of cyclonic shear contributes to frontolysis by diminishing the magnitude of $\partial\theta/\partial y'$, as in the case for the open wave cyclone in Figure 13.6. In this typical example, confluence and shear tend to be frontogenetic along the cold front, but the terms have opposite signs along the warm front, where shear acts frontolytically. For the warm front to be well defined in this type of pattern, the magnitude of the confluence must exceed that of the frontolytic effect of shear. Recall that the forcing terms in the quasi-geostrophic omega equation can also have opposite signs. If we reverse the east–west gradient of temperature along the warm front, such that colder air is located north and east of the warm front, the shear deformation is positive. (This is typical of Pacific Ocean cyclones reaching the western coast of North America.)

The tilting effect is more difficult to evaluate because of its dependence on the vertical motion. In the schematic diagram of Figure 13.8, air is sinking on

355

the warm side of the front and rising on the cold side. Thus, $\partial\omega/\partial y'$ is negative. Since $\partial\theta/\partial p$ is also negative, the product of these derivatives is positive and the tilting effect is frontogenetic. In this example, the circulation is indirect, cold air rising and warm air sinking. The situation is nevertheless frontogenetic because the total derivative of the horizontal temperature gradient gains a positive contribution by the tilting term. The more typical case is one in which there is a direct circulation, such as that analyzed by Sanders in Figure 13.2c, where warm air is rising and cold air is sinking. Tilting often dominates the process of frontogenesis in upper-tropospheric fronts (in conjunction with jet streams or tropopause folds). Tilting is usually a passive effect, responding to forcing by the dynamics, but nevertheless responsible for the rapid weakening with height of lower-tropospheric fronts. Thus, ascent in the warm air and descent in the cold air generate kinetic energy, but act to reduce the horizontal temperature gradient (the available potential energy) by adiabatic cooling and warming. The magnitude of the tilting term in (13.2) increases with height due to the increase with height from the surface of the magnitude of the vertical motion.

The role of diabatic heating is less obvious than that of confluence and shear in frontogenesis. For fronts in which the diabatic heating rate $(d\theta/dt)$ decreases toward colder air, $\partial(d\theta/dt)/\partial y'$ is negative and, consequently, differential heating contributes to frontogenesis. This is illustrated in Figure 13.9a. Here, heating of the surface occurs as the result of more insolation occurring on the warm side of the front (due to the presence of a stratiform cloud in the cold air). The result of such a distribution of sensible heating in

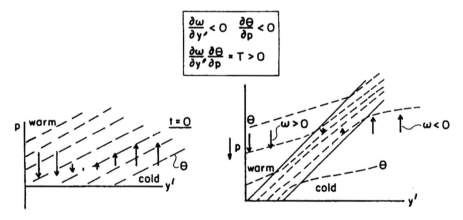

Figure 13.8 Schematic pattern of vertical motion (arrows showing direction and magnitude of vertical motion (ω)) and isentropes (broken lines) in a cross section through a surface front. Full lines are frontal boundaries. Initial conditions are shown at the left. Mathematical symbols enclosed in box refer to signs of derivatives and tilting term in equation (13.2).

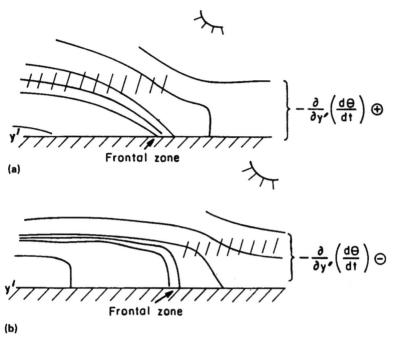

Figure 13.9 Schematic cross section through a front showing isentropes (full curves) and stratus deck (hatching). (a) Here the cloud (stratus) lies in the cold air behind the front and the diabatic heating term (mathematical symbols) in equation (13.2) is positive due to differential warming of the boundary layer imposed by the effect of cloud on insolation. (b) Now the cloud lies in the warm air and the diabatic heating term is negative near the surface.

the planetary boundary layer is to increase the horizontal potential temperature gradient across the lower part of the front. The difference in the strength of the front between Figures 13.9a and b is meant to illustrate differential heating associated with the presence of a dense cloud shield on one side of the cold front.

Even in the frontolytic example (Fig. 13.9b), the isotherm gradient is not everywhere weakened. Mixing due to surface sensible heating behind the cold front increases the vertical potential temperature gradient (the static stability) and the horizontal temperature gradient in the upper part of the front in Figure 13.9b. This reversal of the sign of the diabatic heating term on the horizontal temperature gradient near the upper part of the mixing layer is due to a reversal in the direction of the horizontal temperature gradient near the top of the front and a change of sign with height in the vertical divergence of the turbulent heat flux. Inclusion of precipitating clouds further complicates the effects of diabatic heating on frontogenesis,

357

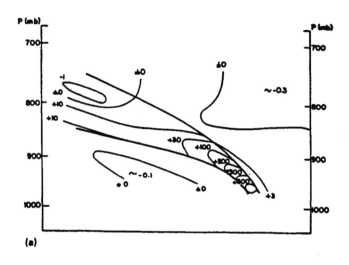

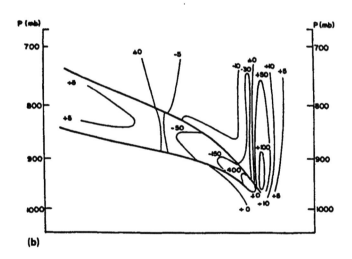

Figure 13.10 Cross sections of frontogenetic effects, showing part of cross section (E–N) in Figure 13.2. Units of tendency term are K $(100 \text{ km})^{-1}$ $(3 \text{ h})^{-1}$, and positive values indicate frontogenesis: (a) confluence term $(\partial v/\partial y)(\partial\theta/\partial y)$; (b) tilting term $(\partial w/\partial y)(\partial\theta/\partial z)$.

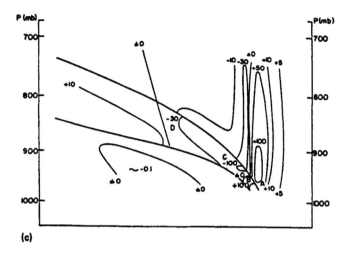

(c)

Figure 13.10 (c) the sum of (a) and (b). (From Sanders, 1955.)

especially in the case of convective clouds, which can produce an effective diabatic heating in their surroundings (Ch. 9).

In the case analyzed by Sanders (Fig. 13.2), confluence was strongly frontogenetic in the lower part of the zone (Fig. 13.10a) and it decreased in magnitude with height. The tilting term (Fig. 13.10b) was strongly frontogenetic in the warm air ahead of the frontal zone and frontolytic within the zone. The combined effect of confluence and tilting (Fig. 13.10c) indicates that the tilting term is of great importance in the warm air ahead of the front and at levels well away from the ground, whereas confluence dominates in the frontal zone near the surface.

It is also evident from examination of the vertical motion pattern that rising warm air and sinking cold air not only serve to reduce the horizontal temperature gradient (and therefore the influence of confluence and shear) but also contribute to a decrease in the slope of the front, thereby maintaining its shallowness.

13.3 The deformation vector

Frontogenesis takes place when the winds increase the temperature gradient by moving isotherms closer together. It can be seen in Figure 13.11 that this will occur along an *axis of deformation (or dilatation)* (A–A'), which is oriented at an angle α with the isotherms. The axis of the dilatation constitutes the center of a zone in which isotherms are being collected and the temperature gradient is increasing with time. It is clear that, if the angle α is

359

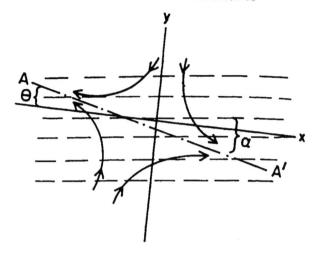

Figure 13.11 Illustration of temperature deformation by confluent wind pattern (streamlines) in a field of isotherms (broken lines). The axis of dilatation is along A–A', and the angles α and θ are those between the axis of dilatation and the isotherms and between the axis of dilatation and the x axis, respectively. Note that the x-y coordinate system is placed at an arbitrary angle with respect to the direction of the isotherms.

less than 45°, frontogenesis occurs along the axis A–A'. If, however, the angle were greater than 45° (but less than 90°), the axis A–A' would become an axis of frontolysis and the perpendicular axis would eventually collect the isotherms and constitute an axis of dilatation and frontogenesis. Indeed, isotherms will always tend to collect along an axis of dilatation.

Earlier in this chapter two types of deformation were discussed: stretching (or confluence) and shear deformation. By themselves, the deformation components describe the instantaneous stretching and shear of the momentum fields. In equation (13.2), deformation refers to shear or stretching of the isotherm pattern along an axis of dilatation. Pure stretching deformation for momentum is defined kinematically as

$$CD = \partial u / \partial x - \partial v / \partial y$$

and shear deformation as

$$SD = \partial v / \partial x + \partial u / \partial y.$$

Both types of deformation are really aspects of one another (coordinate-dependent). The magnitude of the deformation vector is

$$Def = \left[\left(\frac{\partial v}{\partial x} + \frac{\partial u}{\partial y} \right)^2 + \left(\frac{\partial u}{\partial x} - \frac{\partial v}{\partial y} \right)^2 \right]^{1/2} = (SD^2 + CD^2)^{1/2} \qquad (13.3)$$

and the angle (θ) between the x axis and the axis of dilatation (measured counterclockwise) is

$$\theta = \tfrac{1}{2} \tan^{-1}(SD/CD). \qquad (13.4)$$

Figure 13.11 resembles the distribution of isobars and relative wind motion along a limiting streamline on an isentropic surface (Ch. 12). Since the limiting streamline corresponds to a dilatation axis, it is not surprising that fronts in the lower and middle troposphere tend to form on the eastern sides of troughs, where a deformation pattern, resembling that in Figure 13.11, is imposed by confluent air streams of differing origins.

13.4 Deficiencies in the kinematic explanation of frontogenesis

Figure 13.11 suggests that, given a sufficient time, deformation of the temperature field will cause a front to form along an axis of dilatation. Purely kinematic deformation, in which there is no relationship between the isotherm deformation and the v component of the wind responsible for compressing the isotherms, implies a rate of frontogenesis that is too slow. Consider only the confluence term in (13.2) and the schematic confluent temperature pattern shown in Figure 13.11. Let us set the rate of change of the horizontal temperature gradient $d(- \partial\theta/\partial y')/dt$ equal to the confluence term (C), which is equal to $(\partial\theta/\partial y')(\partial v/\partial y')$. If one postulates that $- \partial v/\partial y'$ remains constant (whose value is k), the solution to this differential equation is

$$- \partial\theta/\partial y' = - (\partial\theta/\partial y')_0 \, e^{kt}$$

where the subscript zero refers to temperature gradient at time $t = 0$. A reasonably large value for $- \partial v/\partial y'$ (e.g. $10 \text{ m s}^{-1} (1000 \text{ km})^{-1}$) yields a time constant for this equation ($1/k$, the e-folding time) of about one day. This equation tells us that fronts will increase their intensity by the natural log factor (e = 2.72: approximately doubling) in one day.

It is quite clear from inspection of Figure 13.1 that fronts can double their intensity in a matter of a few hours or less, an order of magnitude not easily explained by a simple kinematic approach. Thus, the e-folding time for an initial horizontal temperature gradient is more rapid than the simple logarithmic function for the static wind gradient model.

Temperature deformation is large in the vicinity of fronts, but so are values of the divergence and vorticity. Indeed, peak vorticity and divergence values in the frontal zone can be an order of magnitude greater than those customarily found on the synoptic scale! Sanders' analysis of the case shown in Figure 13.10 indicates that the maxima in convergence, confluence and vorticity occur almost in the same place and that the narrow zone of very large values of convergence and vorticity corresponds to the frontal zone. According to the vorticity equation, one would expect to find that large values of convergence are associated with large values of cyclonic vorticity. Kinematic theory fails to explain the presence of elevated values of vorticity because it does not take account of divergence or changes in momentum (the dynamics). Large values of divergence and vorticity (and large Rossby numbers) exist in frontal zones on scales of 100 km or less, suggesting that the winds there are highly ageostrophic.

Kinematic theory of fronts fails because temperature is treated as a passive scalar, e.g a neutral pollutant that does not affect the dynamics. The unrealistically long time for frontogenesis to occur by purely kinematic processes arises because $\partial v / \partial y'$ is fixed as a constant in this analysis and, in fact, its magnitude is much larger in the frontal zone than on the synoptic scale.

Thus, the normal wind component to the front must increase with time. In order for a frontal theory to be complete, it must account for this increase, the observed rate of frontogenesis, for a cascade of energy from synoptic to subsynoptic scales and for the presence of subsynoptic-scale convergence, vorticity and ageostrophic motion within the frontal zone. Since fronts are driven by synoptic-scale motions, the theory must also include a mechanism for relating geostrophic forcing to ageostrophic response in the cross-front direction. As is shown in the next chapter, the ageostrophic component of the wind is responsible for the changes in divergence and vorticity that occur within the frontal zone and, indeed, for the continual adjustment of the winds to geostrophic balance.

Problems

13.1 Starting with equation (13.2), derive an expression for the horizontal temperature gradient $(- \partial\theta / \partial y')$ as a function of time for the case of pure confluence, allowing $\partial v / \partial y'$ to be a function of time $(= k_0 + k_1 t)$, where the constant k_0 is the initial value of $\partial v / \partial y'$. Let $k_0 = 10 \text{ m s}^{-1} (1000 \text{ km})^{-1}$ and show that, for any positive value of k_1, the doubling rate for $- \partial\theta / \partial y'$ is larger than for the case of an unchanging momentum field, $k_1 = 0$.

13.2 Derive a frontogenesis equation for the rate of change of $(- \partial\theta / \partial p)$, analogous to equation (13.2).

Further reading

Ballentine, R. J. 1980. A numerical investigation of New England coastal frontogenesis. *Mon. Wea. Rev.* **108**,1479–97.

Bergeron, T. 1937. On the physics of fronts. *Bull. Am. Met. Soc.* **18**, 265–75.

Bosart, L. F. 1975. New England coastal frontogenesis. *Q. J. R. Met. Soc.* **101**, 957–78.

Bosart, L. F., C. J. Vaudo and J. H. Helsdon Jr 1972. Coastal frontogenesis. *J. Appl. Meteor.* **11**, 1236–58.

Miller, J. E. 1948. On the concept of frontogenesis. *J.Meteor.* **5**, 169–71.

Newton, C. W. 1954. Frontogenesis and frontolysis as a three-dimensional process. *J. Meteor* **13**, 449–61.

Reed, R. J. and F. Sanders 1953. An investigation of the development of a mid-tropospheric frontal zone and its associated vorticity field. *J. Meteor.* **10**, 338–49.

Sanders, F. 1955. An investigation of the structure and dynamics of an intense surface frontal zone. *J. Meteor* **12**, 542–52.

Shapiro, M. A. 1981. Frontogenesis and geostrophically forced secondary circulations in the vicinity of jet stream-frontal zone systems. *J. Atmos. Sci.* **38**, 954–73.

Shapiro, M. A. 1982. Mesoscale weather systems of the Central United States. CIRES Report, NOAA/University of Colorado, Boulder, CO. 78 pp.

14

Ageostrophic motion and the dynamics of fronts

An underlying premise of this text is that the physical systems that are represented on weather charts obey quasi-geostrophic constraints. A paradox in quasi-geostrophic theory is that geostrophic forcing can be used to explain ageostrophic motion. For instance, the omega equation allows one to diagnose, using fields of geostrophic advections, the vertical component of the wind and, therefore, the divergent part of the horizontal flow. Since the divergent part of the wind is almost entirely ageostrophic, the balanced (geostrophic) part of the wind, expressed as geostrophic advections, determines the unbalanced part of the wind. Conversely, the unbalanced part of the winds acts toward restoring geostrophic balance.

Until now, we have assumed that the synoptic-scale processes are always in approximate geostrophic balance. The quasi-geostrophic framework implies that thermal wind balance is always maintained and, therefore, changes in the temperature (mass) and wind (vorticity) fields are not independent. Actually, movement of isotherms and of momentum isopleths implies that balance is continuously being disturbed and that the return to balance requires a finite length of time, which is on the order of $1/f$. A movement toward restoring geostrophic balance is accomplished by the forces on the air and by the ensuing accelerations toward equilibrium.

Motion of the wind implies not only a displacement of the isotherms but a displacement of the momentum field. In the quasi-geostrophic system both temperature and momentum are related, and their geostrophic advections contribute equally to the forcing of ageostrophic motions. *Thus, the balanced part of the flow is responsible for promoting a deviation from the balanced state, while the unbalanced part of the flow acts to restore that state.* In the process of restoring geostrophic balance (which never is actually achieved), the accelerations and their associated ageostrophic motion lead to patterns of divergence, to spin up and spin down of weather systems and to changes in the kinetic energy of the flow, depending on whether the motions are direct or indirect.

Vertical motion, therefore, is intimately related to horizontal ageostrophic motions. Together, the vertical and horizontal ageostrophic motions form circulations in the x, p and y, p planes. These "transverse/vertical" circulations are responsible not only for the spin up of the geostrophic component of the vorticity field, but for the formation of fronts and jet streaks, features that are small-scale in their cross-wind dimension. Fronts and jet streaks are also inherent features of a rotating baroclinic fluid and can be observed to form in an initially uniform temperature gradient in rotating tank models, such as the type referred to in the introduction to Chapter 5.

On the scale of fronts and jet streaks, the cross-wind (transverse) ageostrophic component of these transverse/vertical circulations is apt to be rather significant and may become as large in magnitude as the geostrophic wind velocity. Consequently, cross-front advections by ageostrophic winds cannot be neglected. In order to diagnose the cross-front transverse/vertical circulation, however, it is necessary to modify the quasi-geostrophic assumptions to include ageostrophic motions in the plane normal to the frontal zone. This modification is called the "geostrophic momentum" approximation.

We began this chapter by introducing the concept of ageostrophic motion and its role in the formation of jets and cyclones. Some simple illustrations will be used to show how the ageostrophic motions operate from the standpoint of the basic equations of motion. This introduction is followed by a re-examination of frontogenesis in the context of tranverse/vertical motion in frontal zones. To do this, we present a qualitative description of how a front is created by the operation of dynamic principles, showing where the kinematic ideas give way to the dynamics of frontogenesis. Next, we introduce the modification of quasi-geostrophic theory in order to derive the Q-vector equations. This formalism will be applied to explain the evolution of fronts. We will show how the transverse/vertical circulations can be described by equations similar to the quasi-geostrophic omega and pressure tendency equations, which form a special case of the more general geostrophic momentum formulation. We will show that the method of solenoids can be used to diagnose the transverse/vertical motions and the large-scale quasi-geostrophic forcing of frontogenesis in a manner similar to that for quasi-geostrophic forcing of vertical motions and surface pressure tendency. Illustrations of the use of these equations, with emphasis on surface fronts (following the format of Chapter 13), constitute the concluding part of this chapter. This sets the stage for a description of upper-tropospheric fronts, jet streaks and tropopause folding in the next chapter.

14.1   The four-quadrant model

A practical illustration of transverse/vertical circulations can be made with regard to the jet streak, which we define as being a core of locally maximum wind speeds, usually in the upper troposphere. Figure 14.1 shows isotachs and geopotential height contours for the case where air moves through the wind speed maximum, such as at 300 mb. This *four-quadrant model* of the jet, sometimes referred to as the Riehl model (after H. Riehl), is usually applied to upper-tropospheric wind speed maxima, although the principles apply to any level in the atmosphere.

Let us see how this model operates, starting first with what we know to be an empirical fact: air tends to move through wind speed maxima in the upper troposphere. If the air moves faster than the wind speed in the core of the jet, an air parcel approaching the jet from behind (the chain arrows) experiences an acceleration in its forward motion, the largest acceleration occurring in parcels arriving near the axis of the jet. For convenience we divide the

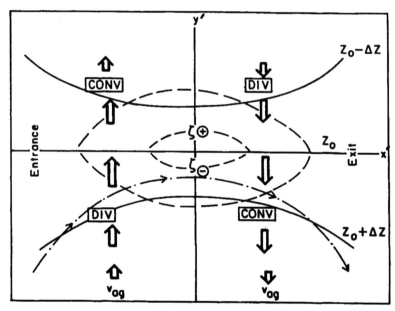

Figure 14.1   Four-quadrant model of a jet streak. Isotachs are broken curves, maximum wind speed is located at the center of the x' and y' axes, geopotential height contours (Z) are full curves, the chain barbed curve is a trajectory and the double-shafted arrows (labeled v_{ag}) show the direction and magnitude of the v component of the ageostrophic wind. The four quadrants are labeled according to whether there is divergence (DIV) or convergence (CONV). Exit and entrance regions are labeled. The symbol ζ refers to negative ($-$) or positive ($+$) relative vorticity right and left of the jet core.

core of maximum winds into four quadrants, two on the downstream (exit) and two on the upstream (entrance) sides of the jet. Right and left quadrants are defined facing downwind.

The equations of motion governing frictionless flow in the x and y directions (equation (1.26)) are restated as

$$\mathrm{d}u/\mathrm{d}t = f v_{ag} \quad \text{and} \quad \mathrm{d}v/\mathrm{d}t = -f u_{ag} \quad \text{ageostrophic/geostrophic system.}$$

Acceleration in the entrance and deceleration in the exit regions therefore cause ageostrophic motions shown in Figure 14.1. As a result of the induced ageostrophic motions, convergence occurs in the left entrance and right exit regions, and divergence in the right entrance and left exit regions. Dines' compensation (see Ch. 2) indicates that convergence (divergence) overlies divergence (convergence) in *straight upper-tropospheric jets*; ascent occurs in the right entrance and left exit quadrants, and descent in the left entrance and right exit quadrants of the jet.

For lower-tropospheric jets, the same ageostrophic and divergence patterns occur (for relative motion through the jet), but the reverse circulations are implied: divergence and descent in the left exit and right entrance quadrants, and convergence and ascent in the left entrance and right exit regions.

Diagnosis of vertical motions around upper-tropospheric jet streaks is useful where there are few radiosonde stations or where the jet streak is so narrow that there are insufficient data to define its vorticity pattern properly. The model, however, is not always correct because the assumptions leading to a diagnosis of divergence and convergence patterns are restrictive.

In the example for 2 December 1983 (Fig. 14.2), cloud cover and radar echoes are shown to be most extensive in the right entrance and left exit quadrants of a 300 mb jet. Some precipitation extends into the right exit quadrant, but there is little cloud and no precipitation in the left entrance quadrant, a sector that is associated in the four-quadrant model with upper-level convergence and descent. In general, however, cloud cover around a jet streak is not clearly explainable in terms of the four-quadrant model because the basic premises are faulty. These assumptions are:

(a)   a relative motion of the trajectory with respect to the jet streak;
(b)   straight flow; and
(c)   synoptic-scale response.

In fact, the vertical motion about the jet may occur over a narrow or curved region, and the jet streak may exhibit several jet cores in a zone of high wind speeds.

A fundamental question is why does the air move through the jet and not

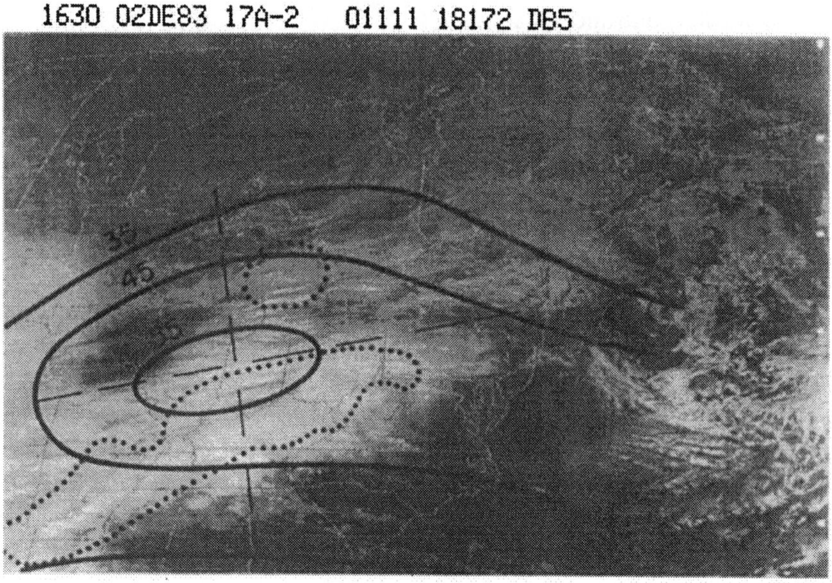

1630 02DE83 17A-2   01111 18172 DB5

Figure 14.2 Satellite (GOES) visible-wavelength photograph, isotachs (kt) and radar echoes (dotted curves) for 1630 GMT 2 December 1983. The broken axes identify the four quadrants of the jet, according to the model prescribed in Figure 14.1. (The isotach pattern has been translated slightly toward the east (right) in order to coincide with the satellite and radar times.)

simply advect the isotach field? The answer is related to the discussion in Chapter 6 concerning the movement of 500 mb troughs relative to the mean flow. In that chapter we showed that the reason why troughs and ridges move as coherent entities is due to the vertical coupling of the vertical motion and divergence patterns. Therefore, let us now avoid the *a priori* assumption that the air motion is relative to the movement of the jet.

In a simple jet streak such as the one shown in Figure 14.1, the vorticity pattern consists of a single couplet in which there is a maximum on the left-hand side and a minimum on the right-hand side of the jet; this vorticity maximum is indicated in the figure by the relative vorticity symbols (ζ) labeled positive and negative. Since both the vorticity and zonal wind component increase with height between the level of the jet and the lower troposphere, positive vorticity advection increases with height in the right entrance and left exit quadrants of the jet. If the isotherms are parallel to the wind direction, vertical increase of absolute vorticity advection is the dominant effect in the omega equation. Thus, rising motion occurs in sectors where there is an increase of positive vorticity advection with height.

Since the magnitude of vorticity advections increases with height from the surface, dynamically forced ascent (descent) must occur in the left (right) exit and right (left) entrance quadrants, *provided that one neglects the effects of temperature advection, which vanishes only when the isotherms are parallel to the jet axis.* We will see in Chapter 15 that different patterns of convergence and divergence around a jet streak are possible when temperature advection is permitted along the jet. Consequently, air parcels in the left entrance or right exit quadrants in the high troposphere experience descent and upper-level convergence and, therefore, a negative change in a parcel's absolute vorticity with time. This arrangement is equivalent to the acceleration and deceleration of the parcel in the front and rear sectors of the jet. Alternatively stated, ageostrophic/vertical circulations are required by the configuration of the geostrophic advections. As discussed in Chapter 6, the result of Dines' compensation and the transverse/vertical circulations is to maintain an approximately uniform phase speed of the jet and the wave with height.

Since jet streaks move slower than the wind speeds at 300 mb, one can speculate that the difference between the migration speed of the jet and its maximum wind speed is related to the strength of the transverse/vertical circulation surrounding the jet. Thus, 300 mb jet streaks that move much slower than the wind speeds at 300 mb are associated with intense transverse/vertical circulations around the core. As stated in Chapter 12, cloudiness tends to favor the right-hand side of jets, with the jet axis corresponding in many cases to a sharp left-hand edge of a cirrus shield. In situations where the jet is curved, the curvature term in the expression for relative vorticity in natural coordinates (2.3) may be more important than shear in indicating the sign of the relative vorticity maximum on the anticyclonic side of the jet. Indeed, it is surprising how little curvature is required in order for the curvature term to be significant when the wind speed is large. Where curvature becomes important, a single positive vorticity maximum is found, implying a two-quadrant (rather than a four-quadrant) model; positive or negative vorticity advection then occurs everywhere in the exit region and negative or positive vorticity advection in the entrance region of the jet. In the case of a trough, advection of positive absolute vorticity (and ascent) would take place downstream from the vorticity center, and advection of negative absolute vorticity (and descent) upstream from the jet core. Nevertheless, ascent may be strongest in the left exit quadrant, even in curved jet situations.

14.2 Isallobaric wind

It should now be evident that the divergent component of the wind field is an inherent property of the ageostrophic motion, although the latter does also possess a significant non-divergent component. Since surface pressure tendency and low-level divergence are interrelated, it is possible to derive an equation relating the surface (or sea-level) pressure tendency to the ageostrophic motion, starting with the quasi-geostrophic approximation to the momentum equations (1.28) with the friction term set to zero. (Recall that this equation is consistent with the derivation of the quasi-ageostrophic vorticity equation (3.6).)

Consider the equations of motion for horizontal, quasi-geostrophic flow (equation (1.27)), but without the friction terms. We write this equation as

$$\frac{\partial u_g}{\partial t} = -\left(u_g \frac{\partial u_g}{\partial x} + v_g \frac{\partial u_g}{\partial y}\right) + f v_{ag}. \qquad (14.1a)$$

The left-hand side of this equation can be transformed using the definition $u_g = (-g/f_0) \, \partial Z / \partial y$ to become

$$v_i = -\frac{g}{f_0^2} \frac{\partial}{\partial y}\left(\frac{\partial Z}{\partial t}\right) = -\frac{1}{f_0}\left(u_g \frac{\partial u_g}{\partial x} + v_g \frac{\partial u_g}{\partial y}\right) + v_{ag} \qquad (14.1b)$$

where the term labeled v_i has the units of speed and is defined as the component of the *isallobaric wind* in the y direction. Note that the geopotential height tendency can be equated with a pressure tendency on a constant-elevation surface. The first term on the right-hand sides of (14.1a) and (14.1b) is the advection of u momentum and the second term is the ageostrophic component of the wind. Similarly, the v momentum equation can be transformed using the geostrophic wind definition to arrive at an expression for the u component of the isallobaric wind. Neglecting horizontal advection:

$$u_i = -\frac{g}{f_0^2} \frac{\partial}{\partial x}\left(\frac{\partial Z}{\partial t}\right). \qquad (14.1c)$$

In general, therefore,

$$V_i = -\frac{g}{f_0^2} \nabla_p \frac{\partial Z}{\partial t} \qquad \nabla_p \cdot V_i = -\frac{g}{f_0^2} \nabla_p^2 \frac{\partial Z}{\partial t}. \qquad (14.1d)$$

In the absence of friction and advection of momentum, the isallobaric wind is equal to the ageostrophic component of the wind. For straight flow, the total wind is composed of a geostrophic component that blows parallel to the isobars and an ageostrophic component blowing normal to the isallobaric field toward more negative values of pressure tendency, as shown by the isallobaric wind in Figure 14.3. Maximum surface-level divergence or convergence is in regions where the Laplacian of the surface isallobaric field is maximized or minimized, respectively.

In accordance with arguments posed in regard to the vorticity equation,

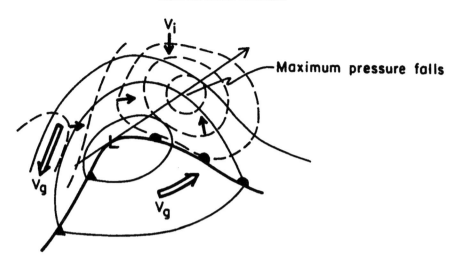

Figure 14.3 Schematic surface isallobaric pattern (broken curves) poleward of a warm front and cyclone. Isobars (full curves) and fronts are shown for surface pattern. Full arrows labeled V_i represent direction of isallobaric (and ageostrophic) wind velocity. Double-shafted arrows show the direction of the geostrophic wind velocity. The long thin arrow indicates the direction of motion of the low.

the magnitude of the advection of vorticity is small compared to the divergence term at the surface; consequently, the advection terms in (14.1) tend to be small near the surface compared with the ageostrophic terms. Thus, we may consider the isallobaric wind at the surface to be a close approximation of the non-frictional part of the surface ageostrophic wind. It is easy to see that the pattern of convergence associated with pressure falls and spin up of cyclonic vorticity at the surface is also compatible with a circulation in which air moves across isobars and converges in regions of pressure falls. Ascent takes place in the vicinity of the negative isallobars with compensating divergence at upper levels and with sinking motion occurring elsewhere in the vicinity of positive sea-level isallobars. In this respect, the pattern of ageostrophic (isallobaric) wind in Figure 14.3 closely resembles that of the Q-vector in the area north of the warm front in Figure 14.16 (see Sec. 14.7).

14.3 How a front is made

The last chapter concludes by demonstrating that frontogenesis is not merely the passive concentration of isotherms by an advecting wind. We now show that it is a type of instability in which the differential advection of temperature and momentum feed back, via ageostrophic motions, to produce a

371

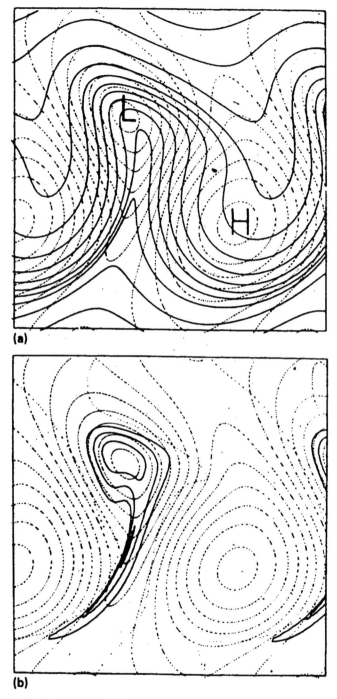

Figure 14.4 Surface maps of (a) potential temperature (full curves) and geopotential height field (thin broken contours) and (b) the relative vorticity and height field at day 5.5 of simulation. The height pattern (broken contours) is at intervals of 50 m and the vorticity isopleths (full curves) are at $-0.5f$, 0, $0.5f$, $2f$, $4f$ and $8f$, where f is the Coriolis parameter. Isotherms are given every 4 °C. (From Hoskins and West, 1979.)

singularity known as a front. A remarkable property of the synoptic-scale motion is that it is able to create such frontal boundaries. Results of numerical experiments show that exceedingly narrow gradients of temperature and vorticity can form in a short period of time starting with a typical pattern of isotherms and geopotential height contours. Even if the initial flow is in geostrophic balance and highly smoothed, the forcing is sufficient to form an intense temperature gradient in just a few hours after an initial period of organization. The results of Hoskins and West shown in Figure 14.4 were obtained from computer simulations that began with smoothed initial conditions, yet were able to produce exceptionally narrow and intense gradients of temperature and vorticity within comma-shaped zones. Although these simulations required five days before the large gradients appeared, the changes were explosive after an initial period of organization. In order to understand how synoptic-scale forcing can produce a front, subject to the dynamics, the individual processes are now dissected in the form of a series of steps starting from a geostrophic flow leading to the formation of a front.

Consider an initial geostrophic pattern of isotherms, isotachs and geopotential height contours. Figure 14.5, easily recognizable as corresponding to the pure confluence case described in Figure 13.7, is characterized by a straight frontal zone parallel to the x' axis and by decreasing geopotential height (streamlines) and temperatures (thin broken lines) toward the positive

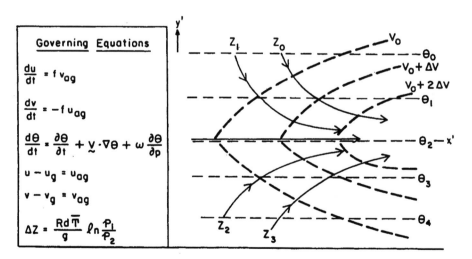

Figure 14.5 Isotherms (thin broken lines labeled θ, colder toward the top of the figure) and geopotential height contours (full curves representing geostrophic wind streamlines) for pure confluence. Geostrophic streamlines are also contours of the geopotential height (labeled Z). Heavy broken curves are isotachs of the wind speed (V). Equations governing the dynamics of frontogenesis are shown (see text).

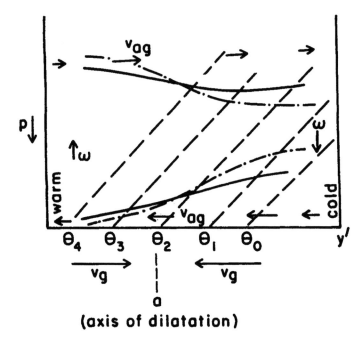

Figure 14.6 Cross section illustrating how a front is formed during the first instant following forcing by confluence for 700 mb pattern shown in Figure 14.5. Broken lines are isentropes (warm to left, cold to right), full curves are the initial geopotential surfaces near the surface and at an upper level (700 mb) and chain curves are the geopotential height surfaces an instant after confluence has begun. Arrows within the frame of the cross section are proportional to the direction and magnitude of the ageostrophic or vertical motion; arrows below the y' axis show the direction of the geostrophic wind normal to the frontal zone; and the vertical broken line labeled "a" is the initial axis of deformation.

y' axis. In accordance with the decrease in spacing of the geostrophic wind streamlines toward the right, the geostrophic wind speeds (heavy broken curves are almost identical to isotachs of u_g) increase toward the right.

For convenience, the governing equations, reproduced from Chapter 1, are written to the left of the figure. The cross section in Figure 14.6 lies along the y' axis, the initial isentropes are shown by the broken lines and the initial heights of two pressure surfaces (one near the surface and the other at 700 mb) by full curves; the initial geostrophic wind speed along y' is given by the arrows below the y' axis. The following discussion outlines the physical processes that occur to make a front, starting with an initially geostrophic flow. The steps that follow, however, are only loosely chronological, without necessarily implying cause and effect between successive steps.

Step 0

Geostrophic balance ($V = V_g$) exists everywhere and there are uniform gradients and straight isotherms. (Confluent deformation is assumed to exist only within the figure; outside of the domain there is no deformation.) Figure 14.5 represents the pattern at the upper level (700 mb) in Figure 14.6. Note that the slope of the upper geopotential surface is consistent with Figure 14.5, in that lower heights are situated on the negative y' side and higher heights on the positive y' side. The lower geopotential surface slopes in the opposite direction, but this slope is unimportant for the following arguments except to illustrate unambiguously that temperature decreases in the positive y' direction.

Step 1

Confluent deformation of the temperature field by the v component of the wind begins. Since v_g (and therefore v) decreases toward positive y', $\partial v / \partial y'$ and $\partial \theta / \partial y'$ are both negative and therefore the confluence term (C) in equation (13.2) is positive. In effect, there is advection of cold air toward negative y' and warm air toward positive y', on either side of the axis of deformation that lies along isentrope θ_2. Changes occur only within the domain of Figures 14.5 and 14.6.

Step 2

Hydrostatics demands that since $-\partial \theta / \partial y'$ increases due to advection of θ, the thickness gradient increases in the first instant of time. Accordingly, let us agree that heights fall aloft and rise below where there is cold air advection (north of the axis of deformation), and rise aloft and fall below where there is warm air advection. As we will see, this arrangement constitutes the only change in the geopotential height pattern that is consistent with the evolution of the ageostrophic wind field. The new configuration of geopotential heights is given by the chain contours in Figure 14.6.

Step 3

Consider *only* temperature changes for the present. Let only the geopotential heights and geostrophic winds vary as the result of these changes. Since the temperature gradients are changing along the y' direction, the geopotential height gradients are changed along the y' axis. Therefore, u_g changes and the flow is no longer geostrophic. Because the geopotential height changes, u_g becomes more negative at the surface than u, increasing from an easterly direction. At the upper level u_g becomes greater than u. Therefore, u_{ag} becomes positive and dv/dt negative at the lower level, and u_{ag} becomes negative and dv/dt positive at the upper level. Since dv/dt is negative at low levels and positive at upper levels, the v component of the wind gains an initial ageostrophic component in these directions (toward positive y' above

375

and toward negative y' near below) as the result of the acceleration of this wind component from an initially geostrophic balance.

Step 4

Now, let us hold the heights constant and allow the momentum to be advected. At the upper level, isotachs of u_g (and therefore u) increase in magnitude toward positive x' and, consequently, the winds advect lower values of u toward positive x'; the winds also advect u momentum meridionally. *Consequently, from the first instant, u becomes less than u_g and dv/dt is positive* due to advection of u momentum along the x axis. The reverse occurs at the lower level. The component of u_{ag} produced by advection of momentum is exactly equal to that for temperature advection, *the net effect being double that of the individual geostrophic advections of either temperature or momentum.*

Step 5

In accordance with the establishment of meridional accelerations (dv/dt non-zero), a movement of the wind in v_{ag} is created (the full arrows near the geopotential height surfaces in Figure 14.6), and there is a component of ageostrophic motion toward the left (negative y') at low levels and toward the right (positive y') above. Consequently, convergence below and divergence above (ascending motion) takes place on the far left, and divergence below and convergence above (descending motion) takes place on the far right. Note that the convergence pattern augments the effect of confluence in compressing the isotherms by increasing the horizontal temperature gradient on the left at low levels and on the right above.

Effectively, a zone of convergence is established that extends from the lower left to the upper right of Figure 14.6. This zone corresponds to one of enhanced vorticity generation (because of the convergence) and enhanced temperature gradient (the frontal zone). The configuration of ageostrophic/vertical motions constitutes a direct circulation (clockwise in the plane of Figure 14.6), which begins to tilt the isotherms toward the horizontal while increasing the horizontal temperature gradient within the zone of convergence. The imposition of the ageostrophic wind component in the negative y' direction at low levels starts to shift the axis of net (geostrophic and ageostrophic) dilatation toward the left.

Step 6

Changes in the gradient of geopotential height in the y' direction imply changes in u_g, the geostrophic component of the wind normal to the cross section. In this example, u_g becomes more positive (out of the plane of the figure) aloft and more negative near the surface. At the same time du/dt increases (toward positive x') aloft (because v_{ag} is positive) and decreases (increases toward negative x') near the surface (where v_{ag} is negative). Thus, *the ageostrophic motions act to force the actual winds toward maintaining geostrophic (thermal wind) balance.* How quickly this *geostrophic adjustment* occurs depends on latitude and on the scale of motion. (In reality, changes in the u component of the wind are much less profound at the surface than aloft due to friction.) The essential result of this is that changes in $\partial\theta/\partial y'$ are complemented by changes in $-\partial u_g/\partial p$, thereby maintaining an approximate thermal wind balance.

Step 7

Convergence augments the confluence, and rotates the axis of dilatation, shifting it toward negative y' below and toward positive y' aloft. This process forms a frontal zone that slopes backward toward colder air, as in Figure 14.7. Large values of convergence $(\partial v_{ag}/\partial y')$ in the frontal zone

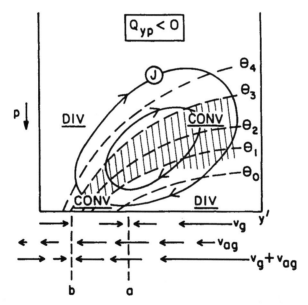

Figure 14.7 Cross section through a front formed by confluence. Broken curves are isentropes (labeled θ, increasing upward), thin full streamlines depict the streamfunction of the transverse ageostrophic circulation, and arrows below the y' axis are proportional to the magnitude and direction of the geostrophic, ageostrophic and total wind components normal to the frontal zone at the surface (vectors labeled accordingly). DIV and CONV refer to regions of divergence and convergence, respectively. Shading shows the region of largest cyclonic vorticity and J is the location of the principal wind speed maximum (jet) normal to the plane of the figure. The inequality $(Q_{yp} < 0)$ in the box above the figure indicates the sign convention for Q_{yp}.

contribute to a spin up of cyclonic vorticity (greatly exceeding the magnitude of the Coriolis parameter) in the frontal zone. The combination of $\partial v_{ag}/\partial y'$ and $\partial v_g/\partial y'$ contributes to a further enhancement with time of the confluence and the temperature deformation, to an acceleration of the u wind component at upper levels over the temperature gradient and to a progressive narrowing of the frontal zone. The cycle of steps 1–7 continue at an increasing rate as long as the synoptic-scale forcing is maintained. The process forms an increasingly intense and narrow front with time. In this respect, frontogenesis resembles cyclogenesis, except that the scale of intensification is subsynoptic along the cross-front direction and becomes increasingly smaller with time.

If frontogenesis is allowed to continue, the cross section in Figure 14.6 comes to resemble that in Figure 14.7. The axis of dilatation at the surface is displaced from point "a" to "b" and the frontal zone has tilted more toward the horizontal. Confluence in the frontal zone consists of both geostrophic and ageostrophic motion. Acceleration of the u component of the wind leads to the formation of a jet aloft (toward positive x', symbolized by the circled J) and to stronger winds toward negative x' at low levels. Within the frontal zone, there are large values of convergence and cyclonic vorticity (hatching). The convergence weakens with height due to the increasing importance of the vertical motion, which acts to weaken the frontal gradients via the tilting effect (warm air rising and cold air sinking). Both the ageostrophic and vertical components of the wind (and therefore the tilting effect) are viewed as *responses* to the forcing by the synoptic-scale motions rather than separate forcing mechanisms.

14.4 The Sawyer–Eliassen Q-vector: equations governing transverse/vertical motion

We now derive equations describing the transverse (ageostrophic)/vertical circulations normal to frontal zones and jet streaks (the circulation depicted in Figure 14.7). Like the quasi-geostrophic omega equation, the Q-vector equations are expressed in terms of a quasi-geostrophic forcing and an ageostrophic response. Diagnosis of the forcing is made using conventional geopotential height and temperature patterns, specifically the basis of the distribution and number density of solenoids, as in the case of the omega equation. A fundamental difference in the equations developed in this chapter and the quasi-geostrophic equations of Chapter 4 is that we include advections by the ageostrophic component of the wind *normal* to the frontal zone or jet streak and the vertical advection of geostrophic momentum.

In 1962, A. Eliassen, following the work of J. S. Sawyer in the 1950s, published a classic paper on fronts in which he performed a scale analysis on the equations of motion and found that the cross-frontal ageostrophic advection terms ($v_{ag} \partial u_g / \partial y'$, $v_{ag} \partial \theta / \partial y'$ and $v_{ag} \partial v_g / \partial y'$) are of first-order importance in the vicinity of fronts (Eliassen, 1962). He explained that advections of geostrophic wind components by the ageostrophic or vertical wind components are important in vertical planes normal to the frontal zone but they are small along the direction of the front. Terms consisting of products of ageostrophic winds and their derivatives are generally small, normal or along the front, although accelerations along the direction of the front constitute an important manifestation of the geostrophic response to changing temperature gradients normal to the frontal zone.

The reason for the difference in the magnitude of the cross-frontal versus the along-frontal terms is the difference in scale, the latter being much larger than the former. Consequently, it is correct to assume that wind components parallel to the front maintain an approximate geostrophic balance whereas the cross-frontal ageostrophic motion is equally important as that of the cross-frontal geostrophic wind. This simplification of the primitive equations is a special case, which formally defines the geostrophic momentum approximation.

In keeping with these assumptions and using the rotated coordinate system of x' and y' parallel and normal to the front, the basic geostrophic momentum and thermodynamic equations (1.12) and (1.13) are written as follows:

$$\left(\frac{du}{dt} \right)_{gm} = \frac{\partial u_g}{\partial t} + u_g \frac{\partial u_g}{\partial x'} + \left[(v_g + v_{ag}) \frac{\partial u_g}{\partial y'} + \omega \frac{\partial u_g}{\partial p} \right] = f v_{ag} \qquad (14.2)$$

and

$$\left(\frac{d\theta}{dt} \right)_{gm} = \frac{\partial \theta}{\partial t} + u_g \frac{\partial \theta}{\partial x'} + \left[(v_g + v_{ag}) \frac{\partial \theta}{\partial y'} \right] + \omega \frac{\partial \theta}{\partial p} = \left(\frac{p_0}{p} \right)^{\kappa} \frac{\dot{Q}_{db}}{c_p}. \qquad (14.3)$$

These equations constitute an expansion of the quasi-geostrophic system expressed in (1.28) and (1.29). Acceleration terms are subscripted gm to denote the geostrophic momentum approximation. Terms not included in the quasi-geostrophic momentum equations are the ageostrophic and ω terms enclosed in large square brackets. These additional terms account for advection of the geostrophic momentum or temperature by the ageostrophic component of the wind normal to the front; they are not contained in the quasi-geostrophic version of the Q-vector or in the quasi-geostrophic vorticity equation (3.6). Here, the vertical advection of momentum and the diabatic heating effects are retained for the present.

Derivation of the Sawyer–Eliassen Q-vector equations for the geostrophic momentum equations is similar to that for the omega equation (Ch. 4). The geostrophic wind is defined as in (1.24) and the thermal wind components are $\partial u_g / \partial p = \gamma \, \partial \theta / \partial y'$ and $\partial v_g / \partial p = - \gamma \, \partial \theta / \partial x'$. The "Boussinesq" approximation is made to γ, requiring its vertical and time derivatives be neglected; the horizontal variation in the Coriolis parameter is also neglected. The divergence of the geostrophic wind is set to zero, requiring that

$$\partial u_g / \partial x' = - \partial v_g / \partial y'.$$

Continuity is expressed as:

$$\frac{\partial u_{ag}}{\partial x'} + \frac{\partial v_{ag}}{\partial y'} = - \frac{\partial \omega}{\partial p} \approx \frac{\partial v_{ag}}{\partial y'}$$

as is consistent with the geostrophic momentum equations.

Derivation of the transverse/vertical circulation function in the geostrophic momentum system proceeds with the temperature equation (14.3), which is differentiated with respect to y', and the u momentum equation (14.2), which is differentiated with respect to p. Using the thermal wind identity and continuity, these manipulations lead to the following equations for temperature and momentum:

$$\frac{d}{dt}\left(\frac{\partial \theta}{\partial y'} \right) = \left[- \frac{\partial u_g}{\partial y'} \frac{\partial \theta}{\partial x'} - \frac{\partial v_g}{\partial y'} \frac{\partial \theta}{\partial y'} \right] - \left(\frac{\partial v_{ag}}{\partial y'} \frac{\partial \theta}{\partial y'} - \frac{\partial \omega}{\partial y'} \frac{\partial \theta}{\partial p} \right) + \frac{\partial \dot{\theta}}{\partial y'}$$

$$= \frac{1}{\gamma} \left[\frac{\partial u_g}{\partial y'} \frac{\partial v_g}{\partial p} - \frac{\partial v_g}{\partial y'} \frac{\partial u_g}{\partial p} \right] + \frac{\partial \omega}{\partial p} \frac{\partial \theta}{\partial y'} - \frac{\partial \omega}{\partial y'} \frac{\partial \theta}{\partial p} + \frac{\partial \dot{\theta}}{\partial y'} \qquad (14.4)$$

and

$$\frac{d}{dt}\left(\frac{\partial u_g}{\partial p} \right) = - \left[\frac{\partial u_g}{\partial y'} \frac{\partial v_g}{\partial p} - \frac{\partial v_g}{\partial y'} \frac{\partial u_g}{\partial p} \right] - \frac{\partial \omega}{\partial p} \frac{\partial u_g}{\partial p} - \frac{\partial v_{ag}}{\partial p} \left(\frac{\partial u_g}{\partial y'} - f_0 \right)$$

$$= - \left[\frac{\partial u_g}{\partial p} \frac{\partial u_g}{\partial x'} + \frac{\partial v_g}{\partial p} \frac{\partial u_g}{\partial y'} \right] + \frac{\partial v_{ag}}{\partial y'} \frac{\partial u_g}{\partial p} - \frac{\partial v_{ag}}{\partial p} \left(\frac{\partial}{\partial y'} (u_g - f_0 y') \right)$$

$$(14.5)$$

where the first term (in square brackets) on the right-hand side of each equation is the forcing term and the remainder is the response. Note that the first terms on the right-hand sides of (14.4) and (14.5) resemble the shear and confluence terms in the frontogenetic equation (13.2), except that these terms contain only geostrophic winds. Equation (14.5), representing the momentum equation, is not included in the kinematic theory. Like (14.4), it is also separated into geostrophic and ageostrophic terms; *the forcing terms, however, are identical in both equations.* Thus, the forcing term describes geostrophic motions (except for the diabatic heating term) and the response terms describe the vertical and ageostrophic motions.

The momentum equation can be rewritten in a slightly different form for variable f_0 $(=\beta y')$, but this will not be done here. Allowing that $\partial(f_0 y')/\partial p = 0$, and letting $m \equiv u_g - f_0 y'$, one obtains the result that

$$
\frac{d}{dt}\left(\frac{\partial u_g}{\partial p}\right) = -\left(\frac{\partial u_g}{\partial y'}\frac{\partial v_g}{\partial p} - \frac{\partial v_g}{\partial y'}\frac{\partial u_g}{\partial p}\right) + \frac{\partial v_{ag}}{\partial y'}\frac{\partial}{\partial p}(u_g - f_0 y')
$$

$$
- \frac{\partial v_{ag}}{\partial p}\frac{\partial}{\partial y'}(u_g - f_0 y'). \tag{14.6}
$$

To eliminate time derivatives, the thermodynamic equation (14.4) is multiplied by γ, the thermal wind constant. Since γ is not made a function of time or height, the total time derivatives in both the thermodynamic and momentum equations are identical,

$$
\frac{d}{dt}\left(\gamma\frac{\partial\theta}{\partial y'}\right) = \frac{d}{dt}\left(\frac{\partial u_g}{\partial p}\right).
$$

The geostrophic forcing terms are also identical, but of opposite sign. Eliminating the total derivative by subtraction of γ times (14.4) from either (14.5) or (14.6) yields the result

$$
-\frac{\partial v_{ag}}{\partial y'}\frac{\partial(u_g - f_0 y')}{\partial p} + \frac{\partial v_{ag}}{\partial p}\frac{\partial}{\partial y'}(u_g - f_0 y') + \gamma\frac{\partial\omega}{\partial p}\frac{\partial\theta}{\partial y'} - \gamma\frac{\partial\omega}{\partial y'}\frac{\partial\theta}{\partial p}
$$

$$
= -\frac{\partial v_{ag}}{\partial y'}\frac{\partial m}{\partial p} + \frac{\partial v_{ag}}{\partial p}\frac{\partial m}{\partial y'} + \frac{\partial\omega}{\partial p}\frac{\partial m}{\partial p} - \gamma\frac{\partial\omega}{\partial y'}\frac{\partial\theta}{\partial p}
$$

$$
= -2\left(\frac{\partial u_g}{\partial y'}\frac{\partial v_g}{\partial p} - \frac{\partial v_g}{\partial y'}\frac{\partial u_g}{\partial p}\right) - \gamma\frac{\partial\theta}{\partial y'} \tag{14.7}
$$

where the parameter m, representing the absolute momentum, is here defined as $u_g - f_0 y'$ and $-\partial m/\partial y'$ is the absolute vorticity. The geostrophic forcing terms for the geostrophic momentum equations are identical to those which would be derived from quasi-geostrophic theory. Only the response differs from that for quasi-geostrophic theory.

The expression in large parentheses on the right-hand side of (14.7) is defined in terms of the y component of a vector, the Q-vector, as

$$2\gamma Q_{yp} = 2\left(\frac{\partial u_g}{\partial y'}\frac{\partial v_g}{\partial p} - \frac{\partial v_g}{\partial y'}\frac{\partial u_g}{\partial p}\right) = -2J_{yp}(v_g, u_g). \tag{14.8a}$$

Substitution of $\partial u_g/\partial x' = -\partial v_g/\partial y'$ from the geostrophic continuity equations yields an alternative expression for Q_{yp}:

$$2\gamma Q_{yp} = 2\left(\frac{\partial u_g}{\partial p}\frac{\partial u_g}{\partial x'} + \frac{\partial v_g}{\partial p}\frac{\partial u_g}{\partial y'}\right). \tag{14.8b}$$

Substitution of the thermal wind relationships for the vertical (p) derivatives yields

$$2\gamma Q_{yp} = -2\gamma\left(\frac{\partial \theta}{\partial x'}\frac{\partial u_g}{\partial y'} - \frac{\partial \theta}{\partial y'}\frac{\partial u_g}{\partial x'}\right). \tag{14.8c}$$

Using geostrophic continuity one obtains

$$2\gamma Q_{yp} = -2\gamma\left(\frac{\partial \theta}{\partial x'}\frac{\partial u_g}{\partial y'} + \frac{\partial \theta}{\partial y'}\frac{\partial v_g}{\partial y'}\right). \tag{14.8d}$$

The y' component of the Q-vector is the forcing for the transverse/vertical circulation in the y', p plane. Forcing is defined above according to Hoskins and Pedder, rather than Eliassen, in which Q(Eliassen) $= -2Q\gamma$. The most useful forms of (14.8) are (14.8a) and (14.8c) or (14.8d).

For the purpose of consistency, the definition of the Q-vector is essentially identical to that of Hoskins and Pedder except for minor differences in defining the constants $\tilde{s}$ and γ. (Note, however, that the sign differs from that in Eliassen's original derivation of Q.) In order to represent the circulation in the y', p plane, a *streamfunction* (ψ) is defined such that

$$v_{ag} = -\partial\psi/\partial p \tag{14.9a}$$

and

$$\omega = \partial\psi/\partial y'. \tag{14.9b}$$

Figure 14.8 illustrates the sense of the transverse/vertical circulation, the equivalent signs of ψ and Q_{yp} and the sign of geostrophic forcing. This convention has no particular significance with regard to whether something is positive or negative. Signs could be reversed and the forcing called negative instead of positive; the results would be identical.

Using the streamfunction relationships (14.9), equation (14.7) becomes

$$- \gamma \frac{\partial^2 \psi}{\partial y'^2} \frac{\partial \theta}{\partial p} + 2 \frac{\partial m}{\partial p} \frac{\partial^2 \psi}{\partial y' \partial p} - \frac{\partial^2 \psi}{\partial p^2} \frac{\partial m}{\partial y'}$$

$$= -2 \left(\frac{\partial u_g}{\partial y'} \frac{\partial v_g}{\partial p} - \frac{\partial v_g}{\partial y'} \frac{\partial u_g}{\partial p} \right) - \gamma \frac{\partial \dot{\theta}}{\partial y'} = -2\gamma Q_{yp} - \gamma \frac{\partial \dot{\theta}}{\partial y'}. \quad (14.10)$$

Equation (14.10) resembles the quasi-geostrophic omega equation, except that the temperature equation is differentiated only once with respect to pressure rather than twice (with respect to x' or y' and then with respect to pressure). Written in terms of p and m, equation (14.10) is known as the *semi-geostrophic* version of the Q-vector equation. Semi-geostrophic coordi-

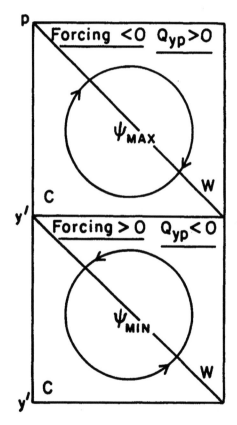

Figure 14.8 Schematic illustration of the sense of the Q_{yp} component, the forcing ($L(\psi)$) and the streamfunction for the direction of circulation shown by the thin circular streamlines. The sloping diagonals represent sloping isentropes in a frontal zone, and the letters C and W, respectively, refer to the locations of the cold and warm air.

nates (m, p) are conceptually more difficult to diagnose than y', p coordinates, but are mathematically more elegant in their simplicity. Those equations reduce to the quasi-geostrophic form if the quasi-geostrophic momentum and temperature equations are used. *The factor of 2 in the forcing term
originates because identical contributions are made to the circulation by the
advection of temperature and momentum.*

Equation (14.10) can be written

$$S \frac{\partial^2 \psi}{\partial y'^2} - 2B \frac{\partial^2 \psi}{\partial y' \partial p} + \zeta_a \frac{\partial^2 \psi}{\partial p^2} = -2\gamma Q_{yp} - \gamma \frac{\partial \dot{\theta}}{\partial y'} \qquad (14.11a)$$

where the symbols S, B and ζ represent the static stability, baroclinicity and
absolute vorticity (inertial stability) effects, defined as

$$S = -\gamma \partial\theta/\partial p$$

$$B = -\gamma \partial\theta/\partial y' = -\partial m/\partial p$$

$$\zeta_a = -\partial m/\partial y' = f_0 - \partial u_g/\partial y'.$$

Eliassen showed that, for a given point source of geostrophic forcing, the
circulation cells take the shape of an ellipse (as in Fig. 14.7), whose major and
minor axes are determined by the relative magnitudes of the thermal stability
$S(-\partial\theta/\partial p)$ and the absolute geostrophic vorticity $\zeta_a(-\partial m/\partial y')$, and whose
vertical slope depends upon the baroclinicity B $(\partial\theta/\partial y'$ or $\partial m/\partial p)$.
Further, if the potential vorticity is large and the absolute vorticity is small
(large static stability), the circulation ellipses are oriented horizontally. If the
potential vorticity is small and the absolute vorticity is large (small static stability), the circulation ellipses are oriented vertically. Higher static stability by
itself compresses the circulation cells and tends to weaken them.

The left-hand side of (14.11a) can be reduced to two response terms.
Defining

$$\delta \equiv \gamma \left(\frac{\partial\theta}{\partial p} \frac{\partial m}{\partial y'} - \frac{\partial\theta}{\partial y'} \frac{\partial m}{\partial p} \right)$$

the equation is transformed to become

$$\left[\frac{\partial}{\partial m} \delta \left(\frac{\partial \psi}{\partial m} \right) \right]_p + \left(\frac{\partial^2 \psi}{\partial p^2} \right)_m = -\frac{2\gamma Q_{yp} + \gamma \partial\dot{\theta}/\partial y'}{f_0 - \partial u_g/\partial y'}. \qquad (14.11b)$$

This semi-geostrophic equation (14.11b) now contains no mixed derivatives
and one variable coefficient (δ), which is equivalent to the potential vorticity;

therefore, it must always be positive for the baroclinic current to be stable and, therefore, to obtain a solution. On the right-hand side are terms determined from the geopotential height and temperature fields. Written thus, the numerator of the right-hand side is sometimes referred to as the "source strength" of the transverse/vertical circulation. Q_{yp} is a measure of the geostrophic source strength in the y, p plane. On the left is δ, the response to the forcing, which can be determined from a numerical solution of the differential equation. Further discussion concerns itself solely with the representation of the circulation equation in y', p or x', y' coordinates.

There are certain advantages in representing the equation in m and p coordinates, however. Not only is (14.11b) simpler than (14.11a), but the latter seems to give better resolution in frontal zones, where gradients of wind speed are large. Air circulates around areas where Q_{yp} is a minimum or maximum; conversely, no closed streamlines exist where Q_{yp} is equal to zero. Elliptical streamlines are found close to the source. Nearer the boundaries they change shape to conform to the boundary conditions, which impose the restriction that there is no transverse circulation across the lateral and vertical boundaries.

It is clear that the forcing and the coefficients of the response terms are governed by the geostrophic processes. As in quasi-geostrophic theory, the assumption governing (14.10) is that the ageostrophic response to geostrophic, diabatic or frictional forcing is effectively instantaneous. This assumption is supportable insofar as the forcing by the geostrophic wind does not totally destroy the balance between vertical wind shear and the horizontal temperature gradient, a situation that could arise on small time and space scales. This equation also breaks down in sharply curved flow in which the acceleration of the ageostrophic velocity is of equal magnitude to the geostrophic acceleration due to the importance in the centripetal component of the motion. A more sophisticated theory could include gradient wind (instead of geostrophic) balance, or make use of the primitive equations of motion without simplifications. Numerical solutions of the quasi-geostrophic and the geostrophic momentum versions of the Q-vector circulation equations show, however, that the latter is able to capture the details of frontal circulations much more completely, but that the former is, nevertheless, capable of resolving the overall features of the circulation. Geostrophic momentum theory is most accurate in describing the transverse/vertical circulations in straight flow.

Interpretation of the Q-vector equations for two-dimensional circulations

The geostrophic Q-vector (source strength) terms resemble the geostrophic advection terms on the right-hand side of the omega equation. The magnitude and distribution of the source strength in the y', p plane can be diagnosed in a similar manner to that of geostrophic advection by applying convections that are now to be established.

It is convenient to write (14.11a) as

$$L(\psi) = -2\gamma Q_{yp} - \gamma \partial\dot{\theta}/\partial y' = L(\psi_d) + L(\psi_{db}). \quad (14.12)$$

Terms in large parentheses in (14.8) determine the geostrophic forcing of the circulation function $L(\psi)$. We will call the operator $L(\psi)$ the "geostrophic forcing" and ψ the response. If Q_{yp} is negative (positive forcing), its contribution to $L(\psi)$ is positive. Since $L(\psi)$ represents a kind of Laplacian, it follows that a positive value of $L(\psi)$ corresponds to a negative value of ψ. A minimum in Q_{yp}, therefore, corresponds to a minimum in ψ and to a circulation operating in the sense of the streamfunction lines in the lower part of Figure 14.8, rising on the negative y' and sinking on the positive y' side of the figure. This circulation is direct because warm air (W) rises and cold air (C) sinks. *It must be emphasized, however, that the sign of Q_{yp} does not define whether a circulation is direct or indirect because the choice of the positive y' axis in the direction of cold air is arbitrary and the direction of the cold air may be ambiguous in some cases.* The equations are, of course, valid, regardless of the direction of y'; and their interpretation is equally consistent for any y' direction provided that the other components are defined with respect to this system; the x' axis is orthogonal to y', with positive values on the right.

Physical interpretation of (14.12) can be made by evaluating the geostrophic forcing terms in (14.8). Consider (14.12) with the definition of Q_{yp} given by (14.8d). This is written

$$L(\psi) = -2\gamma Q_{yp} - \gamma \frac{\partial\dot{\theta}}{\partial y'} = 2\gamma \left(\frac{\partial\theta}{\partial y'} \frac{\partial v_g}{\partial y'} + \frac{\partial\theta}{\partial x'} \frac{\partial u_g}{\partial y'} \right) - \gamma \frac{\partial\dot{\theta}}{\partial y'}$$

$$= 2\gamma(C_g + S_g) + DB$$

$$= -2\gamma(Q_{yp} + Q_{yp}^d) \quad (14.13)$$

where the term in large parentheses consists of two parts, one representing the effect of confluence (analogous to the term labeled C in (13.2)) by the geostrophic component of the wind (C_g), and the other the effect of shearing by the geostrophic component of the winds (S_g) (analogous to term S in (13.2)). $L(\psi)$ is also expressed as the sum of three terms, C_g, S_g and DB, where S_g is the geostrophic shear term, C_g the geostrophic confluence term and DB the diabatic term.

The last term on the right-hand side of the equation represents the diabatic heating effect expressed in terms of the Q-vector, Q_{yp}^d, which is equal to

$\frac{1}{2}\partial\dot{\theta}/\partial y'$. Note that this equation is similar to (13.2), except that the forcing is geostrophic and the tilting effect is contained in the response term ($L(\psi)$).

As expressed in (14.13), geostrophic confluence and shearing produce tranverse/vertical circulations in the y', p plane. Similar equations can be derived for a circulation in the x', p plane by simply rotating the y' axis by 90° and relabeling the coordinates accordingly. Evaluation of the three-dimensional dynamically driven circulation (Q_{xp}, Q_{yp}) is somewhat more complicated than a simple transformation to arrive at the other Q component because the lateral ageostrophic advections must be retained in both x and y momentum and temperature equations when both are treated as a combined system of equations.

Positive forcing (S_g and/or C_g greater than zero) is frontogenetic with respect to the effect of geostrophic deformation of the isotherm pattern. Compression of the isotherms into an increasingly larger horizontal gradient is aided, however, by the convergence, that is by the action of $\partial v_{ag}/\partial y$, which acts to compress the isotherms further. Thus, the effect of the transverse component of the circulation is to accelerate the process of frontogenesis in the frontal zone when the circulation is direct. Note, however, that the vertical component of the circulation (which vanishes at the surface) acts to reduce the horizontal temperature gradient in the sense described by the tilting term in (13.2). The same argument applies to vorticity, which is increased in the frontal zone due to the effect of convergence ($\partial v_{ag}/\partial y > 0$). The latter is partially offset by the tilting effect in the vorticity equation.

Let us now evaluate (14.13) algebraically, beginning with the geostrophic shearing (S_g). Note the similarity between Figures 14.9 and 13.5, the differ-

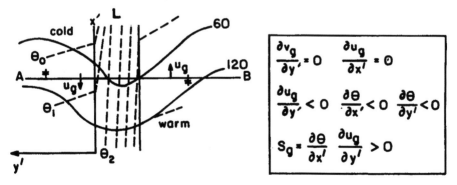

Figure 14.9 Shearing by geostrophic wind. Broken lines are isotherms (labeled θ, cold toward upper left), full curves are the geopotential height contours (labeled in dam), arrows (labeled accordingly) show the direction of the u_g component and the full lines denote the boundaries of the frontal zone. The double cross symbols indicate the boundary between cyclonic and anticyclonic shear of v_g. The signs of the relevant terms in the Q_{yp}-forcing equation are shown. Segment A–B refers to the cross section in Figure 14.10.

ence being that the geopotential height contours determine the forcing by shearing, rather than the total wind components. The front is drawn north to south in Figure 14.9 in the manner of a cold front, with cold air to the west (positive y'). The trough is centered on the front and is symmetric, so that the variation of the normal component of the geostrophic wind (v_g) along y' (and therefore the confluence term) is everywhere equal to zero. Since there is no confluence or diabatic heating, only shear operates to produce this front. Note, however, that in equation (14.10) the shear forcing leads to a component $\partial v_{ag}/\partial y'$ along the front.

It is instructive to note from Figure 14.9 that u_g varies from plus to minus across the front and that the magnitude of this wind component first increases and then decreases with increasing distance from the middle of the frontal zone. The shear term is therefore positive near and within the frontal zone because $\partial u_g/\partial y'$ and $\partial \theta/\partial x'$ are both negative, the shear is positive and Q_{yp} is negative. By the convention of (14.13), $L(\psi)$ is positive and ψ negative. Accordingly, the direction of the circulation ellipse in the y', p plane is the same as in Figure 14.7 (and in Figure 14.8 for positive forcing, $Q_{yp} < 0$), rising on the negative y' and sinking on the positive y' sides of the plane through the front. This circulation ellipse is shown in Figure 14.10, which represents a cross section through Figure 14.9 along segment A–B. Note, however, that the magnitude of Q_{yp} is greatest in the frontal zone, where the circulation is direct, but that there are two weaker satellite cells on either side (both are indirect).

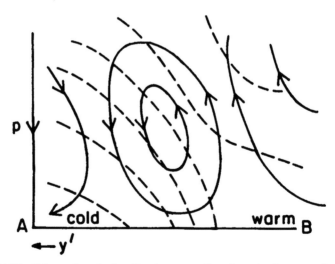

Figure 14.10 Schematic solution for the streamfunction obeying equation (14.11) for geostrophic shearing along the segment A–B in Figure 14.9. Full streamlines represent ψ and broken curves are the isentropes, increasing upward with cold air on the left.

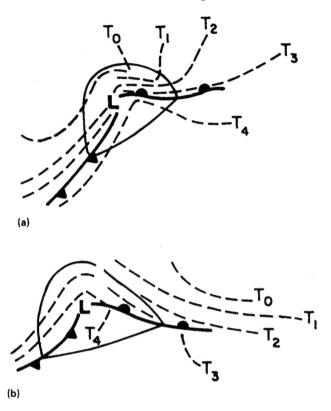

Figure 14.11 Typical surface warm front patterns with isotherms (broken curves, cold air toward top left in (a) and toward the right in (b)). Full contours represent isobars.

These two satellite cells exist because the sign of $\partial u_g/\partial y'$ changes further away from the frontal zone; the sign changes at the small double cross signs in Figure 14.9. The magnitude of $\partial u_g/\partial y'$ is nevertheless weak outside of the double cross signs and therefore the indirect satellite cells are weak. *Note that the transverse/vertical circulation in the frontal zone, shown in Figure 14.10, contributes a component of $\partial v_{ag}/\partial y'$ along the leading edge of the front, thereby augmenting the geostrophic confluence (which in this case is zero), resulting in a further contraction of the isotherms and intensification of the circulation.*

Figure 14.9 resembles a cold front, although the figure can be rotated sideways to look like a warm front. As discussed in regard to equation (13.2), the sign of the shear term in a typical warm front, such as shown in Figure 14.11, depends on where the cold air is located. In Figure 14.11a, $\partial\theta/\partial x'$ is positive and $\partial u_g/\partial y'$ is negative; Q_{yp} due to shear is positive and ψ is positive, implying an indirect circulation, opposite to that in Figure 14.10. In the other case (Fig. 14.11b) the sign of $\partial\theta/\partial x'$ is negative and $\partial u_g/\partial y'$ is negative; Q_{yp} is negative and ψ is negative, and the circulation is direct as in Figure 14.10. According to thermal wind arguments, the indirect case conforms to an increasing component of the wind with height across the frontal zone from the warm air side of the front. Therefore, negative shearing in the vicinity of warm fronts is compatible with the front being anabatic, whereby air ascends slantwise toward the colder air. This does not mean that the warm frontal circulation is indirect because, as stated earlier, confluence in the direct sense tends to dominate along warm fronts.

In the case of pure confluence shown in Figure 14.12, the deformation pattern resembles that of Figure 13.7, except that the geostrophic components of the wind replace the kinematic ones. Kinematic frontogenesis due to confluence was first postulated by T. Bergeron in 1928. The geostrophic wind pattern shows a high to the south and a low to the north of the frontal zone, which produces frontogenesis due to geostrophic forcing by confluence. Figure 14.12 shows that $\partial v_g/\partial y'$ and $\partial\theta/\partial y'$ are both negative and, by consequence, Q_{yp} is negative and the forcing positive. The circulation here is direct normal to the front. This is shown in Figure 14.13 for the case of Figure 14.12, which shows the temperature and geopotential height contours at 700 mb. The circulation in Figure 14.7 operates to move the surface front toward the warm air and the front aloft toward the cold air, thereby retarding the translation of the front aloft. (The same principles apply to a coordinate system moving with the mean speed of the front.)

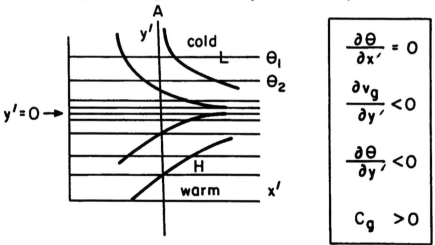

Figure 14.12 Geopotential height contours (full curves, lower values toward the top) and isotherms (full lines, colder toward the top) for the case of forcing by geostrophic confluence. The signs of the relevant terms in the Q_{yp}-forcing equation are shown.

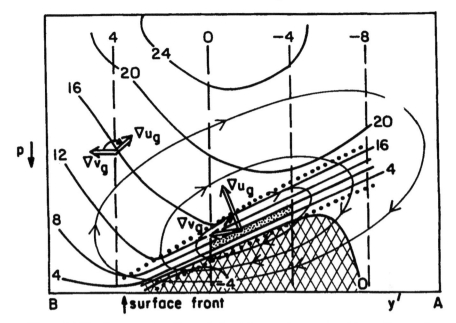

Figure 14.13 Cross section through frontal zone for geostrophic confluence (e.g. Fig. 14.12). The full curves are isotachs of u_g, broken lines are isotachs of v_g and thin full streamlines are the streamfunction obeying (14.11). Short bold arrows labeled accordingly show the direction and magnitude of the gradients of v_g and u_g; the rotation angle for the left-hand rule is indicated by short curved arrows between the gradient arrows. The shaded solenoid in the frontal zone is proportional to the magnitude of the forcing by confluence; the cross-hatched region is the area where the u_g wind component has the opposite sign from that above the front (into page). The location of the surface front is indicated.

Acceleration of the wind that occurs to the right of v_{ug} is into the plane of the figure near the surface and out of the plane of the figure above the frontal zone. Accordingly, if the process continues, the speeds increase toward positive x' aloft and toward negative x' near the surface, i.e. in the easterly sense (the cross-hatched region for a northward-oriented y' axis). Note that the transverse circulation itself operates to contract isotherms in the frontal zone and thereby aid the frontogenesis by confluence. *This secondary effect of the transverse circulations in enhancing the initial geostrophic confluence in the frontal zone constitutes an essential difference between the dynamic and kinematic analyses of frontogenesis.*

It should be noted that the tilting terms do not enter explicitly into the geostrophic forcing terms; their effect is felt indirectly as a weakening of the frontal zone with height. This can be seen because the direct circulation involves an adiabatic cooling of warm air by ascent and an adiabatic warming of cool air by descent. This process reduces the horizontal temperature gradients. Thus, tilting often acts to counter the frontogenetic effects of confluence and shear, although it nevertheless serves to increase kinetic energy at the expense of available potential energy. In fact, the decrease in the potential energy is the result of the direct circulation's weakening of the horizontal temperature gradient. This weakening in the horizontal temperature gradient is reflected in a decrease with height of the horizontal vorticity gradient and even in the forcing itself. Consequently, there is a broadening with height of the circulation ellipse, as drawn in Figure 14.13. Of course, the tilting term exerts no such effect in modifying the circulation at the surface, with the result that lower-tropospheric fronts tend to be strongest at the surface and weaken rapidly with height. Therefore, the vertical motion pattern within frontal zones may not conform to the bowstring model (Fig. 2.8), although the magnitude of ω is nevertheless relatively small at the surface and increases to a maximum within the frontal zone.

The diabatic term in (14.13) is similar to that in (13.2), except that it produces a diabatic component of the circulation ($Q^d_{yp} \equiv \frac{1}{2} \partial\dot{\theta}/\partial y'$) in the y', p plane. Thus, on the left of Figure 14.14, $\partial\dot{\theta}/\partial y'$, is positive, Q^d_{yp} is positive and ψ is positive, yielding a circulation ellipse in which air rises on the

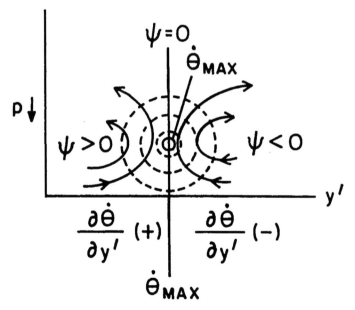

Figure 14.14 Cross section showing streamfunction (full curves) forced by diabatic heating (broken contours, with maximum value of $\dot{\theta}$ at center). The streamfunction for the solution of (14.11) is positive at the left, negative at the right and zero at the central axis. The sign of the y' derivative of the heating rate ($\dot{\theta}$) is shown below the y'-axis.

positive y' and sinks on the negative y' side of the figure. On the right of the figure, $- \partial \dot{\theta} / \partial y'$ is negative, Q_{yp} is negative and ψ is negative, and the sense of the diabatic circulation cell is clockwise in the $y - p$ plane.

The convention of orienting the y' axis toward the cold air may appear ambiguous in some circumstances, although careful application of the sign conventions of Figure 14.8 can resolve the direction of the circulation without ambiguity. Consider the situation depicted in Figure 14.14, in which the temperature gradient is not known, but a heat source is placed in the middle of the figure. The circulation is in the opposite direction on either side of Figure 14.14 because the sign of $\partial \dot{\theta} / \partial y'$ changes at the center. This double circulation cell is arranged such that diabatic forcing produces upward vertical motion near the source, a result previously discussed with regard to diabatic heating in the omega equation (Ch. 9). If the diabatic heating were to produce a warming of the temperature pattern near the center of the figure, both circulation ellipses would become direct, although Q_{yp}^{d} possesses different signs on either side of the heating source. The point is that the choice of the y' axis is immaterial, as long as mathematical consistency is maintained.

14.5   Graphical interpretation: the left-hand rule

A diagnosis of the transverse/vertical circulation can be done without having to evaluate mathematical terms in a complex differential equation. Let us now codify some rules for graphically evaluating the sense and magnitude of the transverse/vertical circulation in the y', p plane. First, let us note that, written as in (14.8a) or (14.8c), Q_{yp} can be expressed as a Jacobian and as the scalar component of a two-dimensional vector cross product. If ψ_{d} is the dynamically forced part of the transverse/vertical circulation (that forced by the geostrophic part of the flow)

$$L(\psi_d) = - 2\gamma Q_{yp} = - 2 \left(\frac{\partial v_g}{\partial y'} \frac{\partial u_g}{\partial p} - \frac{\partial u_g}{\partial y'} \frac{\partial v_g}{\partial p} \right)$$

$$= - 2 J_{yp}(v_g, u_g)$$

$$= - 2 i \cdot \begin{vmatrix} i & j & k \\ \partial v_g / \partial x' & \partial v_g / \partial y' & \partial v_g / \partial p \\ \partial u_g / \partial x' & \partial u_g / \partial y' & \partial u_g / \partial p \end{vmatrix}. \qquad (14.14)$$

First note that the cross product is normal to the plane of the two vectors forming the product; the sign of $L(\psi_d)$ is determined by the direction of the unit vector i along the positive or negative x axis. (For convenience, we will henceforth drop the subscript d.) The two gradient vectors in the y', p plane of Figure 14.13, ∇v_g and ∇u_g, are drawn (not to scale) from a point located near the center of the circulation ellipse. The sign of Q_{yp} is negative, $L(\psi)$ is positive and ψ is negative. Gradient vectors, respectively pertaining to the two arguments in the Jacobian $-(J_{yp}(v_g, u_g))$, are directed perpendicular to isopleths of v_g and u_g and toward higher values. According to (14.14), the sign of Q_{yp} is equal to that of the Jacobian in these equations.

In order to evaluate both the strength *and* sense of the transverse/vertical circulation using (14.12), we employ the *left-hand rule*. Standard mathematical treatment of vectors makes use of a right-hand rule. The left hand is employed here rather than the right hand because the direction of the fingers denotes the direction of the streamlines in the y', p plane and the direction of the. thumb along the x' axis denotes the sign of forcing ($L(\psi)$). Not only are we able to avoid complex equations, but the left-hand rule makes it unnecessary to confront the issue of positive or negative signs in the equations. Further, the left-hand rule can also be applied in evaluating circulations in the y', p plane by inspection of u_g (or v_g) and θ isopleths in the x', y' plane. Finally, we will also show that the strength of the y', p circulation can be evaluated qualitatively from inspection of solenoids in either the y', p or x', y' planes.

Let us now see how the left-hand rule operates. Since the sense of the transverse/vertical circulation is given by the direction of the cross product (and therefore the sign of the Jacobian), we obtain the following rules for evaluating $L(\psi)$:

(1) Find the direction of the gradients of v_g and u_g.
(2) With the left hand, point the fingers along the gradient of v_g (first argument of the Jacobian) toward higher values.
(3) Rotate the fingers in the direction of the gradient of u_g (second argument of the Jacobian). The angle between the two vectors must be less than 180°.
(4) The direction of the thumb along the positive or negative x' axis indicates the sign of $L(\psi)$, and the curling of the fingers is the sense of the transverse/ageostrophic circulation.

Thus, one can imagine that the fingers represent lines of the streamfunction and the transverse/vertical circulation flows along the fingers. Note that the curling of the fingers is still in the same sense if the right hand is used, but the thumb does not give the correct sign of $L(\psi)$. Conventional vector evaluation is usually given by the "right-hand" rule. The left thumb determines the sign of $L(\psi)$ and the forcing by its direction along the x' axis. For the case of $Q_{yp} < 0$, $L(\psi)$ is positive, ψ is negative (the bottom part of

Figure 14.8) and the thumb points along the positive x axis. For $Q_{yp} > 0$, $L(\psi)$ is negative, ψ is positive (the top part of Figure 14.8) and the thumb points along the negative x' axis.

Let us see how the left-hand rule operates in Figure 14.13. The left-hand rule, applied at the two locations where the two gradient vectors (∇v_g and ∇u_g) are drawn (one inside the frontal zone and one above it), demonstrates that the circulation is clockwise in both cases. The smaller the angle between the vectors, the smaller the value of the cross product for a given pair of vectors. Vectors $180°$ or $0°$ apart imply no geostrophic forcing and therefore a zero value for $L(\psi)$. By contrast, vectors pointing at right angles maximize $L(\psi)$ for a given magnitude of ∇v_g and ∇u_g. Consequently, the circulation cell changes sign at the level of the jet where the gradients of v_g and u_g are parallel.

The strength of the circulation cell can be evaluated by realizing that *the magnitude of a Jacobian is inversely proportional to the area of solenoids formed by the intersection of isopleths of the components in the argument of the Jacobian.* Gradient vectors that are parallel form no solenoids, while those at right angles to one another maximize the number of solenoids. This is exactly analogous to the geostrophic advection (e.g. equation (2.20)), which can thus be written as a Jacobian. Recall that geostrophic temperature advection is expressed as the number of unit solenoids formed by intersection of the geopotential height contours (Z) and the isotherms (T). Restating (1.30),

$$- V_g \cdot \nabla_p T = - \left(u_g \frac{\partial T}{\partial x} + v_g \frac{\partial T}{\partial y} \right) = - \frac{g}{f_0} J_{xy}(Z, T).$$

In Figure 14.13, there is only one circulation cell below the level of the wind speed maximum, which is situated at the top of the figure. Because of the configuration of the gradients of v_g and u_g, the left-hand rule requires that the thumb points toward the positive x' axis in this figure. The circulation cell possesses one center in the figure because there is only one minimum in the size of solenoids and therefore in the streamfunction. The v_g isotachs are everywhere vertical and equispaced, signifying that this is a case of pure confluence since $\partial v_g / \partial p$ (which is equivalent to $\gamma \partial \theta / \partial x'$) is everywhere zero. Shearing vanishes where $\partial u_g / \partial y'$ is zero, which occurs where isotachs of u_g are horizontal, for example, directly below the wind speed maximum. Conversely, confluence is zero where isotachs of u_g are vertical ($\partial \theta / \partial y' = 0$) or where isotachs of v_g are horizontal ($\partial v_g / \partial y' = 0$). In Figure 14.13, isotachs of u_g are vertically oriented to the right and left sides of the wind speed maximum. Accordingly, the sense of the circulation changes at the level of the jet streak from that below. (An analysis of circulation patterns in the vicinity of the jet streak using the principles formulated in this chapter is presented in the next chapter.) Mixed shearing and confluent forcing can

395

occur where the isotachs of u_g and v_g are both tilted at a non-zero angle with the horizontal and vertical axes. The total forcing is zero where the isotachs of u_g and v_g are parallel.

The left-hand rule allows one to evaluate the sign of $L(\psi)$ by inspection of u_g (or v_g) and θ fields in the x', y' plane. Since the circulation is still in the y', p plane, the left-hand rule must be evaluated in two steps. First, let us write the equation for Q_{yp} in terms of u_g and θ (14.8c) and rearrange it such that the Jacobian of these two variables has the same form as (14.14), i.e. with a negative sign in front of the Jacobian:

$$L(\psi) = -2\gamma Q_{yp} = 2\gamma \left(\frac{\partial \theta}{\partial x'} \frac{\partial u_g}{\partial y'} - \frac{\partial u_g}{\partial x'} \frac{\partial \theta}{\partial y'} \right)$$

$$= 2\gamma J_{xy}(\theta, u_g) = -2\gamma J_{xy}(u_g, \theta). \tag{14.15}$$

According to (14.15) we can evaluate the sense of $L(\psi)$ by crossing ∇u_g (the vector pointing up the gradient of u_g) toward $\nabla \theta$ (the vector pointing up the gradient of θ) with the left hand, exactly as indicated by the left-hand rule. In this case the fingers do *not* determine the direction of the circulation. Since the thumb does point toward positive or negative z, we will adopt the convention that positive z denotes positive $L(\psi)$ and negative z a negative value of $L(\psi)$. The strength of $L(\psi)$ in the y', p plane at that point is nevertheless inversely proportional to the solenoid areas formed by u_g and θ. Therefore, for *positive* $L(\psi)$ (thumb up) we rotate the thumb to point toward *positive* x'; the circulation in the y', p plane is now given by the direction of the fingers on the left hand. The reverse is true for negative $L(\psi)$ (thumb down in evaluating $-J_{xy}(u_g, \theta)$).

Let us apply the left-hand rule to the confluent case illustrated in Figure 14.12. The fingers point toward positive x' along the gradient of u_g and are rotated toward negative y' in the direction of the θ gradient and the thumb points upward defining the sign of the Jacobian $-J_{xy}(u_g, \theta)$ and the forcing in (14.15) as positive and $L(\psi)$ as positive. *Note that the rotation of the fingers has no significance in the x', y' plane because the circulation is in the y', p plane.* The direction of the thumb expresses the sign of the forcing only, although the strength of the circulation is nevertheless inversely proportional to the area of the solenoids of θ and u_g. Thus, the circulation is strongest in the frontal zone, particularly where the isotherms are crossing isotachs of u_g at right angles. Since $L(\psi)$ is positive (thumb now points toward positive x'), the sense of the circulation given by the direction of the fingers conforms to the lower part of Figure 14.8.

As in the case of advection, a Jacobian can be expressed in natural coordinates. Equation (14.15) can be expressed as

$$- 2\gamma J_{xy}(u_g, \theta) = - 2\gamma \frac{\Delta\theta}{\Delta s}\frac{\Delta u_g}{\Delta n} \qquad (14.16)$$

where the s and n axes are defined as along and at right angles (positive to *left*) to the direction of the temperature gradient. The denominator on the right-hand side of (14.15) is the area of a solenoid, whose magnitude is expressed by one term, rather than by both confluence and shear. This equation is analogous to the Q-vector form of the omega equation discussed later in this chapter, in which the two forcing terms on the right-hand side of the omega equation are combined in a single mathematical expression.

14.6 A numerical simulation of frontogenesis

A simulation of the transverse/vertical circulation, made by D. Keyser using a primitive equation model for geostrophic confluence, is shown in Figure 14.15. The center of the circulation is located in the frontal zone, which is recognizable by the larger horizontal gradient of potential temperature between $+ 800$ km and $- 800$ km from the center line. Vertical motions are strongest on either side of the front at about 500 mb, but (unlike the bowstring model) the direction of the vertical motion changes sign in the frontal zone. Vertical motion is very weak near the surface. Horizontal gradients of temperature are largest in the frontal zone near the surface, as is the horizontal component of the circulation (the ageostrophic wind). Strongest convergence occurs at the surface in the frontal zone, where there is the largest vorticity. Accelerations into and out of the plane of the figure occur to the right of the ageostrophic wind velocity. These accelerations lead to the formation of a wind speed maximum into the plane of the figure at upper levels above the frontal zone and out of the plane of the figure below the frontal zone. In reality, large wind speeds are not usually attained near the surface due to the presence of surface friction, which was not taken into account.

The temperature structure of Figure 14.15 resembles those of Figures 13.2 and 14.7, except that the depth of the simulated circulation extends to the tropopause. The reason for this greater depth of circulation in the computer simulations is due to deeper forcing. Note, however, that the tropopause is higher on the right-hand side of Figure 14.15. Differences in tropopause height are thought to be dynamic in origin, a consequence of the difference in direction of the vertical motion. (Tropopause behavior is treated in more detail in Chapter 15.) The center of circulation in this figure is situated not far from the region of smallest solenoids of v_g and u_g.

Simulations such as the one shown in Figure 14.15 are able to produce realistic fronts, although often they are deeper than observed. Such simu-

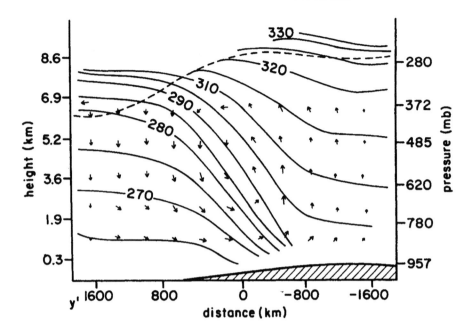

Figure 14.15 Numerical simulation of a front for geostrophic confluent deformation after 48 h of integration time. The isentropes (full curves in K) have begun to form a front. Arrows are proportional to magnitude of vertical/ageostrophic winds. The broken curve denotes the tropopause. (Courtesy of D. Keyser.)

lations eventually produce fronts with extremely large temperature gradients, which approach the limit of grid resolution in the model. Without the ability to generate turbulence, such as would occur in the atmosphere, simulated fronts would attain unrealistically large wind speeds and temperature gradients. Recent theoretical evidence, however, suggests that there is a physical limit of frontogenesis, just as there is self-limitation in cyclogenesis. Actually, the generation of large gradients of wind and temperature lead to small-scale turbulent breakdown of the flow, which prevents the atmosphere from achieving extremely intense large gradients. Frontogenesis, subject to the retardation by turbulent mixing, continues as long as the geostrophic forcing is imposed. This forcing is dictated by the large-scale geopotential height and temperature fields.

14.7 Quasi-geostrophic omega equation

A form of the quasi-geostrophic omega equation, derived by Hoskins and Pedder (1980), combines the two forcing terms in (4.3) in the Q-vector.

Derivation of the Q-vector form of the omega equation proceeds in exactly the same sequence as shown earlier in this chapter, except that the quasi-geostrophic momentum and temperature equations (1.28) and (1.29) for adiabatic flow are used instead of the geostrophic momentum equations.

As earlier, we differentiate (1.29) with respect to y', multiply it by the thermal wind constant and combine the result with the u momentum equation, which has been differentiated with respect to pressure. After making use of the geostrophic continuity equation and substituting the thermal wind relationship in the y', p plane, one obtains

$$2\gamma Q_{yp} = f_0 \frac{\partial v_{ag}}{\partial p} + \gamma \frac{\partial \theta}{\partial p} \frac{\partial \omega}{\partial y'} = 2\gamma Q \cdot j \qquad (14.17a)$$

where the y component of the Q-vector (Q_{yp}) is defined exactly as in (14.8).

Similarly, the quasi-geostrophic versions of the momentum and adiabatic temperature equations are used to obtain the following relationship for Q_{xp}:

$$2\gamma Q_{xp} = f_0 \frac{\partial u_{ag}}{\partial p} + \gamma \frac{\partial \omega}{\partial x'} \frac{\partial \theta}{\partial p} = 2\gamma Q \cdot i. \qquad (14.17b)$$

The components for Q_{xp} are

$$\gamma Q_{xp} = - \left(\frac{\partial u_g}{\partial p} \frac{\partial v_g}{\partial x'} + \frac{\partial v_g}{\partial p} \frac{\partial v_g}{\partial y'} \right) \qquad (14.18a)$$

or

$$\gamma Q_{xp} = - \gamma \left(\frac{\partial \theta}{\partial y'} \frac{\partial v_g}{\partial x'} - \frac{\partial \theta}{\partial x'} \frac{\partial v_g}{\partial y'} \right) \qquad (14.18b)$$

or

$$\gamma Q_{xp} = - \gamma \left(\frac{\partial \theta}{\partial x'} \frac{\partial u_g}{\partial x'} + \frac{\partial \theta}{\partial y'} \frac{\partial v_g}{\partial x'} \right). \qquad (14.18c)$$

Combining the two expressions for the Q-components, (14.17a) and (14.17b), and using the continuity relationship for the ageostrophic wind components,

$$\frac{\partial u_{ag}}{\partial x'} + \frac{\partial v_{ag}}{\partial y'} = - \frac{\partial \omega}{\partial p}$$

results in the following expression:

$$2\gamma \nabla_p \cdot Q = \gamma \frac{\partial \theta}{\partial p} \nabla_p^2 \omega - f_0 \frac{\partial^2 \omega}{\partial p^2}. \tag{14.19}$$

In terms of the omega equation,

$$2\gamma \frac{f_0}{g} \nabla_p \cdot Q = -\frac{f_0}{g} \tilde{s}\gamma \nabla_p^2 \omega - \frac{f_0^2}{g} \frac{\partial^2 \omega}{\partial p^2}. \tag{14.20}$$

Note that (14.20) is virtually identical to (4.3) except for the neglect of advection of f_0 in the forcing; it is exactly identical to (4.3) on an f plane. Here, it is expressed in terms of the horizontal divergence of the Q-vector, which represents the combined forcing of terms F_1 and F_2 in the omega equation (4.3), except for the neglect of the advection of Coriolis parameters. Thus, we see again that the division of the omega equation into two quasi-geostrophic forcing terms is somewhat artificial since both terms are aspects of one another.

The sign of the response term in the Laplacian is determined by the sign of the vertical motion; in this case *rising motion occurs where there is convergence of the Q-vector (the forcing function is positive ($\nabla_p \cdot Q$ negative)) and sinking motion is found where there is divergence of the Q-vector.*

The importance of the Q-vector in diagnosing frontal development can be seen more clearly by expressing the Q-vector in terms of the time rate of change of the horizontal temperature gradient. From the definition of Q, one can write

$$Q = \frac{d_g}{dt} \nabla_p \theta = u_g \frac{\partial}{\partial x'} \nabla_p \theta + v_g \frac{\partial}{\partial y'} \nabla_p \theta + \frac{\partial}{\partial t} \nabla_p \theta \tag{14.21a}$$

and $d_g(\nabla_p \theta)/dt$ is the time rate of change of the temperature gradient moving with the horizontal geostrophic velocity. It can also be shown that

$$(\partial/\partial t + V_g \cdot \nabla_p)(\nabla_p \theta)^2 = 2Q \cdot \nabla_p \theta. \tag{14.21b}$$

To illustrate further the use of the Q-vector, consider the 700 mb geopotential height and isotherm chart of Figure 14.16a. The distribution of the Q-vectors for this case is shown in Figure 14.16b with rising and sinking motion implied, respectively, in regions of Q convergence and divergence. The greatest convergence of Q, corresponding to the strongest rising motion, occurs poleward of the warm fronts, particularly in the region just poleward of the junction between cold and warm fronts near the surface low center. Strong divergence of the Q-vector, and therefore strong descent, is occurring southwest of the surface low and west of the surface cold front.

Although the Q-vector formulation is mathematically elegant, it is more difficult to use subjectively than some other forms of the omega equation, at

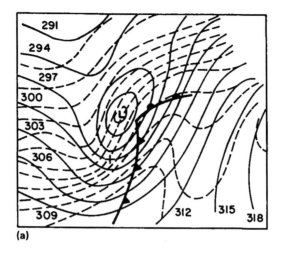

(a)

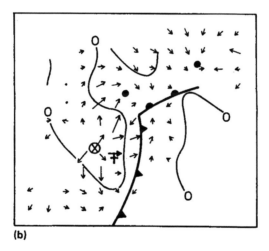

(b)

Figure 14.16 (a) The 700 mb height contours (full curves labeled in dam) and temperature contours at 2 °C intervals for 10 November 1975. Surface fronts are also shown. (b) The Q-vectors (arrows with length proportional to magnitude) for the pattern shown in (a). The thin full curves represent the isopleth of zero Q divergence, and the centers of convergence and divergence, respectively, are denoted by the filled and crossed circles. The cross is referred to in the text. (From Hoskins and Pedder, 1980.)

least without a great deal of experience, because the distribution of Q is difficult to visualize and the divergence of this vector even more so. Rules for interpreting the sign and relative magnitude of the Q-vector and the sense of the transverse ageostrophic circulation can be determined qualitatively from the left-hand rule.

Let us illustrate the fact that fronts are associated with convergence of the Q-vector and therefore with ascent. Consider a horizontal (x, y) plane and the x and y components of Q, given by (14.8c) and (14.18b), at the location of the cross (+) just west of the cold front in Figure 14.16b. Evaluation of the sign of the Q-vector at that point depends on evaluation of the geostrophic wind components and the temperature field. Note that Q_{yp} is vanishingly small at this point because both $\partial u_g/\partial x'$ and $\partial u_g/\partial y'$ are very small along the axis of the cyclone trough. Similarly, Q_{xp} reduces to $-(\partial\theta/\partial y')(\partial v_g/\partial x')$ because $\partial v_g/\partial y'$ is very small. Since $\partial\theta/\partial y'$ is negative and $\partial v_g/\partial x'$ is positive at that point, Q_{xp} is positive. A positive value of Q_{xp} requires that the vector point along the positive x' direction, i.e. toward the east. Further east, in the warm sector, the gradients are weak and, accordingly, the magnitude of Q is small. Thus, $\partial Q_{xp}/\partial x'$ is negative along the front (convergence of Q_{xp}). This diagnosis is in agreement with the pattern of vectors shown in Figure 14.16b in which strong convergence of Q (and implied ascent) is taking place along the cold front. On the cold side of the front, there are zones in which the total rate of change of the temperature gradient following an air parcel increases with time, implying that it is also a region of frontogenesis. Finally, from an inspection of this pattern and from the form of the Q-vector, it is shown that the latter is largest in regions of strong temperature and geopotential height gradients, i.e. where solenoids formed by isopleths of θ and u_g are small.

Problems

14.1 Figure 14.13 depicts a cross section through a lower-tropospheric front. Relabel the v_g isopleths (vertical broken lines) with the opposite signs from those in the figure and answer the following questions:

(a) Indicate with a sketch the streamlines of the ageostrophic circulation in the y', p plane. Indicate the sense of the circulation by using arrows, locating the circulation center.

(b) Is this circulation direct or indirect?

(c) Give a dynamic argument (using the vorticity equation) explaining why the magnitude of the vorticity tends to be so large in the frontal zone, particularly at the surface.

(d) What is the direction of the acceleration of the wind velocity normal to the plane of Figure 14.13 due to ageostrophic motion: at the surface below the frontal zone and at upper levels above the frontal zone?

14.2 Imagine a y', p cross section showing positive values of the geostrophic wind components, v_g and u_g. The gradient of the former points toward the lower left (toward negative y' and increasing p and perpendicular to the x' axis) and the gradient of the latter points toward the top of the diagram (toward decreasing p). What is the sign of the streamfunction and the sense of the circulation at that point? What does this imply about the gradient of potential temperature along the x' axis? What is the sign of the horizontal temperature advection along the x' axis? What is the direction of the acceleration of the total u wind component (along x' axis) normal to the plane of the figure at the level of the u_g maximum (top of figure)?

14.3 Write down an expression for the geostrophic advection of relative thermal vorticity by the thermal wind in the form of a Jacobian.

Further reading

Eliassen, A. 1962. On the vertical circulation in frontal zones. *Geofys. Publ.* **24**, 147–60.

Hoskins, B. J. 1975. The geostrophic momentum approximation and the semigeostrophic equations. *J. Atmos. Sci.* **32**, 233–42.

Hoskins, B. J. and F. P. Bretherton 1972. Atmospheric frontogenesis models: mathematical formulation and solution. *J. Atmos. Sci.* **29**, 11–37.

Hoskins, B. J. and W. A. Heckley 1981. Cold and warm fronts in baroclinic waves. *Q. J. R. Met. Soc.* **107**, 79–90.

Hoskins, B. J. and M. A. Pedder 1980. The diagnosis of middle latitude synoptic development. *Q. J. R. Met. Soc.* **106**, 707–19.

Hoskins, B. J. and N. V. West 1979. Baroclinic waves and frontogenesis. Part II: Uniform potential vorticity jet flows – cold and warm fronts. *J. Atmos. Sci.* **36**, 1663–80.

Keyser, D. and M. A. Shapiro 1986. A review of the structure and dynamics of upper-level frontal zones. *J. Atmos. Sci.* **114**, 452–99.

Sawyer, J. S. 1956. The vertical circulation at meteorological fronts and its relation to frontogenesis. *Proc. R. Soc.* **A234**, 246–62.

Shapiro, M. A. 1981. Frontogenesis and geostrophically forced secondary circulations in the vicinity of jet stream–frontal zone systems. *J. Atmos. Sci.* **38**, 954–73.

15

Upper-tropospheric fronts and jet streaks

Early investigations of upper-tropospheric structure began in the 1920s and 1930s using lightweight retrievable balloons over Europe. This probing of the upper troposphere helped to generalize previous frontal concepts constructed by the Norwegian school a decade earlier. Advent of systematic radiosonde measurements during the 1950s set the stage for understanding the dynamics of fronts and jets. The localized zones of enhanced winds, whose cores of maximum winds are called jets or jet streaks, were found to contribute a substantial fraction of the dynamic forcing associated with synoptic-scale waves. As pointed out in Chapters 5 and 6, jet streaks and frontal zones are intrinsic features in a rotating, baroclinic fluid.

In the lower troposphere, the magnitude of the divergence tends to maximize near the bottom and top boundaries, the latter being less well defined than the former. Divergence associated with these upper-level features and its ageostrophic motion contribute to mass changes in the vertical column and, therefore, to important changes in surface pressure and to the weather in the lower troposphere. Upper fronts have been of particular interest because they are associated with a variety of phenomena not normally referred to as "weather", such as clear-air turbulence, stratospheric–tropospheric interchanges and billow clouds. Therefore, the location and intensity of upper-level fronts are important to aircraft, in the study of cyclogenesis and in understanding stratosophere–troposphere mass exchanges.

Transverse / vertical circulations associated with jet streaks (Fig. 14.1) are often confined to the upper troposphere, except in cases of lower-tropospheric cyclogenesis in which a lowered static stability aids in coupling upper- and lower-level circulations. Upper-tropospheric fronts are more often associated with a strongly sinking branch of the circulation on the western sides of 500 mb troughs (and in the vicinity of wind speed maxima), where there are clear skies below and above the frontal zone. At times, the apparent intensification of jet streaks and upper-tropospheric fronts prior to surface cyclogenesis suggests that these features are a cause, or at least a precursor, of cyclone development. This association of jet streaks with cyclogenesis is most dramatic in cases when the deformation of the tropopause leads to an intrusion of a thin wedge of stratospheric air into the middle and lower troposphere. This process is referred to as *tropopause folding*.

Two types of upper-tropospheric wind speed maxima have been widely investigated, these being the polar front jet, which follows the westerlies at middle latitudes, and the subtropical jet, which is generally confined to latitudes equatorward of about 30°. In both the polar front and subtropical jets, cores of faster-moving winds tend to migrate eastward and become involved in mid-latitude disturbances. Unlike the polar front jet, the subtropical jet is completely confined to the high troposphere and its associated frontal zone joins with the stratosphere at very high levels. Some meteorologists refer to an "arctic front", which may be closely linked to wintertime outbreaks of cold continental air at mid-latitudes. Unlike lower-tropospheric fronts, upper-tropospheric fronts are not associated with outbreaks of cold air from polar latitudes, but function as dynamic entities.

Mechanisms for the formation of upper-troposphere fronts and jet streaks have been proposed since the 1940s, when Namais and Clapp (1949) suggested that the so-called Bergeron confluence in the entrance region of a jet streak (exemplified in Figure 14.12) is marked by transverse/vertical circulations that promote the intensification of the jet streak and the attendant horizontal temperature gradient. Namais and Clapp proposed that upper-tropospheric jets are formed in the confluence near the ridge between an equatorward anticyclonic flow and a cyclonic flow on the poleward side of the ridge. Their model also resembles the example in Figure 14.5, which typically occurs when a ridge builds poleward into a region of zonal or cyclonic flow. In kinematic terms, this results in a confluence pattern in which the isotherms are pushed toward the axis of dilatation, thereby increasing the horizontal temperature gradient. Eliassen showed, however, that changes in both the wind field and the horizontal temperature gradient are intimately related to ageostrophic circulations resulting from force imbalances. These ageostrophic circulations cause an acceleration of the wind speed in the direction of maintaining thermal wind balance.

Upper-tropospheric fronts, jet streaks and the mechanism by which tropopause folding is produced constitute the focus of this chapter. First, jet streaks are re-examined in light of the Sawyer–Eliassen circulation equations. Using rules set forth in Chapter 14 we diagnose the transverse/vertical circulations in the vicinity of jet streaks and upper-tropospheric fronts. These transverse ageostrophic circulations are related to the formation and intensification of the fronts, especially along the tropopause, where the resulting vertical motions promote the intrusion of stratospheric air into the troposphere. Finally, these ideas are tied to the three-dimensional structure of the atmosphere within the framework of the relative-wind model discussed in Chapter 12.

15.1 Transverse/vertical circulations along jet streaks

Jet streaks are often located in the region where the tropopause level slopes sharply upward. The relationship between a sloping tropopause and its horizontal temperature gradient is evident in Figure 12.10c. Clearly, there is an ambiguity in defining the height of the tropopause in the region of large tropopause slope; a break is suggested between the high and low tropopause in the region near the jet. In this "tropopause break" region, the lower-stratosphere tropopause is continuous with the troposphere within a bundle of closely spaced isentropes that slope downward below the jet streak into the middle troposphere. This is the upper-tropospheric front.

The tropopause region itself constitutes a region of strong horizontal temperature gradient in the break where the isentropes slope upward above the higher tropopause over the warm air in the *lower* troposphere and cold air in the high troposphere. In the vicinity of the jet streak, the isentropes are approximately horizontal ($\partial\theta/\partial y' = 0$). Since a jet streak is defined as a region where the vertical gradient of wind speed is zero, the wind speed maximum occurs where the horizontal temperature gradient normal to the jet streak vanishes. This point lies close to where the vertically integrated (or averaged) baroclinicity between the surface (p_b) and the level of the jet streak (p_t) is a maximum. Viewed on a cross section, the jet can be found in the apex of a v-shaped pattern of isentropes, which slope upward above the level of the jet and downward below the level of the jet toward the right-hand side of it. Thus, the wind speed maximum occurs at the flare in the isentropes, near the tropopause break. The strong horizontal temperature gradient in the upper-tropospheric frontal zone below the jet streak comprises a large fraction of the total baroclinicity in the column extending from the surface up to the jet streak. The rapid decrease in wind speed with height just above the jet streak is associated with a reversal with height in the slope of the isentropes in the break region.

Mathematically, the vertically averaged baroclinicity between the surface and the level of the jet streak is written

$$-\frac{R_d}{f_0}\int_{p_b}^{p_t}\frac{\partial T}{\partial y}\frac{dp}{p} = \frac{R_d}{f_0}\frac{\partial \bar{T}}{\partial y}\ln\left(\frac{p_b}{p_t}\right) = u_t - u_b \tag{15.1}$$

where u_t is the velocity component at the top of an arbitrary layer, u_b is the velocity component at the bottom of the layer and $\partial\bar{T}/\partial y$ is the vertically averaged (with respect to the logarithm of pressure) horizontal temperature gradient in the plane of the cross section. If the cross section is taken approximately normal to the wind velocity at the level of the wind speed maximum (which is approximately the case for Figure 12.10c), u_t is almost equal to that in the jet core (whose wind speeds are much greater than u_b).

Thus, the maximum wind speed in the jet is approximately proportional to the average horizontal temperature gradient averaged over the atmospheric column below the jet streak.

15.2   Confluent jets

Transverse/vertical circulations in a plane normal to a jet streak near the tropopause break region, discussed by Eliassen (1962) in his seminal paper, are shown in Figure 15.1. The tropopause is situated close to the level of the jet streak on the positive y' side of the wind speed (u_g) maximum. To the left of the jet streak, the pair of isentropes flare and the tropopause level lies above the higher isentrope. Since the v_g isotachs are vertically oriented $(\partial v_g/\partial p = 0)$, there is no temperature gradient along the direction of the jet

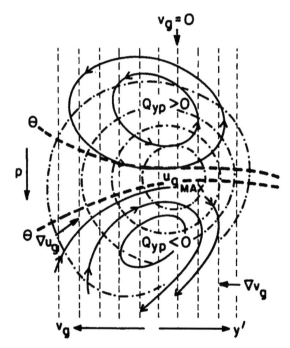

Figure 15.1   Transverse-vertical circulation (streamlines) for the case of pure confluence in the entrance region of a jet streak, $\partial v_g/\partial y'$ negative and $\partial v_g/\partial p = 0$. Chain curves represent the u_g isotachs and broken lines the v_g isotachs increasing toward the left (negative y'). (Note that the location of the $v_g = 0$ isopleth separates positive v_g on its left from negative v_g on its right.) The two thick broken curves are schematic isentropes. The directions of the gradients of u_g and v_g are indicated by the short arrows labeled accordingly, and the sign of Q_{yp} is labeled at the centers of the circulation cells. (Figure derived from Eliassen, 1962.)

streak ($\partial\theta/\partial x' = 0$) and, consequently, no temperature advection along the jet. Although the temperature gradient is not explicitly specified, the distribution of the figure implies that warmer air lies to the left (negative y' direction) below the level of the wind speed maximum and to the right at higher levels. This case is recognizable as one of pure confluence ($\partial\theta/\partial x' = 0$) and corresponds to the entrance region of the jet streak, with an axis of dilatation ($v_g = 0$) coinciding with the jet axis.

Let us now consider the transverse/vertical circulations occurring in the y', p plane of Figure 15.1. Since v_g and u_g isotachs intersect and form solenoids, it is clear that a circulation must exist. The left-hand rule is applied (Ch. 14) by first designating the positive y' axis; this is arbitrarily chosen at the right-hand side of the horizontal axis. Next, one points the fingers of the left hand towards increasing v_g, which is toward the left, and rotates the hand in the direction of the gradient of u_g, which is toward higher values of u_g.

This operation shows that the circulation below the level of the wind speed maximum is everywhere clockwise in the plane of Figure 15.1, i.e. Q_{yp} is less than zero and ψ, the streamfunction, is negative. Minimum ψ is located where the solenoids of u_g and v_g are smallest, which is just below the jet core. This circulation is consistent with the analysis of Figure 14.1, which shows that geostrophic forcing operates to produce a direct circulation in the entrance (rear) quadrants of a jet streak. In this circulation, mass adjustment requires an ageostrophic flow at jet level from the anticyclonic to the cyclonic side of the jet streak in the rear and from the cyclonic to the anticyclonic side of the jet streak in its exit (front) region. At lower levels, the mass adjustment is in the opposite direction from that at jet level. Consequently, maximum surface pressure falls often occur in the left exit (or sometimes the right entrance) region of upper-tropospheric jets (Fig. 15.3).

Above the level of the wind speed maximum, however, the circulation is reversed because the rotation of ∇v_g toward ∇u_g is reversed. *Although Q_{yp} has the opposite sign from that below the jet, the circulation is still direct because of the reversal of the horizontal temperature gradient above the jet core.* Thus, the configuration of v_g and u_g permits two direct-circulation cells to form, one below and one above the jet streak.

For the opposite case, that for the exit region of the jet streak, the winds are diffluent and the gradient of v_g is reversed, as are the circulations, which are indirect. Although consistent with the explanations presented in regard to Figure 14.1, the present argument is more rigorous because it does not require the assumption that air moves through the jet streak.

In reality, the circulations shown in Figure 15.1 are considerably damped in the stratosphere because of the higher static stability. Moreover, actual v_g isotachs are not perfectly vertical or evenly spaced as in this figure. Considering only the tropospheric circulations, as depicted in Figure 15.2, the

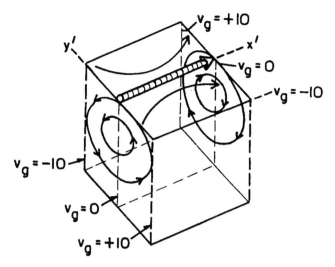

Figure 15.2 Schematic illustration of the four-quadrant model of a jet streak in three dimensions for the case of pure confluence. The tubular arrow represents the jet core, and the two curved arrows the geostrophic wind flow in the horizontal plane intersecting the jet. Vertical broken lines are isotachs of v_g in vertical planes normal to the axis of the jet, values increasing to the right in the entrance region of the jet streak and to the left in the exit region of the jet. The full streamlines are the transverse (ageostrophic)/vertical circulation in the vertical planes normal to the jet axis.

transverse/vertical circulations suggest that the jet streak moves relative to the winds because of divergence patterns produced by geostrophically forced transverse/vertical circulations. Because of the indirect circulation ahead of the jet and the direct one behind it, the horizontal temperature gradient is being created in the exit region of the jet streak, and destroyed in the entrance region. This cancels the effect of diffluence which destroys the horizontal temperature gradient in that region. The net effect is that the temperature gradient ahead of the wind speed maximum increases and behind it decreases, thereby accounting for the progressive increase in vertical wind shear downstream and the movement of the jet, but with a relative movement of air through the jet. Recall from Chapter 6 that troughs and ridges are slowed down by divergence and convergence in the upper troposphere, allowing air to move through the systems from west to east.

The horizontal scale of these circulations is relatively narrow and the depth of the two cells depends on the atmospheric forcing and on the static stability. Sharply defined cloud edges, discussed in Chapter 12, are probably maintained on subsynoptic scales by such narrow zones of enhanced confluence and vertical circulation (descent on the dry side).

It cannot be too strongly emphasized that jet streaks and fronts are not entities that *cause* storms to form or air to descend or ascend, but are consequences of the transverse ageostrophic motions, which are more fundamental processes in producing the weather.

15.3 Variants of the four-quadrant model

In the confluence case (Fig. 15.2), the jet streak is confluent in the entrance and diffluent at the exit. For pure confluence or diffluence, isopleths of v_g are vertical ($\partial v_g / \partial p = 0$) and the circulation assumes the double-cell structure of Figure 15.2, clockwise in the exit and counterclockwise in the entrance region of the jet streak facing downstream.

Vertically oriented isopleths of v_g imply that there is no temperature gradient along the axis of the jet streak ($\partial \theta / \partial x' = 0$) and, therefore, the

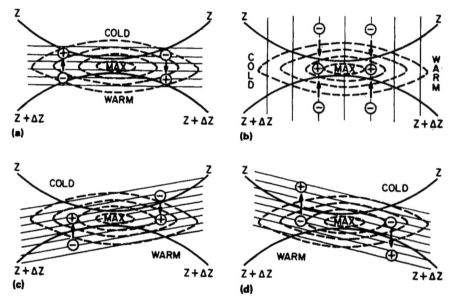

Figure 15.3 Schematic illustration of mid-tropospheric geostrophic deformation fields for a level just below a straight jet stream wind maximum. Heavy full curves are geopotential height contours; heavy broken curves are isotachs; thin full lines are isentropes; arrows indicate the sense of the secondary (ageostrophic) circulation due to the Q-forcing. North- and southward directed Q-vectors indicate thermally direct or indirect circulation forcing, respectively. Plus and minus signs give sense of the vertical motion (ω) for given distributions of Q, with $\omega > 0$ being (+) and $\omega < 0$ being (−). (a) Pure stretching (confluence); (b) pure shear with $\partial \theta / \partial x' > 0$; (c) confluence plus shear with $\partial \theta / \partial x' > 0$; (d) same as (c), but for $\partial \theta / \partial x' < 0$. (Based on a figure by Shapiro, 1982.)

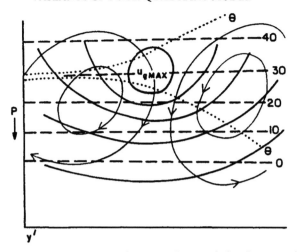

Figure 15.4 Schematic cross section through a jet streak for the case of pure shear (Fig. 15.3b). Thin full curves are streamlines, broken lines are isotachs of v_g, heavy full curves are isotachs of u_g and the two dotted curves are isentropes near the level of the wind speed maximum. The transverse/vertical circulation exhibits two cells, one direct (on right) and one indirect.

geostrophic shear (S_g) is everywhere zero. This is the pure confluence case of Figures 15.1 and 15.2. It is also equivalent to Figure 15.3a. Obviously, other arrangements are possible including the pure shear case in which the isopleths of v_g are horizontal in the plane of the cross section. The two extremes, that of pure geostrophic confluence and that of pure geostrophic shear, are illustrated in Figures 15.3a and b, respectively. Note that the horizontal gradient of temperature implies that a geostrophic vertical wind shear exists and therefore that the maximum wind speed is located above the level of the figure.

The pure shear case, in which isopleths of v_g are horizontal, is shown in Figure 15.4. In this case, cold air upstream and warm air downstream imply cold air advection just below the level of the jet streak. Accordingly, the thermal wind dictates that v_g increases everywhere with height. There is no v_g gradient in the y' direction; thus, $C_g = 0$ because $\partial v_g / \partial y' = 0$. The left-hand rule requires that two cells form on either side of the jet streak, one cell being direct and the other indirect. The direct cell is situated beneath the higher tropopause and in the warmer air. The indirect cell is centered near the lower tropopause. Both cells produce sinking motion in the vicinity of the jet streak. This case conforms to Figure 15.3b.

Figures 15.3c and d are intermediate cases between pure shear and pure confluence. Isotherms lie at an angle with the jet axis, requiring that the v_g isopleths tilt at an angle with the vertical in the y', p plane. (Note that the y' axis is still taken normal to the jet core.) Figure 15.3c shows the temperature gradient from cold to warm along the direction of the jet streak. This implies

411

cold air advection ($\partial\theta/\partial x' > 0$) with $\partial v_g/\partial p$ negative and v_g increasing with height, as is illustrated in Figure 15.5, which is located in the entrance region of the jet streak. The tilted v_g isopleths in the y', p plane, corresponding to a skewed distribution of θ at jet level, permit both confluence and shear. A second circulation cell appears, while the original circulation cell shifts from directly below the axis of the jet streak (Fig. 15.2) to the warm side of the jet streak in the entrance region (Fig. 15.5) and to the cold side of the jet in the exit region.

Figures 15.3c and d correspond, respectively, to cold and warm advection cases; v_g increases with height for cold advection (Fig. 15.3c) and decreases with height for warm advection (Fig. 15.3d). This tilt of v_g isotachs in the vertical and the angle made by the temperature field along the axis of the jet streak are consistent with the presence of both confluence and shear. The principal circulation cell, while displaced from the axis of the jet streak, remains direct in the entrance and indirect in the exit quadrants of the jet

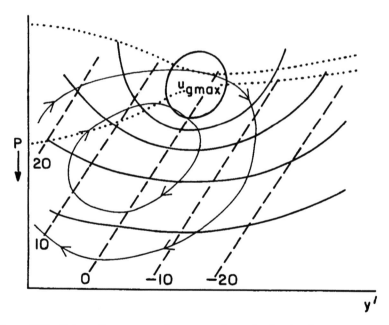

Figure 15.5 Schematic cross section normal to the jet in the entrance region, showing the u_g isotachs (thick full curves), v_g isotachs (broken lines), two isentropes near the level of the wind speed maximum (dotted curves) and the streamlines of the transverse/vertical circulation (thin full curves) for the case of both shear and confluence and cold air advection (corresponding to Figure 15.3c). Here, isotachs of v_g are inclined in the vertical ($\partial v_g/\partial p < 0$), in accordance with the downstream increase in θ ($\partial\theta/\partial x' > 0$); the direct circulation cell is displaced toward negative y'.

streak. With increasing tilt in the v_g isotachs from the vertical, convergence or divergence begins to occur along the jet axis, depending on whether the horizontal distribution of isotherms crosses the jet axis to allow cold or warm advection, as in Figures 15.3c and d, respectively. This also allows increasing descent or ascent to occur just below the jet streak.

15.4 Movement of jet streaks

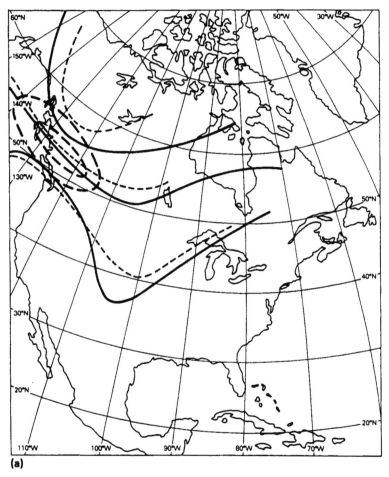

(a)

Figure 15.6 Schematic depiction of the propagation of a mid-tropospheric (~ 400 mb) jet streak through a synoptic-scale wave over a 72 h timespan. Geopotential height contours are shown as heavy full curves, isotachs as heavy broken curves and potential temperatures as thin broken curves. (Based on a figure by Shapiro, 1982.) (a) Formation of initial jet streak in the confluence of mid-latitude and polar latitude currents along the amplifying ridge of a mid-latitude wave; at $t = t_0$.

413

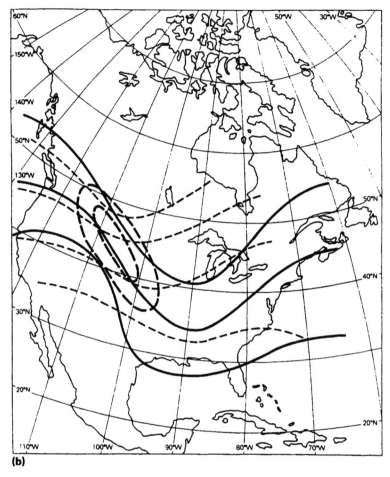

(b)

Figure 15.6 (b) Jet streak situated in the northwesterly flow of an amplifying wave; $t = t_0 + 24$ h.

Because jet streaks are not simply advected with the wind and are subject to rapid changes with time in intensity (depending on the synoptic-scale forcing), the movement of a jet streak is often difficult to follow from one wave to another. Most studies of upper-tropospheric jet structure pertain to the region downstream from the ridge, although some refer to southwesterly flow east of a trough axis near a developing cyclone. The movement of a traveling jet streak and its attendant temperature and geopotential height patterns are presented in Figure 15.6.

Typically, a jet streak and upper-tropospheric front may form at high altitudes in the ridge south of a polar trough, which is precisely the confluent

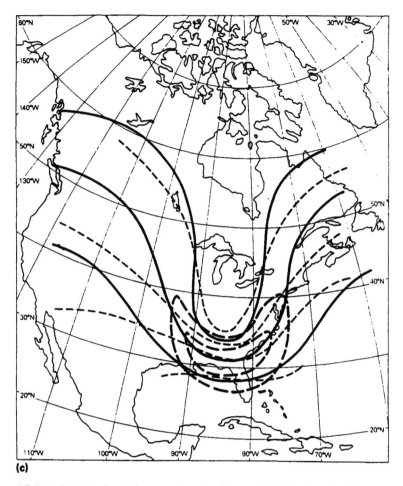

(c)

Figure 15.6 (c) Jet streak at the trough axis of a nearly fully developed wave; at
$t = t_0 + 48$ h.

model proposed by Namias and Clapp (1949) (Fig. 14.5). Formation or
enhancement of the upper-tropospheric jet streak, designated as stage (a)
(pure confluence) in Figure 15.6, occurs along the poleward side of the warm
conveyor belt (Ch. 12), in the region where the latter reaches the high
troposphere. The conceptual model of Namais and Clapp shows confluence
in the rear quadrants of the jet streak between the warm conveyor belt and a
polar trough. Warm advection occurs in the right entrance region of the jet
streak in the middle and upper troposphere. Development of the upper-
tropospheric front accompanies rapid deepening of an upstream trough and
cyclone.

415

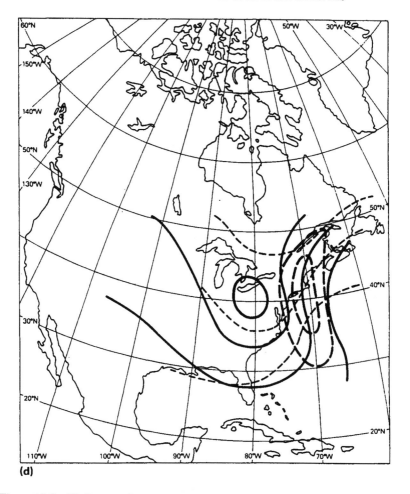

(d)

Figure 15.6 (d) Jet streak situated in the southwesterly flow of wave, which is starting to weaken; at $t = t_0 + 72$ h.

In stage (b) in Figure 15.6, cold advection occurs along the entire length of the jet streak; the configuration of isotherms resembles those in Figure 15.3c, which show cold advection with a mixture of confluence and shear. It follows that this configuration must produce a transverse/vertical circulation in the sense of Figure 15.5. Geostrophic shear is now an important forcing mechanism, coexisting with confluence. Propagation of the jet streak into the downstream trough then accompanies a deepening of the downstream trough (stage (c)). This sequence of events corresponds to the apparent deepening process discussed in Chapter 6 and also to development of the

500 mb vorticity maximum, which is discussed in Chapters 7 and 12. Advection of cold air behind the jet streak into the downstream trough, occurring at stage (b), is consistent with the phase lag between geopotential height and thermal patterns and therefore with wave and surface cyclone development.

In stage (d) of Figure 15.6, the jet streak has propagated into the region between the downstream trough and downstream ridge, and the thermal trough begins to lead the geopotential height trough; the jet streak now lies close to the surface cyclone center, in the region of the dry tongue. Both the trough and the surface cyclone begin to weaken. Warm advection exists below the jet streak. There is shear, but the circulation resembles that for Figure 15.3d.

Figure 15.6 corresponds to the sequence of development described in Chapter 10, in which the jet streak moves to the east side of the trough near the storm center and the wave undergoes barotropic (apparent) weakening and a cessation of the forcing as it moves poleward. The wave also develops a negative tilt in the axis as the cyclone intensifies and subsequently reaches occlusion (e.g. Figs 10.4g & h). (Note that the jet streak in this figure moves through the trough during the intensification phases and arrives at a point just equatorward of the surface cyclone center as the system achieves maximum strength. Simultaneously, jet streak formation or enhancement occurs at high levels in the downstream ridge and the sequence of successive downstream development is re-established.)

15.5 Coupled jet streaks

Consider the exit region of the jet streak in Figure 15.2. At the level of the jet, ageostrophic motion is from the left to the right side of the jet streak, requiring a deceleration in the u component of the wind speed at jet level according to (14.2). Conversely, there is an ageostrophic flow from right to left below the jet, requiring an ageostrophic *acceleration* toward positive x' by air parcels at these lower levels. Thus, at lower levels in the exit region of the jet, air will cross geopotential height lines toward the left of the upper jet (as the result of ageostrophic motion) and accelerate in the direction of the upper-level jet. This may lead to the formation or enhancement of a lower-level jet with a component of motion directed from right to left across the upper jet, as in Figure 15.7. Expressed alternatively in terms of isentropic motion, airflow at lower levels moves up sloping isentropic surfaces beneath the jet streak toward the left in the forward quadrant of the jet. This interaction between lower- and upper-level jet streaks, first proposed by Uccellini and Johnson (1979), is thought to play an important role in the release of potential instability and the production of severe local storms when latently unstable air in the planetary boundary layer is forced to ascend

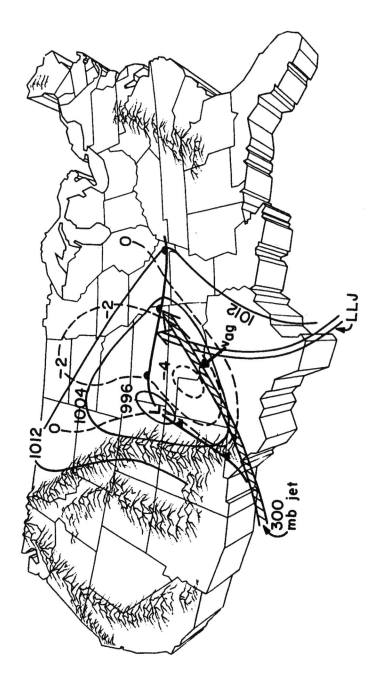

Figure 15.7 Pattern of surface isobars (full lines labeled in mb) and isallobars (broken curves labeled in mb (3 h)$^{-1}$) and fronts around the cyclone over the southern Great Plains of the United States. The hatched double-shafted arrow shows the 300 mb wind speed maximum, the double-shafted arrow shows the position of the low-level jet (LLJ) and the small arrow labeled v_{ag} the direction of the ageostrophic motion at the surface.

in the left front (or sometimes right rear) quadrants of the jet streak.

Whether an upper-level jet streak induces a lower-level one to form probably depends on both strong forcing and lowered static stability beneath the upper jet streak. Thus, the formation of a lower-tropospheric jet streak in response to the transverse/vertical circulations associated with the upper jet streak occurs selectively in regions where the static stability is favorable, such as over the Great Plains of the United States during the spring and summer months (Fig. 9.12).

Figure 15.7 shows a case of an upper-tropospheric jet streak, in which there are lower-tropospheric geopotential height falls (negative isallobars) in its left exit and height rises in its right exit quadrants. A surface isallobaric (ageostrophic) component of motion is directed from the right to the left exit regions. Thus, the surface wind velocity gains a component of acceleration normal and to the right of the ageostrophic velocity. Despite the effects of surface friction, this acceleration may produce a lower-level jet streak. Many cases of cyclogenesis over the central United States exhibit a low-level jet (LLJ), which is centered near 850–900 mb and extends from the Gulf of Mexico toward the north and northwest (Fig. 15.7). There is strong ageostrophic motion at low levels in the vicinity of the low-level jet streak with a velocity component toward surface pressure falls.

When the upper-level jet streak is present, these surface pressure falls tend to occur in its left exit or right entrance quadrants. The low-level jet streak is highly ageostrophic along the direction of the geopotential height contours and cross-isobaric toward lower pressure. The low-level jet is related to orography, low static stability and surface heating, which permit the ageostrophic motions to become very large. In cases where an upper-level jet streak or surface cyclone moves over a region of low static stability due to intense surface heating, the efficiency of the vorticity advections in producing vertical motions and low-level convergence (and therefore a more vigorous isallobaric wind) is enhanced; consequently, the isallobaric minimum intensifies. There is an increase in the ageostrophic component of motion at low levels toward the region of surface pressure falls and an acceleration of the flow at low levels toward the right of the isallobaric wind.

It is not uncommon to find a separate intense negative isallobaric region located southeast of a cyclone over the southern plains of the United States, where, in springtime, there is apt to be low static stability associated with a deep mixing layer (Fig. 9.12). Numerical simulations show that both surface heating and vertical mixing are essential for maintaining and enhancing the low-level jet and that the topography of Mexico (the presence of a high plateau) serves as a block to the tropical easterly flow, causing the latter to turn northward from the Gulf of Mexico (Ch. 16).

15.6   Stratosphere–troposphere exchanges: tropopause folding

Upper-tropospheric frontogenesis, jet streak formation and tropopause distortions are all intimately related. Extrusion of a narrow slice of stratospheric air into the troposphere begins with the formation of a tropopause undulation, in response to strong descent at tropopause level. As shown in Figure 10.8, the development of a tropopause undulation often accompanies rapid upper-level frontogenesis and lower-tropospheric cyclogenesis. In extreme cases, air with stratospheric properties arrives at middle and low levels in the troposphere in these upper-tropospheric fronts.

Intensive investigation of frontal systems, including upper-tropospheric fronts, began during the early 1950s with the work of Reed and Sanders (1953). Later, Danielsen (1968) showed that stratospheric air can be traced by analysis of its particular properties, these being high static stability, high ozone content, high radioactivity, low water vapor content and high potential vorticity. Prior to the 1950s it was thought that the tropopause was a well-defined boundary between the stratosphere and the troposphere and that there was little mass transport across the tropopause, the interchange being slow and diffusive or in the form of synoptic-scale descent. We now know that intrusions of stratospheric air into the tropopause are not haphazard, but occur in narrow regions near upper-tropospheric fronts.

Tropopause folding came to be studied intensively during the 1960s, first in order to describe the transport of stratospheric air into the troposphere and later to determine its connection with upper-tropospheric fronts and jet streaks. Danielsen (1964) noticed that some inversions found at low to middle levels could be traced by isentropic analysis to the tropopause. In one example (Fig. 15.8), he showed that the potential temperatures, within a low-level inversion at Winslow, Arizona (located over the southwestern United States), contained almost exactly the same potential temperatures as in successively higher-level inversions toward the north. Two inversions are found at Ely. The lower one is the bottom of a frontal inversion (and the official tropopause) at 390 mb and the upper one about 70 mb higher. Las Vegas and Winslow also show a 70–100 mb deep layer of stratospheric air, which is characterized by a nearly isothermal lapse rate. These inversions slope downward toward the south, while the true tropopause moves upward from 240 mb at Las Vegas to 220 mb at Winslow and 200 mb at Albuquerque. At Ely, Nevada, the inversion is barely distinguishable from the tropopause, and still further toward the northwest (at Winnemucca), a single inversion constitutes the lower portion of the stratosphere.

Mechanisms for producing tropopause folding, and the relationship of the fold to the dynamics of surface development, became a focus of attention during the 1970s and 1980s. At this same time, various hypotheses con-

420

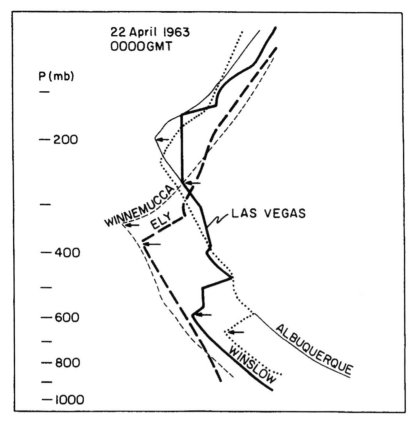

Figure 15.8 Temperature sounding at five locations approximately north to south across an upper-tropospheric front: (From north to south the soundings are at Winnemucca, Ely, Las Vegas, Winslow and Albuquerque, respectively.) Horizontal arrows refer to the base of the frontal inversion or the tropopause. Note that the Albuquerque sounding is discontinued between 600 and 300 mb for clarity. (Based on a figure by Danielsen, 1964.)

cerning the role of potential vorticity and the behavior of the tropopause in the triggering of severe local convection and in cyclogenesis were offered.

Tropopause folding constitutes an intense phase of upper-tropospheric frontal development in which the tropopause undulation collapses. Unlike the surface, the tropopause is not a hard boundary and, consequently, air can be moved into the frontal zone from the stratosphere. Although tropopause undulations are inherent in mid-latitude cyclogenesis, tropopause folds occur less frequently than cyclone development. Most vigorous episodes of tropopause folding take place during winter and spring and are closely associated with strong upper-tropospheric jet streaks. Tropopause folding is also most frequently observed downstream from the ridge, where

421

there is likely to be large-scale descent. Penetration by the stratospheric air is largely within the frontal zone, which is relatively shallow.

Evolution of a tropopause fold is depicted in Figure 15.9, in which an envelope of high static stability, marking the lower boundary of the stratosphere (thick full curve) moves downward with time. A frontal zone is thus formed, which extends from the low tropopause poleward of the jet to low levels on the warm side of the jet approximately within a pair of sloping isentropic surfaces. Descent occurs within and just above the frontal zone. Thermal wind balance demands a progressive increase in the vertical wind shear and in the speed of the jet, as shown by changes in the isotachs.

Upper-tropospheric fronts are marked by high potential vorticity, and potential temperature (implicitly) is the potential *vorticity* (P_θ). The definition of potential vorticity comes from the Ertel vorticity theorem which states that the property

$$P_\theta = \frac{1}{\rho}\, \nabla_p \theta \cdot (\nabla_p \times V + 2\boldsymbol{\Omega})$$

is conserved for frictionless, adiabatic motion. Transforming coordinates, the Ertel theorem can be expressed as a close approximation in the absence of friction as

$$\frac{dP_\theta}{dt} = -\,(\zeta_\theta + f)\, \frac{\partial}{\partial p}\left(\frac{d\theta}{dt}\right) \tag{15.2a}$$

where P_θ, defined in Chapter 12 as

$$P_\theta = -\,(\partial\theta/\partial p)(\zeta_\theta + f) \tag{15.2b}$$

is the potential vorticity (see sec. 12.1). (Note that P_θ is defined in some textbooks as $-g(\partial\theta/\partial p)(\zeta_\theta + f)$.)

Values of P_θ in the stratosphere exceed those in the tropopause by one to three orders of magnitude. Highest values of P_θ are found near the tropopause just above the polar front jet and on the cyclonic side of the jet stream, as shown in Figure 15.10a. The creation of high values of P_θ in the stratosphere can be attributed to diabatic heating due to ozone in the stratosphere and cooling due to long-wave radiation in the troposphere, which is equivalent to a decrease of the diabatic heating term with increasing pressure and therefore a positive value of dP_θ/dt in (15.2a). Observations show that the high values of P_θ are due to the large static stability in the stratosphere. Note, however, that the maximum values in Figure 15.10a occur in the *lower* stratosphere above the wind speed maximum near the entrance to the fold.

The tropopause fold shown in Figure 15.10a developed very rapidly, as

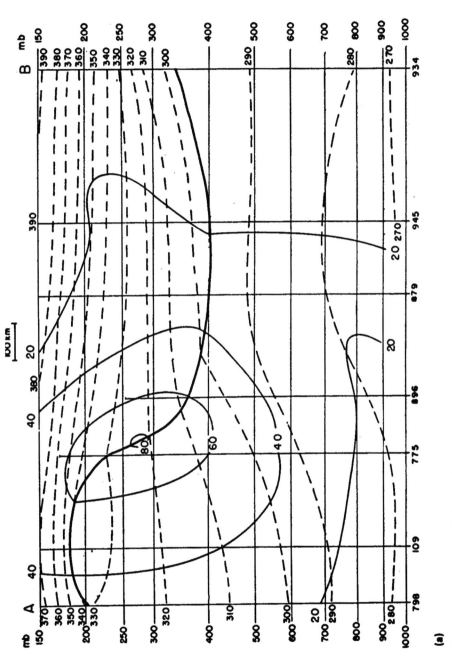

Figure 15.9a (see p. 425 for caption.)

(a)

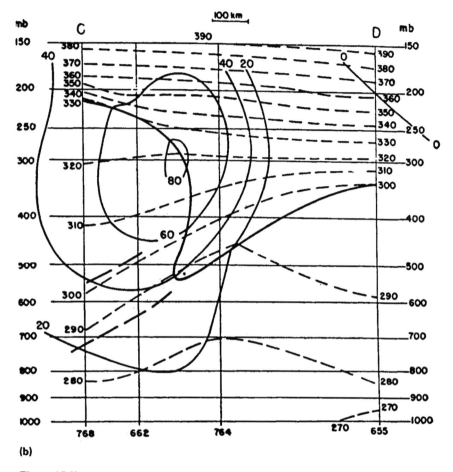

(b)

Figure 15.9b

did the wave, and was marked by strong descent upstream from an expanding cloud shield and in the rear of a jet streak (Fig. 15.10b). High values of potential vorticity moved down the frontal zone into the middle and lower troposphere. Much of the strong descent was concentrated within the tropopause fold. This case is that of the President's Day cyclone (see Ch. 10), which developed explosively along the coast of North America.

Another tracer of stratospheric air is ozone. With the exception of photochemical transformation of industrial and automobile pollutants in the low troposphere, ozone is almost exclusively created in the stratosphere above 20 km. In the situation shown in Figure 15.11, relatively high ozone concentrations were found from these same aircraft measurements in the fold of

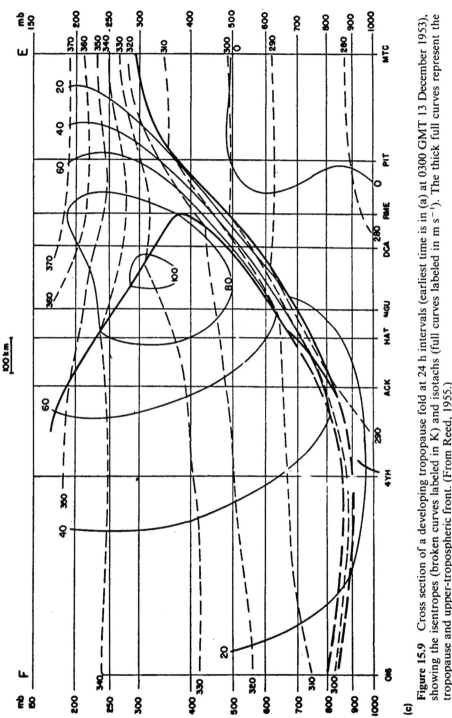

Figure 15.9 Cross section of a developing tropopause fold at 24 h intervals (earliest time is in (a) at 0300 GMT 13 December 1953), showing the isentropes (broken curves labeled in K) and isotachs (full curves labeled in m s⁻¹). The thick full curves represent the tropopause and upper-tropospheric front. (From Reed, 1955.)

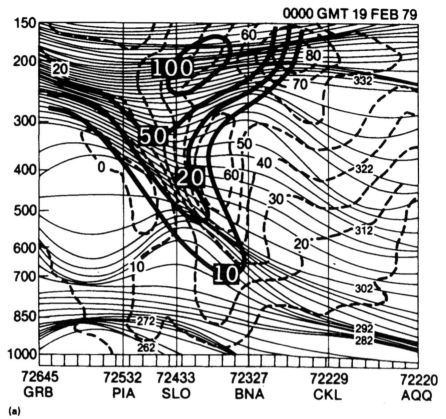

(a)

72645   72532 72433   72327   72229   72220
GRB   PIA   SLO   BNA   CKL   AQQ

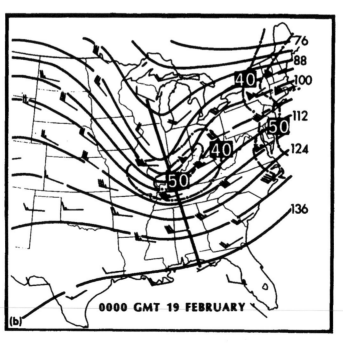

(b)

0000 GMT 19 FEBRUARY

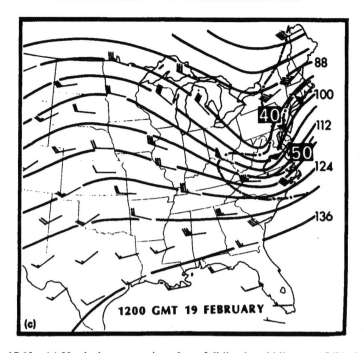

Figure 15.10 (a) Vertical cross section along full line in middle part of (b) showing upper-tropospheric front from Green Bay, WI (GRB) to Apalachicola, FL (AQQ) for 0000 GMT 19 February 1979. Broken curves are geostrophic wind speed (m s^{-1}) computed from the horizontal temperature gradient in the plane of the cross section; thin full curves are isentropes (K) and heavy full curves are potential vorticity, $-(\zeta_\theta + f)(\partial\theta/\partial p)$ (units of 10×10^{-6} K mb^{-1} s^{-1}), shown only for upper portion of frontal zone and stratosphere. (b) Analysis of isotachs (chain curves labelled in m s^{-1}) and Montgomery streamfunction contours (full curves; $100 = 3.100 \times 10^5$ m^2 s^{-1}) for 18–19 February 1979. Wind barbs represent observed speeds (whole barbs denote 10 m s^{-1}). The full line in the middle panel extending northwest to southeast through the entrance region of the jet streak denotes the location of the cross section in (a). (From Uccellini *et al.*, 1985.)

the tropopause, where high values of potential vorticity were also found. The intrusion of ozone-rich air into the troposphere has some importance in that ozone is highly reactive and therefore toxic to humans and animals. In commercial aircraft flights through regions of tropopause folding, air in aircraft cabins can become mixed with ozone to the extent that some passengers maintain that they can sense the presence of the elevated ozone concentrations.

Still another tracer of stratospheric air is radioactivity resulting from the decay of radioactive materials discharged during atomic bomb blasts. During the early 1960s, when testing of nuclear weapons in the atmosphere was commonplace, there was a serious concern about the effects of fallout from

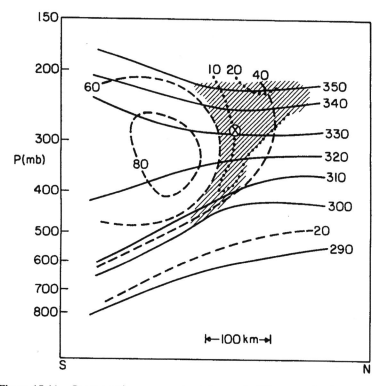

Figure 15.11 Cross section, approximately north (N) to south (S), through a jet streak and tropopause fold. Full curves are isentropes (K) and broken curves are isotachs (m s^{-1}). Shading is region in which potential vorticity exceeds 90×10 K s^{-1} mb^{-1}. Dotted curves represent isopleths of ozone concentration (in units of pphm vol^{-1}). The circled cross denotes the location of the potential vorticity maximum (approximately 120×10^{-5} K s^{-1} mb^{-1}). (Based on a figure by Shapiro, 1978.)

nuclear weapons testing, particularly high-yield bombs that injected large amounts of radioactive debris into the stratosphere. These materials exist in the form of fine dust whose radioactivity decays gradually with time, producing charged particles during the process of decay of primary and secondary isotopes. The most notorious of these radioactive products is strontium-90, which can enter the human biosystem via the ingestion of contaminated grass by dairy cows. The residence time for such atomic debris in the stratosphere tends to be very long, although the most immediate effects of individual atomic explosions may be felt anywhere in the same hemisphere within several days of the event. Aircraft measurements of radioactive decay (beta particles) reveal similar patterns of concentration to those of ozone or potential vorticity, with highest values of beta activity occurring in the lower stratosphere and in the fold region of the tropopause. Radioactivity amounts greatly exceeding normal background levels have also been found at middle levels within the tropopause fold and the upper-tropospheric front.

Some notable episodes of radioactive fallout reaching levels of concern have occurred in the United States as the result of nuclear weapons testing in Asia a week or two prior to the fallout. In one of the most exhaustive studies of its kind, Reiter and his associates were able to show, with the aid of isentropic analysis and a survey of radioactivity measurements made at the surface over the United States (Fig. 15.12), that a dramatic episode of radioactive fallout, which occurred over the southeastern part of the United States (principally Georgia (GA) and the Gulf of Mexico coastal states), was

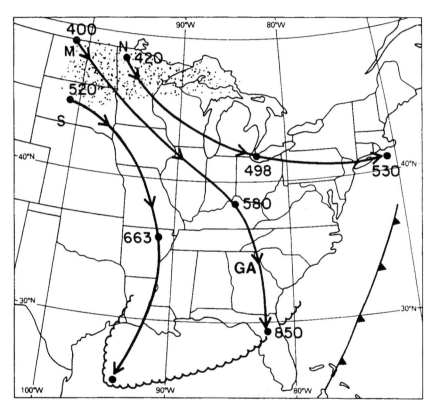

Figure 15.12   Trajectories (ground coordinates) of the 295 K isentropic surface for a 24 h period starting 1200 GMT 22 November 1962. Altitudes of the trajectories (in mb) are shown at 12 h intervals by the three-digit numbers along the respective paths. Stippling denotes region where the trajectories lie above the tropopause and the serrated border is the boundary of the region of elevated surface radioactivity. The location of the cold front at the terminus time is drawn. (Figure adapted from Reiter and Mahlman, 1965.)

the result of the transport of air from near the tropopause to the lower troposphere about a week after the radioactive cloud entered the atmosphere over Siberia. This intrusion of radioactive stratospheric debris took place over a period of about two days, originating in the tropopause fold. Radioactive particles were carried to low levels during a period of strong subsidence to the rear of a major trough in the westerlies.

Reiter's analysis of a 1962 fallout situation, illustrated in Figure 15.12, shows three 24 h trajectories marked by the streamlines and labeled (from north to south) N, M and S, corresponding to an envelope of air that originates west of a trough in the westerlies near the base of the stratosphere on the cyclonic side of an upper-tropospheric jet. Radioactive air descended into the lower and middle troposphere in one day and was subsequently brought to the surface by turbulent exchange within the planetary boundary layer.

Surface measurements showed large concentrations of radioactivity over a relatively small region near the terminus of trajectory M, which originated near the tropopause and reached a level near the top of the planetary boundary layer. (Earlier in the period some high surface concentrations of radioactivity may have been washed down by precipitation, which accompanied the passage of a cold front through the area.) It is possible that high radioactive concentrations were also contained in the trajectory labeled N, but this contamination would have been unable to reach the surface by either large-scale descent or vertical mixing.

A similar fallout incident occurred in 1976 over Pennsylvania: in that case greatly elevated radioactivity was measured at the surface following a period of precipitation on the western side of a cold front. This may have been due to entrainment and fallout of radioactive particles from the dry air into a cold conveyor belt cloud mass.

Such transient and localized episodes of stratospheric intrusions into the lower troposphere, as dramatically illustrated by these radioactive fallout events, are tracers of the organized large-scale meridional and vertical exchange of air streams depicted in the air stream model of Figure 12.20. Thus, air originates in the lower stratosphere west of the trough axis on the poleward (cyclonic) side of the jet and descends anticyclonically to the lower troposphere east of the surface high or crosses the trough and ascends. The origin of this air is generally between 300 and 400 mb within the upper-tropospheric front on the left-hand side of the upper jet streak. Fallout reaches the surface either by intersection of the descending anticyclonic air stream with the planetary boundary layer (S in Fig. 15.12) or by fallout and rain-out on the eastern side of the trough axis.

Although the overall structure of the tropopause fold seems to be preserved following trajectories moving along isentropic surfaces, measurements show that the potential vorticity, ozone and radioactivity diminish by a factor of 3–10 within the stratospheric air as it descends to the lower troposphere; concentrations within this air stream nevertheless remain well above normal values for the lower-troposphere background. This degradation in conservative properties is undoubtedly due to turbulent mixing.

15.7 Tropopause folding and cyclogenesis

The shallow depth of the stratospheric intrusion and the katabatic nature of the upper-tropospheric front are of some meteorological significance, not only because the upper-tropospheric front serves as a conduit for stratospheric air, and thus for radioactive debris and high concentrations of ozone, but also because the folding process is thought to be related to downstream development of cyclones. In such cases (such as the President's Day storm referred to in Chapter 10), tropopause folding may be initiated just downstream from the upstream ridge in the vicinity of a strong jet streak; the jet and the developing tropopause fold may then migrate to the east side of the trough within a day or so.

Some meteorologists speculate that the presence of high potential vorticity in the middle troposphere is intimately related to cyclogenesis on the eastern side of the trough. Development of the fold is arrested and probably reversed by ascent east of the trough. Advection of high potential vorticity at middle levels toward the cyclone center is consistent with the advection of absolute vorticity at 500 mb and of warm air advection near the tropopause (Fig. 10.8) over the storm; thus, tropopause folding may be an aspect of downstream development discussed in Chapter 12. At the present time, operational prediction models are not particularly skillful in capturing either the folding process or downstream development.

Reed's (1955) analysis of tropopause folding, shown in Figure 15.9, remains today a classic example of this phenomenon. Typically, the folding process first occurs in the high troposphere on the cold side of the jet just downstream from the ridge. Reed's example closely parallels the sequence of events depicted in Figure 15.6. Figure 15.13 illustrates the rapid deepening of a cyclone during a tropopause folding episode in Figure 15.9.

Initially, the system is quite weak. After 24 h, however, the folding process and upper-tropospheric frontogenesis and wave amplification are dramatically evident. By the end of the 48 h period (Fig. 15.13c; see Fig. 15.9c), the cyclone is very deep. Reed was able to determine fairly conclusively that the air within this particular frontal zone (Fig. 15.9) was of stratospheric origin because of the large values of potential vorticity within the front. It is worth emphasizing that the tropopause undergoes folding with the amplification of a tropopause undulation and in response to strong descent in the upper-tropospheric front. Neither the front nor the tropopause correspond to a

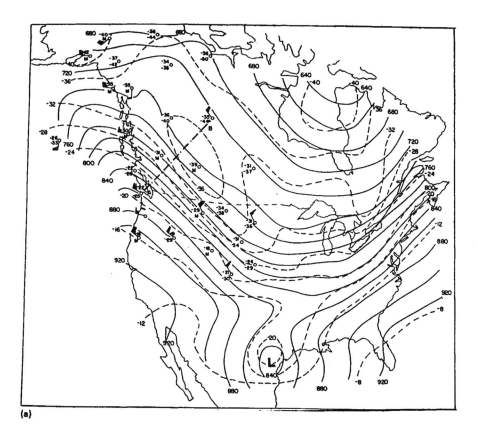

(a)

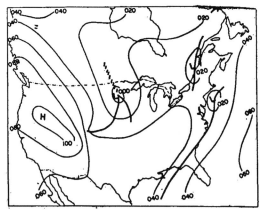

Figure 15.13 (a) Analysis for the 500 mb (top) and 1000 mb (bottom) pressure
surfaces corresponding to Figure 15.9a, for 0300 GMT 13 December 1953. Thin full
curves are height contours (ft); thin broken curves in the top figure are isotherms (°C).
Chain line (top) denotes the position of the cross section appearing in Figure 15.9a.

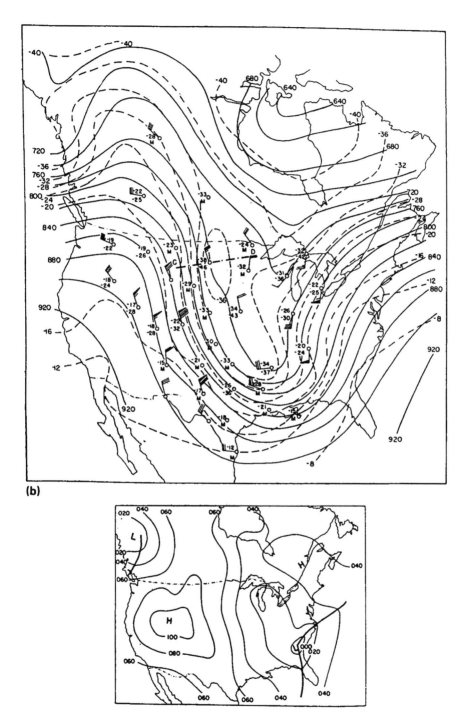

Figure 15.13 (b) Same as (a), except for 0300 GMT 14 December 1953 and corresponding to Figure 15.9b.

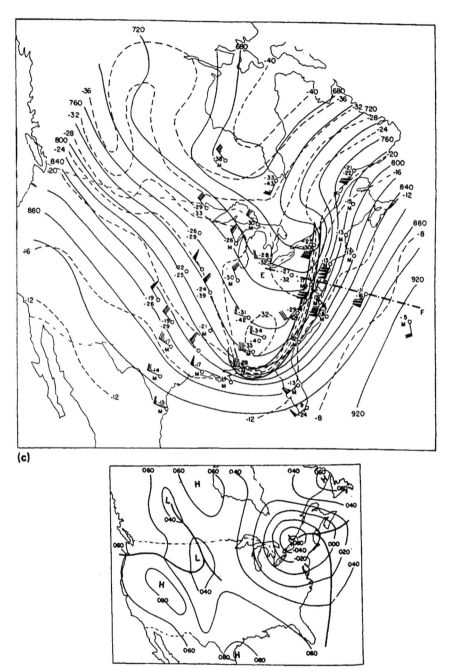

Figure 15.13 (c) Same as (a), except for 0300 GMT 15 December 1953 and correspond-ing to Figure 15.9c. The location of the upper-tropospheric frontal zone is shown by full curves enclosing the region of large horizontal temperature gradient in the top figure. (From Reed, 1955.)

unique isentropic value, nor do the isentropes become distorted to the extent that the static stability becomes negative. The fold and the front are defined in terms of discontinuities in the *gradient* of potential temperature, static stability and wind speed and not in the orientation of the isentropes, which always slope upward toward the cold air. The tropopause level itself is ambiguous in the vicinity of the fold.

Because the stratosphere is a region of large static stability, vertical motions tend to be highly damped above (but may be rather large near) the tropopause. Accordingly, when the air is descending, large vertical gradients in vertical motion can occur. Kinematically, this produces a compression of the isentropes leading to the formation of a frontal zone. This descent tends

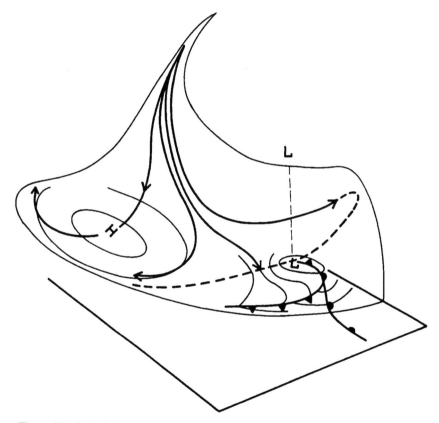

Figure 15.14   Relative trajectories of extruded stratospheric air during tropopause folding (full streamlines within plane of isentropic surface are reaching broken curve). Thin full curves denote surface isobars and the symbols L and H show the location of the surface or 500 mb cyclone or anticyclone centers. (Based on a figure by Danielsen, 1964.)

to concentrate the temperature gradients (both vertical and horizontal) in the upper troposphere below the jet streak. The fact that the tropopause is not a hard boundary allows it to become distorted in response to the dynamic forcing of descent. The tropopause moves downward and is subsequently extruded into a narrow frontal zone.

The relative wind isentropic model suggests that high values of potential vorticity are brought from the lower stratosphere poleward of the jet in the ridge to the lower and middle troposphere in the upper-tropospheric front. Movement of stratospheric air along isentropic surfaces from near the tropopause break to the vicinity of the surface cyclone (Fig. 15.14) results in the intrusion of high values of absolute vorticity upstream from the flow. Conversion of high values of potential vorticity to high values of absolute vorticity occur when the vertical stretching of the frontal surfaces by vertical motion causes the static stability to diminish according to the law of conservation of potential vorticity equations (15.2)). Consequently, the development of the 500 mb wave, the enhancement of the jet streak and the subsequent development of the surface cyclone are all interrelated processes. The time lag between initiation of strong 500 mb vorticity advection and tropopause folding and the formation of an intense surface cyclone may be on the order of hours to a day or more, thereby providing the forecaster with some ability to predict surface cyclogenesis by inspection of tropopause structure.

15.8 Models of tropopause folding and upper-tropospheric frontogenesis

We now turn our attentions to the dynamics of the tropopause fold. The main idea is that strong descent near the jet is a prerequisite for tropopause folding. Danielsen (1964) was among the first to propose a theory of upper tropospheric folding completely consistent with frontal dynamics and with observations. Shapiro (1981) later modified this theory to include a dynamic foundation based on the approach of Sawyer and Eliassen. Danielsen postulated that the circulation accompanying the fold resembled that in Figure 15.15. This figure shows two transverse/vertical circulations cells, one direct and the other indirect, respectively centered on the warm and cold sides of the tropopause break. The two cells produce a confluence zone in the vertical plane. Note that the differential vertical motion requires that isentropic surfaces become compressed to form a frontal zone. Subsidence is maximized on the warm side of the frontal zone in the middle and upper troposphere. The warm side of the upper-tropospheric front is located directly beneath the jet. *The configuration of vertical motion results in a frontogenetic tilting term in (13.2), rather than a frontolytic one, which is the normal case for surface fronts.*

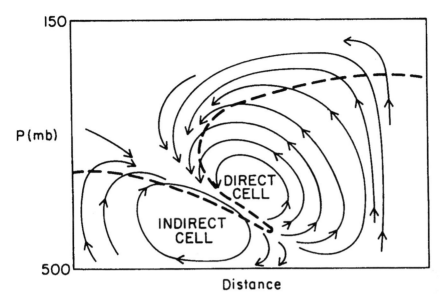

Figure 15.15 Schematic illustration of transverse/vertical circulations conducive to upper-level frontogenesis and tropopause folding. The broken curve represents tropopause and boundary of upper tropospheric front. (Based on a figure by Danielsen, 1968.)

The reason for the extrusion of stratospheric air directly into the upper-tropospheric front beneath the jet streak is explainable in terms of the development of transverse/vertical circulations in that region. Consider Figure 15.4. In this extreme case of pure shear, isopleths of v_g are horizontal, implying a gradient of temperature along the jet (cold upstream) and, therefore, geostrophic cold air advection. The left-hand rule shows that geostrophic shear produces a transverse/vertical circulation with sinking motion beneath the jet streak. This is the case for Figure 15.3b.

More typical, however, is a mixture of both shear and confluence (Fig. 15.5), which maximizes the descent of stratospheric air poleward of the jet into the troposphere. In fact, the frontogenesis process concentrates v_g and u_g solenoids within a narrow circulation ellipse inside the frontal zone. *Both shear and cold air advection along the jet streak are consistent with the extrusion of stratospheric air into the front poleward of the tropopause break and subsequently toward lower levels beneath the jet axis.*

For pure confluence (Fig. 15.3a), isotherms lie parallel to the x axis and the transverse/vertical circulation consists only of a single cell. This arrangement, however, does not allow air to be forced into the front from the stratosphere between the high and low tropopause. Danielsen's model permits a distribution of isotherms approximately parallel to the jet axis, but

437

with a component of the temperature gradient along the jet. This arrangement, which implies a shearing component and cold air advection, permits the formation of two circulation cells. The vertical cross section of u_g and v_g for a case of pure shear is shown in Figure 15.4. Note that the shearing pattern requires cold air advection along the jet for the distribution of isotherms shown in Figure 15.3c, and that the circulation cells are situated such that air is forced downward in the region between the high and low tropopause.

Numerical simulations of upper-tropospheric frontogenesis with confluence alone, as proposed by Namais and Clapp (1949), can generate only the initial stages of folding. Figure 15.16 shows the results of mathematical simulations of tropopause folding, starting with an initial baroclinic atmosphere and allowing only confluence to take place. By varying the geometry of the tropopause, Hoskins and his associates (Hoskins and Bretherton 1972) were able to show that this type of confluence pattern can produce a structure resembling the early stages of a tropopause fold. Absence of a distinct tropopause fold and upper-level frontogenesis in Figure 15.16 is related to the location of the transverse/vertical circulation, which is centered about the mid-tropospheric frontal zone.

The main shortcoming of these early numerical simulations is that they do not allow the upper-tropospheric front to descend below the mid-troposphere. When cold advection is allowed to occur (corresponding to non-zero shearing), the circulation cells are shifted such that the strongest descent occurs below the jet and in the warm air. The average circulation over the domain may be direct, although one cell is indirect and one cell is direct. Models in which cold advection (shear) and confluence are allowed show that the fold descends to the lower troposphere, and an upper-tropospheric front slopes downward toward the warm air below the jet.

A decade after Hoskins' pioneering paper, Shapiro (1981) extended the hypothesis of Danielsen (1968) and proposed that cold air advection along the jet is essential for tropopause folding. Shapiro's model is essentially that in Figure 15.5, except that it substantiates the Danielsen proposition with a set of numerical simulations based on a real case of tropopause folding. The basic premise is that confluence is insufficient to create a real fold, which would occur when there is cold advection and shear along the jet axis.

Shapiro shows that tropopause folding occurs when both confluence and shear are permitted along the jet. The tilt in the v_g isopleths in Figure 15.5 permits both confluence and shear according to the left-hand rule, which requires two cells that join to form a zone of strong subsidence just below the jet. Numerical solutions for this pattern show two cells, one direct and one indirect, which join in the low stratosphere just below the jet. Keyser and colleagues offered a detailed numerical investigation in the mid-1980s, verifying the Danielsen–Shapiro hypotheses that shear and cold air advection

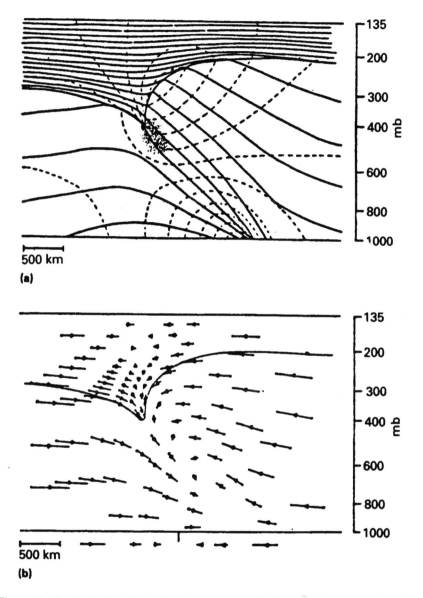

Figure 15.16 Analytical simulation of tropopause folding and frontogenesis at the tropopause and at the ground. (a) Wind speed every 10 m s^{-1}, broken curves; potential temperature every 7.8 K, full curves; tropopause, heavy full curves; small region of shading in the figure denotes a Richardson number less than 1.0. (b) Cross-front motions (geostrophic plus ageostrophic) for (a). Confluence is shown below the lower surface. (Based on a figure by Hoskins and Bretherton, 1972.)

along the jet are necessary to produce two circulation cells at the correct location. The thermodynamically indirect cell enables the tilting effect to be frontogenetic, but the thermodynamically direct cell generates the kinetic energy associated with the acceleration of the wind component parallel to the front.

Normally, both confluence and shear are present along the jet. In fact, initiation of a deep tropopause fold appears to coincide with a change from pure confluence in the ridge of the long-wave system, in which the jet is being formed along the poleward side of a conveyor belt, to a combination of confluence and shear downstream from the ridge. This change in deformation corresponds to the propagation of the jet from the ridge, and to the initiation of a deeper thermal and geopotential trough further downstream,

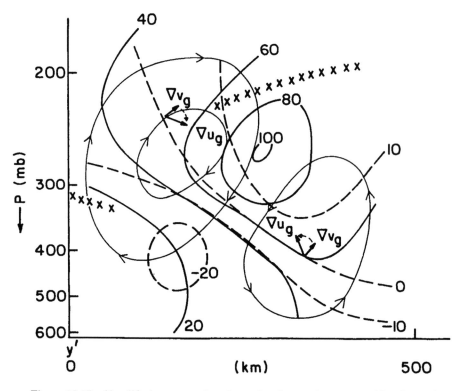

Figure 15.17 Simplified cross section through a jet on the eastern side of a major trough for a case analyzed by Shapiro (1981). Isotachs of v_g are shown as broken curves, those of u_g as full curves and labeled in m s^{-1}. The tranverse/vertical circulation is indicated by the thin full streamlines. The direction and approximate magnitude of the gradients of u_g and v_g and the sense of rotation using the left-hand rule are indicated at selected locations. The tropopause is marked by the line of crosses. (Based on a figure in Shapiro, 1981.)

an idea that is consistent with the discussion of downstream development introduced in Chapter 12 and with the progression of events depicted in Figure 15.6.

Keyser also shows that the geostrophic momentum and even quasi-geostrophic versions of the circulation equations are capable of describing the two-cell circulations, although they are best delineated by the primitive equations. Similarly, they are also better represented and more intense in the geostrophic momentum than the quasi-geostrophic equations. Moreover, although curvature of the front is explicitly absent from the numerical investigations using the geostrophic momentum theory, the results to date suggest that the essential elements in tropopause folding and upper-level frontogenesis are contained within the two-dimensional straight-flow models. Deviations between two-dimensional straight-front analyses and that which would be obtained from a three-dimensional analysis for curved fronts appear to be due to the importance of the along-front super-geostrophic wind component, which augments the cross-front circulation and the displacement of the circulation cell toward the warm air. Another aspect not considered in the diagnostic analyses are the feedbacks between the vertical motions and vorticity tendency. In particular, the tilting effect appears to enhance the vorticity field substantially, which contributes in turn to an enhancement of the ageostrophic circulation.

Let us once more return to the four-quadrant model discussed first in Chapter 14 and depicted in Figure 15.2. This arrangement of vertical motion is valid only for straight jets in which the isotherms lie parallel to the jet. Generally, however, the isotherm distribution is rather more complex, especially during tropopause folding, which demands the presence of shear and cold advection along the jet. Moreover, solutions to the circulation equation (14.11) show that the zone of important transverse/vertical motion is relatively narrow, extending over just a few hundred kilometers (e.g. Fig. 15.17), which is the same scale as the width of the jet streak. Thus, application of the four-quadrant model in diagnosing the vertical motion patterns is limited by scale and by knowledge of the temperature field surrounding the jet.

15.9 Turbulence and mixing in upper-level frontal zones

It is well known that upper-tropospheric fronts above and below jets are preferred regions for turbulence. Typically, turbulence is encountered in the region just below and just above the level of the jet. The likelihood of turbulence in vertically sheared flow is expressed by the Richardson number. When the Richardson number, defined as $Ri = (g/\theta)(\partial\theta/\partial z)/(\partial V/\partial z)^2$, is small (large shear and small buoyancy), the destabilizing effect of vertical wind shear is better able to overcome the restoring effect of buoyancy, resulting in a turbulent exchange of air in the vertical.

Figure 15.16 shows that turbulence (associated with relatively low Richardson numbers) is possible near the tropopause fold in this simulation. Flights through jet streaks for the purpose of measuring their dynamic properties and chemical constituents show that the scales of the lateral and vertical wind shears are exceedingly narrow, the former being on the order of 20 m s^{-1} over less than 100 km and the latter corresponding to roughly half the speed of the jet over a distance of about 100 mb (Fig. 15.11). These favorable zones for clear-air turbulence correspond to strata where the Richardson number is small. In strong jets (speeds exceeding 60 m s^{-1}), turbulence can be quite severe.

The impact of turbulence on the structure of frontal zones can be considerable. It has been noted, for example, that potential vorticity exhibits a maximum, and therefore a source, in the lower stratosphere near the entrance of the upper-tropospheric front (Fig. 15.11). On the other hand, ozone, which is almost exclusively created at altitudes of 20–40 km, shows no such localized maximum in the neck of the front, suggesting a continuous downward transport of ozone from its source higher in the stratosphere. Moreover, although potential vorticity is a conservative property in the absence of diabatic heating or cooling, the fact that potential vorticity decreases following the movement of parcels from the low stratosphere to the middle troposphere on isentropic surfaces demonstrates that the potential vorticity is not conserved in reality. This degradation in potential vorticity with time is due to vertical and horizontal mixing.

The maximum of potential vorticity at the base of the stratosphere suggests that its high values are created locally. Equation (15.2) states that potential vorticity can be created or destroyed if there is a vertical variation in the diabatic heating rate ($d\theta/dt$). Turbulent mixing can alter the vertical distribution of θ and therefore produce a non-zero heating rate at any given level. Consider a vertical distribution of potential temperature, as illustrated in Figure 15.18. The figure illustrates that an initially stable lapse rate is modified by vertical mixing within the layer of mixing. These layers eventually become well mixed (adiabatic lapse rate). Since the potential vorticity is equal to $-(\partial\theta/\partial p)(\zeta_\theta + f)$, and $(\zeta_\theta + f)$ is always positive, P_θ increases with time following a parcel where $(\partial\theta/\partial p)(d\theta/dt)$ is negative or where the diabatic heating rate increases with height. Figures 15.18 and 15.19 illustrate that turbulence brought about by the mixing within the narrow zone of strong vertical wind shear above and below the jet creates high values of potential vorticity. Breakdown of a smooth current into turbulent eddies occurs in the presence of strong vertical wind shear just below and just above the jet axis, the enclosed stippled region in Figure 15.19. Low Richardson numbers are an inherent property of upper-tropospheric fronts and their attendant jets, and the mixing is most intense in the frontal zone. Such mixing, however, serves to entrain tropospheric air into the frontal zone and, accordingly, the stratospheric properties of the air within the frontal zone are constantly being diluted.

Because the jet axis constitutes a region of zero vertical wind shear, the vertical mixing across the jet itself is inhibited. It is not surprising in view of Figure 15.18 that mixing can produce multiple strata of high and low potential vorticity and stability. The breakdown of the current into turbulent eddies is manifested by billow clouds and by gravity waves known as Kelvin–Helmholtz waves, which are generated at the point where the Richardson number falls below 0.25. The occurrence of mechanical mixing by turbulence tends to slow the vertical wind speeds in general, although locally the process can create larger vertical wind shears that will not be rapidly destroyed by turbulence where mixing simultaneously creates a barrier in the form of an inversion. Mixing, however, does not destroy kinetic energy but transfers a certain amount of that kinetic energy from the smooth synoptic-scale motion to scales of turbulence where the energy is dispersed in the form of gravity waves with wavelengths less than 1 km.

It seems quite likely that vertical mixing can serve to sharpen the tropopause inversion or to create multiple tropopause inversions and also to

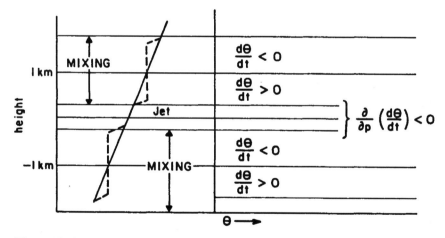

Figure 15.18 Schematic representation of the effects of turbulent mixing on the vertical profile of potential temperature and the tendency of potential vorticity in the vicinity of a jet. On the left is the distribution of potential temperature before (full line) and after (broken lines) mixing has occurred. (The mixing takes place above and below the jet within the vertical segments labeled 'mixing', in conformity with Fig. 15.9.) Indicated on the right is the sign of the vertical distribution of diabatic heating at the jet level (negative) and that of the diabatic contribution to the tendency of potential vorticity in equation (15.2). (After Shapiro, 1976.)

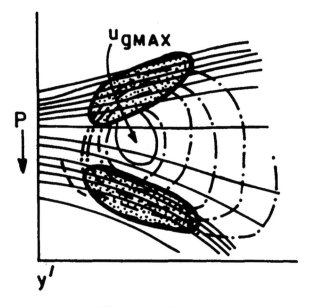

Figure 15.19 Isentropes (thin full curves) and isotachs of u_g (chain curves) through a jet showing region in which there is turbulence (stippled region enclosed in full curves). (After Shapiro, 1976.)

sharpen the upper-tropospheric front. In the vicinity of the tropopause fold, dust, water vapor, anthropogenic pollutants and other surface-derived constituents are introduced into the stratosphere by vertical mixing. What is not yet clear is how significant is the effect of turbulent mixing on tropopause folding.

Problem

15.1 Consider the distribution of v_g and u_g in the vertical (y', p) cross section of Figure 15.20.

(a) Show that this configuration of geostrophic wind components is compatible with tropopause folding by drawing in some schematic streamlines of the ageostrophic/vertical circulation in this plane. Locate the centers of the circulation cells, show their sense of rotation and indicate the strength of this circulation by spacing the circulation lines at a distance roughly inversely proportional to the strength of the ageostrophic/vertical motions.

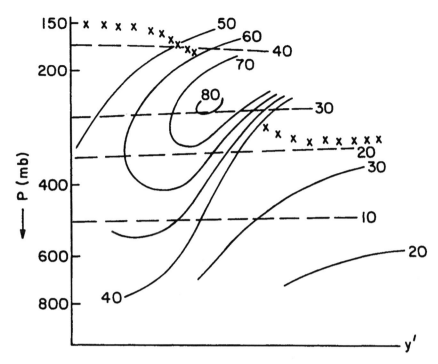

Figure 15.20 Cross section showing isotachs of u_g (full curves labeled in m s^{-1}) and v_g (broken lines) in the vicinity of a jet. The lines of crosses represent the tropopause.

(b) Sketch (perhaps on a piece of tracing paper overlying the figure) some isentropes, showing the probable location of the upper-tropospheric front. Are the circulation cells direct or indirect?

(c) Indicate a region below 400 mb where the potential vorticity is likely to be a maximum and where the Richardson number is likely to be less than 1.0 (i.e. where turbulence is favored).

(d) Demonstrate with an equation or argument that there is geostrophic cold air advection along the x' direction.

Further reading

Brill, K. F., L. W. Uccellini, R. P. Burkhart, T. T. Warner and R. A. Anthes 1985. Numerical simulations of a transverse indirect circulation and low-level jet in the exit region of an upper level jet. *J. Atmos. Sci.* **42**, 1306–20.

Danielsen, E. F. 1964. *Report on Project Springfield*, DASA 1517, HQ, Defense Atomic Support Agency. Washington, D.C. 20301, 97 pp.

Danielsen, E. F. 1968. Stratospheric–tropospheric exchange based on radioactivity, ozone and potential vorticity. *J. Atmos. Sci.* **25**, 502–18.

Eliassen, A. 1962. On the vertical circulation in frontal zones. *Geofys. Publ.* **24**, 147–60.

Hoskins, B. J. 1971. Atmospheric frontogenesis models: some solutions. *Q. J. R. Met. Soc.* **97**, 139–53.

Hoskins, B. J. 1972. Non-Boussinesq effects and further development in a model of upper tropospheric frontogenesis. *Q. J. R. Met. Soc.* **29**, 11–37.

Hoskins, B. J. and F. P. Bretherton 1972. Atmospheric frontogenesis models: mathematical formulation and solution. *J. Atmos. Sci.* **29**, 11–37.

Keyser, D. and M. J. Pecnick 1984. Diagnosis of ageostrophic circulations in a two-dimensional primitive equation model of frontogenesis. *J. Atmos. Sci.* **41**, 1283–305.

Keyser, D. and M. J. Pecnick 1985. A two-dimensional primitive equation model of frontogenesis forced by confluence and horizontal shear. *J. Atmos. Sci.* **42**, 1259–82.

Keyser, D. and M. J. Pecnick 1987. The effect of along-front temperature variation in a two-dimensional primitive equation model of surface frontogenesis. *J. Atmos. Sci.* **44**, 577–604.

Keyser, D., M. J. Pecnik and M. A. Shapiro 1986. Diagnosis of the role of vertical deformation in a two-dimensional primitive equation model of upper-level frontogenesis. *J. Atmos. Sci.* **43**, 839–50.

Kocin, P. J., L. W. Uccellini and R. A. Petersen 1988. Rapid evolution of a jet streak circulation in a pre-convective environment. *Meteor. Atmos. Phys.* **35**, 103–38.

Namais, J. and P. J. Clapp 1949. Confluence theory of the high tropospheric jet stream. *J. Meteor.* **6**, 330–6.

Newton, C. W. and A. Trevisan 1984. Clinogenesis and frontogenesis in jet-stream waves. Part I: Analytical relations to wave structure. *J. Atmos. Sci.* **41**, 2717–34.

Reed, R. J. 1955. A study of a characteristic type of upper-level frontogenesis. *J. Meteor.* **12**, 226–37.

Reed, R. J. and E. F. Danielsen 1959. Fronts in the vicinity of the tropopause. *Arch. Meteor. Geophys. Bioklim.* **A11**, 1–17.

Reed, R. J. and F. Sanders 1953. An investigation of the development of a mid-tropospheric frontal zone and its associated vorticity field. *J. Meteor.* **10**, 338–49.

Reeder, M. and R. K. Smith 1987. A study of frontal dynamics with applications to the Australian summertime "Cool Change". *J. Atmos. Sci.* **44**, 687–705.

Reiter, E. R. 1975. Stratospheric–tropospheric exchange processes. *Rev. Geophys. Space Phys.* **13**, 459–74.

Reiter, E. R. and J. D. Mahlman 1965. Heavy radioactive fallout in the southern United States, November, 1962. *J. Geophys. Res.* **70**, 4501–20.

Shapiro, M. A. 1970. On the applicability of the geostrophic approximation to upper-level, frontal-scale motions. *J. Atmos. Sci.* **27**, 409–20.

Shapiro, M. A. 1976. The role of turbulent heat flux in the generation of potential vorticity in the vicinity of upper-level jet stream systems. *Mon. Wea. Rev.* **104**, 892–906.

Shapiro, M. A. 1978. Further evidence of the mesoscale and turbulent structure of upper-level jet stream frontal systems. *Mon. Wea. Rev.* **106**, 1100–11.

Shapiro, M. A. 1980. Turbulent mixing within tropopause folds as a mechanism for the exchange of chemical constituents between the stratosophere and troposphere. *J. Atmos. Sci.* **37**, 994–1004.

Shapiro, M. A. 1981. Frontogenesis and geostrophically forced secondary circulations in the vicinity of jet stream–frontal zone systems. *J. Atmos. Sci.* **38**, 954–73.

Shapiro, M. A. 1982. Mesoscale weather systems of the Central United States. CIRES, NOAA/University of Colorado, Boulder, CO 78 pp.

Uccellini, L. W. 1986. The possible influence of upstream upper-level baroclinic processes on the development of the QE II storm. *Mon. Wea. Rev.* 114, 1019–27.

Uccellini, L. W., K. F. Brill, R. A. Petersen, D. Keyser, R. Aune, P. J. Kocin and M. des Jardins 1988. A report on the upper-level wind conditions preceding and during the shuttle Challenger (STS 51L) explosion. *Bull. Am. Met. Soc.* 1248–65.

Uccellini, L. W. and D. R. Johnson 1979. The coupling of upper and lower tropospheric jet streaks and implications for the development of severe convective storms. *Mon. Wea. Rev.* 107, 682–701.

Uccellini, L. W., D. Keyser, K. F. Brill and C. H. Wash 1985. The President's Day cyclone of 18–19 February 1979: influence of upstream trough amplification and associated tropopause folding on rapid cyclogenesis. *Mon. Wea. Rev.* 113, 941–61.

Uccellini, L. W. and P. J. Kocin 1987. The interaction of jet streak circulations during heavy snow events along the east coast of the United States. *Wea. Forecast.* 1, 289–308.

Uccellini, L. W., P. J. Kocin, R. A. Petersen, C. H. Wash and K. F. Brill 1984. The President's Day cyclone of 18–19 February, 1979: synoptic overview and analysis of the subtropical jet streak influencing the pre-cyclogenetic period. *Mon. Wea. Rev.* 112, 31–55.

Uccellini, L. W., R. A. Petersen, K. F. Brill, P. J. Kocin and J. J. Tuccillo 1987. Synergistic interactions between an upper-level jet streak and diabatic processes that influence the development of a low-level jet and a secondary coastal cyclone. *Mon. Wea. Rev.* 115, 2227–61.

447

16

Mid-tropospheric fronts, elevated mixed layers and the severe storm environment

Two general classes of fronts, those which form near the surface and those which develop in association with deformation of the tropopause in the upper troposphere, have been discussed in the previous three chapters. Surface fronts are usually most intense not far from the surface; upper-tropospheric fronts are strongest not far below the troposphere. In general, both types of front are weaker in the middle troposphere than near their respective boundaries (the surface or the stratosphere). The decreases in temperature, wind and vorticity gradients with height from the surface are related to the tilting effect (13.2), which weakens horizontal temperature gradients. Both surface and upper-tropospheric frontal circulations exert a net direct effect on kinetic-energy generation, although both may also contain locally indirect circulation cells. In upper-tropospheric fronts, however, the descending vertical motion in clear air just below the jet and near the tropopause are essential factors governing frontogenesis and tropopause folding, whereas lower-tropospheric fronts seem to be closely associated with ascent and precipitation. Nevertheless, both types of fronts usually become quite weak at middle levels.

Unlike the lower and upper troposphere, the middle troposphere is not a region marked by strong fronts, except those which extend into the middle or upper troposphere from below or above and which are rather more intense either below or above the middle levels. Preferred regions for the formation of lower-tropospheric fronts, viewed in the relative wind isentropic system discussed in Chapter 12, favor the eastern sides of troughs, where confluence and shear are strong along the lateral boundaries of ascending and descending air streams. Upper-tropospheric fronts tend to form in the confluence along the poleward edge of the conveyor belt near the downstream ridge.

It is not surprising, therefore, that scant reference has been made to middle-tropospheric fronts as a separate dynamic entity divorced from surface or jet stream influences. One type of mid-level front that has been treated in some detail because of its effect on severe convective storms is the so-called elevated mixed layer front. The formation of this type of front is associated with the differential advection of a deep mixing layer from the planetary boundary layer in a hot, arid (and often elevated) region over a shallow, cooler and more moist mixing layer. The process thereby creates an inversion over the cooler air, which is capped by the warm, elevated, mixed layer. A front forms along one lateral boundary of the plume of the dry, elevated mixed layer whose origins lie within an arid planetary boundary layer. Large vertical potential temperature gradients are formed along the base of the elevated mixed layer. A surface temperature or front or an abrupt change in terrain elevation may mark the boundary along which the surface mixing layer becomes an elevated mixed layer.

16.1 The elevated mixed layer

The lower boundary of the elevated mixed layer plume rises very quickly downstream from the source region and in so doing resembles the warm conveyor belt except for its dryness. The absence of layer cloud and the location of the plume's lateral boundary (effectively, its limiting streamline) are determined, in part, by topography and by the distribution of surface heating. The lateral boundary of the elevated mixed layer resembles the limiting streamline in the conveyor belt model except that its origins remain relatively fixed by its source region. Ventilation of the arid source region, in which the air from the arid boundary layer is transported from its origin, occurs in response to migrating troughs in the westerlies. The elevated mixed layer is carried poleward to follow a path similar to that of the warm conveyor belt.

Deep elevated mixed layers are most often found along or somewhat poleward from the edges of the subtropical deserts, e.g. the southern plains of the United States, southwestern Europe, the tropical Atlantic Ocean, northern India, Australia, southeastern Brazil and northern Argentina, northern South Africa and China. In some regions (the southern Great Plains of the United States and northern India), sharp gradients in soil moisture, which accompany sharp horizontal gradients in surface sensible heating, are manifested at the surface by a moisture front. These surface heating and moisture fronts constitute boundaries along which the warm dry air is able to flow over a cooler and moister surface mixing layer.

The elevated mixed layer front and its surface dry front ("dry line") are intimately associated with outbreaks of severe thunderstorms. Transport of natural particulates from semi-arid to temperate regions is accomplished to a great extent by airflow within elevated mixed layers. Since such particulates are radiatively active at solar and thermal infrared wavelengths, elevated mixed layers and their attendant aerosols may exert some effect on regional, if not global, climate. Consequently, it is worthwhile and interesting to devote a chapter to a description of mid-level fronts and elevated mixed layer plumes.

The chapter contains a description of the origins and migration of elevated mixed layers and describes the mid-level baroclinic zone (front) that characterizes the lateral boundaries of the elevated mixed layer. An analysis of the transverse/vertical circulation associated with mid-level fronts is presented within the context of the Sawyer–Eliassen Q-vector, discussed in previous chapters. Finally, the role of soil moisture in shaping the distribution of surface heating is discussed to show how it affects large-scale atmospheric forcing.

16.2 Origins of elevated mixed layers

Consider a temperature distribution over the lower troposphere, labeled T in the schematic thermodynamic diagram in Figure 16.1. Imagine that surface sensible heating occurs as the result of insolation over a period of one or more days. If the soil is saturated, evapotranspiration is almost that for a wet surface (called "potential evaporation") and the Bowen ratio (the ratio of sensible to latent-heat flux) is very small, typically between 0.1 and 0.2. Growth of the mixing layer is gradual, and after a period of time the sounding in the moist region shows a layer with nearly constant θ extending from the ground over a depth h_m with a temperature profile given by the thin full line labeled M. Up to some height, which is denoted by the horizontal dotted lines, the air is warmed by sensible heating from below and from above. In the moist sounding, the total vertical convergence of sensible heat flux is represented by the area labeled W in Figure 16.1. Above the dotted line and as far as the top of the mixed layer, the air is cooled as the result of downward flux of sensible heat. Thus, warming occurs over the area labeled W and cooling over the area labeled c on the sounding. Both observational and theoretical studies suggest that the ratio of downward to surface sensible heat flux (in proportion to the ratio of the areas W to c) is about 0.2, although the exact value may depend somewhat on the magnitude of the surface sensible heating. It should also be pointed out that vertical mixing affects the vertical distribution of momentum. Although mixing does not produce a vertically constant distribution of the wind components (due to the restoring effect of the balance between Coriolis and pressure gradient forces), the effect of vertical mixing is often to bring down faster-moving air from above. In situations where a jet streak moves over a deep mixing layer,

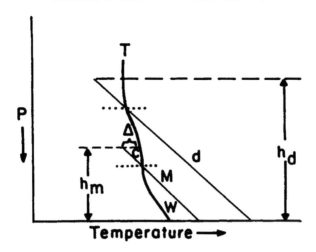

Figure 16.1 Schematic distribution of temperature on a skew T–log p thermodynamic diagram. The initial temperature profile (full curve labeled T), modified by surface sensible heating, forms an isentropic layer between the ground and the top of the mixing layer. Mixing layer heights are at h_m (for sounding M) and h_d (for sounding d), respectively, for a moist and a dry land surface. Overshooting of dry thermals produces a layer of downward heat flux and cooling near the top of the mixing layer above, between the horizontal dotted line and height h_d (for the dry sounding) and the horizontal dotted line and h_m (for the moist sounding). The layers experiencing warming and cooling due to the process of mixing are illustrated with regard to sounding M by the labels c and W, respectively. $\varDelta$ is the potential temperature jump, which is a stable cap at the top of the moist mixing layer.

the downward transfer of horizontal momentum may result in a marked increase in surface wind speeds and in the removal of soil particles in the form of dust storms.

The net result of the surface sensible heating is the formation of a well-mixed layer with a fairly uniform vertical distribution of potential temperature θ, mixing ratio r, soil particulates and gaseous contaminants. There is generally a slight increase with height of potential temperature and a slight decrease with height in the water vapor mixing ratio, particularly near the upper part of the mixing layer; the latter is due to entrainment of lower specific humidity and higher values of θ. The top of the mixing layer tends to be capped by a stable layer or inversion, the equivalent of 1–2 °C difference in potential temperature between that in the isentropic layer below and that at the top of the inversion. The strength of this inversion (labeled $\varDelta$ in Fig. 16.1) tends to depend on the surface sensible heat flux. This cap on the mixing layer is frequently visible in the form of a haze layer top.

Over the arid region, the Bowen ratio may be quite large; indeed, in the absence of available moisture at the surface, a large fraction of the available radiant energy at the ground returns to the atmosphere as sensible heat flux. Accordingly, the growth of the mixing layer over the arid region is much more rapid and extensive than over the moist region. This arid sounding, whose mixing layer depth is h_d, is labeled d in Figure 16.1. If the arid region is sufficiently large and the rate of movement of the air column is relatively slow, an air parcel traversing a homogeneous surface may remain in convective contact with the surface for several days. The result may be the eventual formation of a very deep mixing layer. Over the Sahara or northern India, where an air column may remain over the arid surface for several days, the mixing layer may grow to 5–7 km in depth. It is not uncommon to observe mixing layers over Mexico or over the desert regions of the United States and Mexico that reach elevations of 4–5 km. Over moist surfaces the depth of the mixing layer is typically less than 3 km.

Over regions where there is a large horizontal gradient in soil moisture, horizontal gradients in potential temperature and mixing ratio are large. Figure 16.2 shows a schematic cross section between a desert and a cool, moist region. Because of the large horizontal variation in sensible heating,

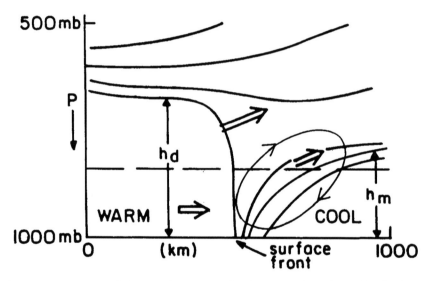

Figure 16.2 Schematic isentropic cross section between a hot, dry region having a deep mixed layer and a cool, moist region with a shallow mixing layer. Differences in surface sensible heating are responsible for differences in the depths and potential temperature in the isentropic mixing layers. The bold-faced arrow shows the direction of the air motion in the plane of the figure. Full curves are isentropes and the depth of the mixed layers over moist and dry regions are labeled, respectively, h_m and h_d. The thin streamline depicts the sense of the circulation ellipse imposed by confluence and shear in the baroclinic zone between the two regimes (see sec. 16.9). The broken horizontal line is referred to in the text (see sec. 16.3).

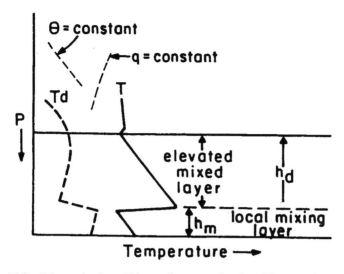

Figure 16.3 Schematic skew T–log p diagram related to Figures 16.1 and 16.2. In this figure the warm, dry mixing layer from sounding d in Figure 16.1 has folded over the moist layer (M) in that figure, forming an inversion at the base of the elevated mixed layer, whose depth is h_d. (For reference, a dry adiabat and an isopleth of constant specific humidity are denoted by dashed lines above.)

there is a baroclinic zone manifested by a front-like structure in the potential-temperature isotherms between the two regions. If the wind blows from dry to moist, the dry air moves over the moist layer, as an elevated *mixed* layer. The base of the arid, elevated mixed layer slides upward in the direction of the advecting wind (toward the right) along the sloping isentropes above the moist air, whose top is at an altitude h_m. The result of the advection of a dry, warm mixed layer over a cool, moist mixing layer is the sounding shown in Figure 16.3. This sounding contains two layers of nearly constant potential temperature derived from different boundary layers. There is a large vertical gradient in potential temperature between the lower, moist mixing layer and the elevated, dry mixed layer.

Note that the terms "dry" or "moist" are relative. Although the mixing ratio tends to decrease with height even in well-mixed boundary layers, the small decrease in mixing ratio with height in a layer with nearly constant potential temperature implies that the relative humidity increases rapidly with height. In effect, the lifting condensation level of the air within the mixed layer (called the "mixing condensation level") is approached as the mixing layer grows with time. Consequently, deep mixing layers may achieve saturation near the top, effectively preventing the further growth of a layer of constant dry potential temperature. In these situations, the top of the mixing layer may be marked by thin layer cloud, often in the form of altocumulus, or altocumulus castellanus if the value of θ_w decreases with height. Above the top of the mixing layer, the relative humidity decreases very rapidly with height to values more typical of cloud-free conditions. Conversely, the relative humidity at the bottom of the well-mixed layer is a minimum.

The elevated mixed layer inversion over the moist air constitutes a brake or "lid" on cumulus convection. Its stratification, shown in Figure 16.3, exhibits a very rapid increase in θ, a very rapid decrease in r and a decrease in θ_w at the inversion. Air in the boundary layer may be latently and potentially unstable with respect to the capping layer. The ability of the capping inversion to impede cumulus convection in the latently unstable air in the boundary layer is expressed as an inversion strength, which is defined as the difference between θ_w in the moist layer and the saturation wet-bulb potential temperature (θ_{sw}) in the warmest part of the inversion. This measure of "lid strength" is very nearly proportional to the difference in potential temperature below and above the inversion, which is often as large as 10–15°C in regions immediately downwind of extensive deserts. Such soundings have been referred to as "type 1" by Fawbush and Miller, who during the 1950s classified soundings on the basis of their potential for severe convective storms. The capping inversion differs significantly from that produced by subsidence, since the layer above the lid is relatively dry and well-mixed in terms of mixing ratio and potential temperature. The elevated mixed layers are frequently recognizable on conventional meteorological soundings hundreds (and sometimes thousands) of kilometers downstream from the arid source region.

In addition to differential soil moisture, a difference in terrain elevation promotes the formation of the elevated mixed layer. Consider a sounding as depicted in Figure 16.1 in which sounding d forms over a plateau, significantly above the elevation of the moist region. Since the potential temperature increases with height in a statically stable atmosphere, mixing layers that form over high terrain will tend to possess a higher potential temperature than at sea level. Even in the absence of differential soil moisture, the movement of air from a plateau over a valley may result in a sounding over the valley that exhibits a very shallow, weak inversion at the level of the plateau, the result of the air's contact with the surface over the plateau. The air in the mixing layer over the valley is also subject to surface heating but starting from a lower potential temperature.

Since the potential temperature decreases with height over the lowest few hundred meters above the ground during conditions of upward flux of sensible heat, the mean θ in the mixing layer is likely to be slightly lower than the afternoon potential temperature measured at screen level, although values of θ and θ_w in an elevated mixed layer are similar to those measured at screen level during the afternoon over the source region. Since the source

Table 16.1 Climatological means of surface dry- and wet-bulb potential temperature at screen level for half-month periods and for various locations during springtime. Units are in °C.

Location		Early April	Late April	Early May	Late May	Early June
Trade wind, Caribbean	θ_w	22.2	21.7	21.9	22.5	22.4
Texas (under lid)	θ_w	18.2	21.6	22.0	23.1	22.0
Mexico	θ	44.5	45.5	45.5	46.5	45.5
	θ_w	16.4	17.5	17.6	17.2	16.8
Continental SW U.S.	θ	33.0	37.5	37.5	39.0	42.0
	θ_w	13.4	14.6	14.6	17.2	17.6

regions for elevated mixed layers are often found at subtropical latitudes, they tend to be relatively barotropic, and values of θ and θ_w in elevated mixed layers are comparable with climatological averages of these quantities at the surface over the source region.

One important source region for elevated mixed layers over the United States is the Mexican plateau, where a combination of dryness and elevation contribute to producing very warm elevated mixed layers during the spring and summer months over the southern and central United States. When there is large-scale southerly flow at middle levels, the Mexican mixing layer is advected over moist, trade wind air from the Gulf of Mexico. Both the air within the capping inversion and the moist, tropical air possess values of potential temperature and mixing ratio that are characteristic of a uniform tropical source region. Table 16.1 lists the climatological values of dry- and wet-bulb potential temperature over Mexico, the wet-bulb potential temperature beneath the capping inversion over eastern Texas, the wet-bulb potential temperature in the trade wind mixing layer over the Gulf of Mexico, and the dry- and wet-bulb potential temperatures over the desert regions of western Texas and New Mexico.

The strength of the capping inversion is reflected by the difference between θ in the capping Mexican air stream and that in the moist trade wind air over the southern United States. Inspection of Table 16.1 shows that the potential temperature in the Mexican elevated mixed layer over the southern United States is between 44 and 46°C, the warmest occurring during late May. Since the potential temperature in the tropical mixing layer corresponding to a wet-bulb potential temperature of about 22°C (typical of the late spring months) is about 28°C, θ increases by about 17°C between the moist mixing layer and the base of the elevated mixed layer. Obviously, the desert air over the southwestern part of the United States cannot produce as strong a

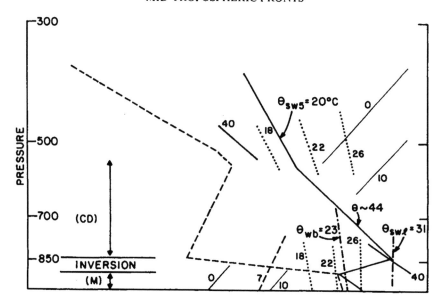

Figure 16.4 Schematic lid sounding on a skew T–$\log p$ diagram. The vertical temperature profile is indicated by the full lines and the dew-point profile is indicated by the broken lines. The thin full line segments slanting upward and toward the right are lines of constant temperature. The broken line segment slanting toward the right at the bottom of the figure is an isopleth of saturation mixing ratio. The full line slanting upward and to the left is a segment of a dry adiabat. The slightly curved dotted lines slanting upward toward the left are moist adiabats. The heavy chain lines are moist adiabats representing, respectively, the mean wet-bulb potential temperature of the surface air (θ_{wb}) and the saturation wet-bulb potential temperature (θ_{swl}) at the base of the elevated mixed layer. The average potential temperature ($\theta = 44°C$) in the elevated mixed layer air is also indicated. The low-level moist layer (air stream M in Fig. 16.9), the lid (labeled INVERSION) and the elevated mixed layer (air stream CD in Fig. 16.9) are indicated at the left.

capping inversion on the moist air since the mean θ in that source region only slightly exceeds 40°C in June.

A useful concept for the analysis of latent instability on a thermodynamic sounding is the saturation wet-bulb potential temperature (θ_{sw}), which is a measure of the actual temperature read on the scale of the moist adiabats. To illustrate this concept, Figure 16.4 shows an elevated mixed layer between about 850 and 550 mb. The saturation wet-bulb potential temperature in the elevated mixed layer is a minimum at the top of the layer (about 20°C), because of its nearly dry adiabatic lapse rate; θ_{sw} is almost constant between 550 mb and the top of the sounding. This minimum θ_{sw} is established by the wet-bulb potential temperature in the source region of the elevated mixed layer and by the amount of lift experienced by the elevated mixed layer; at saturation the minimum θ_{sw} is equal to θ_w in the source region of the elevated mixed layer. Thus, a measure of the maximum possible latent instability in the presence of a lid can be expressed by the difference in θ_w between that in the capped moist layer and that in the elevated mixed layer, which is that in the boundary layer over the source region. According to Table 16.1, this difference is greatest during late May for Mexican inversions capping the trade wind air streams, and during early May for the continental desert air stream capping the trade wind flow.

The strength of the capping inversion, given by the difference between θ in the planetary boundary layer and θ_{sw} at the base of the inversion (which we call θ_{swl}), can be quite large, as illustrated in Figure 16.4. The "negative area" on the sounding (the triangular area between the temperature profile and the $\theta_w = 23°C$ moist adiabat) is a measure of the energy needed by an air parcel in the boundary layer to overcome negative buoyancy and reach its level of free convection, which is the level at which the parcel's virtual temperature is exactly equal to that in the environment. The level of free convection in Figure 16.4 occurs (for a parcel of air lifted from the planetary boundary layer with a value of θ_w equal to 23°C) close to 700 mb. This negative area is roughly proportional to the difference between θ_{swl} and θ_w in the moist layer below. Lifting of the sounding (for example, as the result of synoptic-scale forcing) leads to:

(a) a weakening of the inversion (a lower θ_{swl}) and a reduction of the negative area;

(b) a reduction of the minimum θ_{sw} at the top of the elevated mixed layer; and

(c) an increase in the positive area.

Solar radiation serves to raise both the dry- and wet-bulb potential temperatures of the moist air beneath the capping inversion, even though the moist layer depth is increased due to lifting of the capping inversion. Insolation increases the positive and decreases the negative areas on the sounding, thereby weakening the lid and increasing the latent instability. A combination of insolation and large-scale lifting can greatly increase the latent instability and remove the lid. Latent instability, however, cannot be realized as long as the capping inversion is sufficiently strong to impede convection. In delaying the onset of convection, while allowing latent instability to increase with time, the lid constitutes an agent for the promotion of severe thunderstorms once the lid is removed. Therefore, the lid enables high latent instability to be achieved, whereas, in the absence of a lid, the atmosphere would gradually release the latent instability in more gentle convection.

16.3 Illustrations of airflow within the below elevated mixed layers: a case study

We now illustrate the role of the elevated mixed layer in the evolution of the large-scale severe storm environment. Severe convective storms (the Red River outbreak) occurred over the southern plains of the United States (specifically northern Texas and Oklahoma) during 11 April 1979. Intense convection manifested by tornadoes and large hail moved in a narrow swath, the shaded area in Figure 16.5, causing great damage. The period of very severe weather lasted only about 6–9 h but the storms themselves persisted well into the next day, producing heavy rain, hail and wind damage both south and north of the warm front. The most intense and destructive convective events (heavy shading) were confined to a relatively small region south of the warm front (Fig. 16.5), east of the surface dry front (chain curve) and west of an elevated mixed layer of Mexican origin. There is a strongly confluent pattern between the Mexican elevated mixed layer (whose limiting streamline is the scalloped border) and a dry air stream that originated west of a trough.

The edge of this dry air, denoted by the surface dry front, constitutes a limiting streamline between the moist air from the Gulf of Mexico and the dry air, which has been heated over the southwestern desert region after having descended from the west side of a trough in the westerlies (the "continental" air referred to in Table 16.1). In the case of Figure 16.6, there was a strong jet at upper levels over Mexico and western Texas. Downward mixing of the air within the deep surface mixing layer produced high wind speeds near the surface and considerable blowing of dust along the cold front.

The base of the elevated mixed layer (full isopleths in Figure 16.5) is near 900 mb over southern Texas but rises toward the west and north; near the warm front the elevated mixed layer is above about 700 mb. This large-scale ascending vertical motion is illustrated using relative isentropic analysis (Fig. 16.6). Capping inversions generally ascend as they are advected from the source. In order to understand the large-scale dynamics of this ascent, let us consider the horizontal distribution of temperature associated with this capping inversion. The rise in the base of the diversion toward the north in Figure 16.5 is consistent with a poleward gradient in θ_{swl} and, therefore, in the temperature on isobaric surfaces. Accordingly, large-scale ascent is consistent with the implied warm air advection in the flow coming from Mexico and middle levels.

Consider the potential temperatures within the elevated mixed layer over southeastern Texas (Fig. 16.7). Mean potential temperatures in the mixing layer west of the dry front ($\theta = 30$–$40°C$) are much lower than those in the Mexican air stream ($\theta = 40$–$49°C$). This cooler mixing layer west of the dry

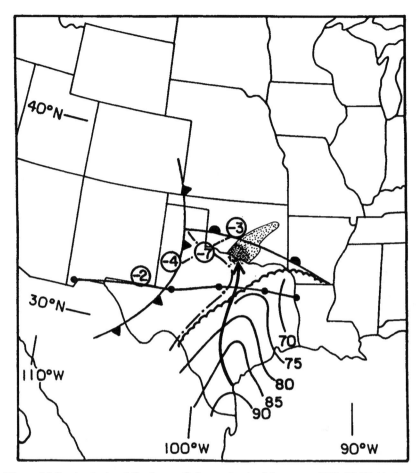

Figure 16.5 Analysis of the base of elevated mixed layers at 0000 GMT 11 April 1979 (the "Red River" tornado outbreak). Full curves represent isopleths of lid base (in tens of mb). The scalloped border and chain curves represent, respectively, the western edge of the Mexican air stream (lid) and the surface position of the dry front. The series of small filled circles along the thin full curve oriented west to east denote stations in the cross section of Figure 16.9. Shading represents the primary region of tornadic storms over a 12 h period centered at 0000 GMT; heavy shading shows the region of most violent convection. Numbers located in circles indicate the 3 h pressure tendencies at the isallobaric minimum at 1500, 1800, 2100 and 0000 GMT. Surface fronts are represented in conventional symbols. (Adapted from Carlson *et al.*, 1983.)

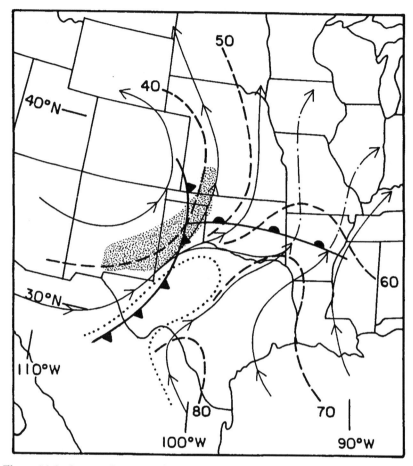

Figure 16.6 Isentropic analysis at 0000 GMT 11 April 1979 on the $\theta = 38°C$ surface (intersection of ground along dotted curve). Relative wind flow (with respect to the migrating trough to the west) is indicated by the streamlines. Broken curves are isobars on the potential temperature surface in tens of mb. The area enclosing wind speeds in excess of 35 m s^{-1} is shown by shading. (From Carlson *et al.*, 1980.)

front moves over the triangular area north of the dry front as a weak elevated mixed layer. Mexican air lies east of the scalloped border. Within the triangular area between the surface dry front, the Mexican air and the warm front, there is almost no negative area on the sounding. Air within the planetary boundary layer within this triangle is latently very unstable, having arrived from underneath the lid with high values of θ_w. Thus, latent instability (positive area on the sounding) was being maximized in this triangular area. *The advection of air with high values of θ_w from beneath the lid is referred to as "underrunning".* Not surprisingly, the latent instability was

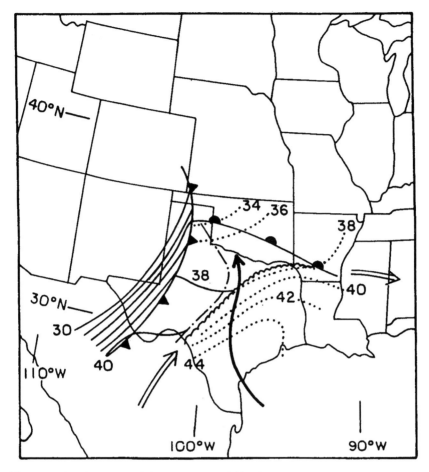

Figure 16.7 Mean potential temperature within the surface mixing layer (full curves labeled in °C) or elevated mixed layer (dotted curves labeled in °C). The edge of the Mexican air over Texas is indicated by the scalloped border and the surface dry front by the chain curve. Double-shafted arrows refer to the relative air motion along the edge of the Mexican lid. The heavy full arrow denotes the movement of moist surface-level air. (From Carlson *et al.*, 1980.)

released rather violently by underrunning, which involves ascent and removal of the negative area.

16.4 Climatology of the lid

Elevated mixed layers sometimes travel for considerable distances at the same latitude (e.g. from the Sahara to the Caribbean Sea). Poleward move-

461

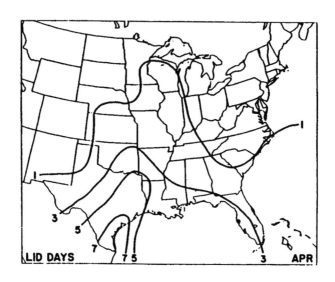

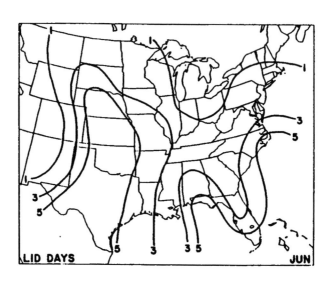

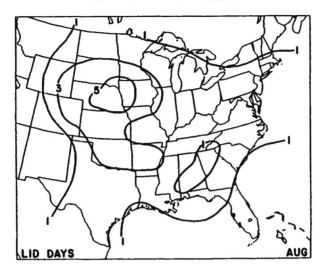

Figure 16.8 Frequency of lid soundings by month, based on a sample of all available rawinsonde data from 1983 to 1986. (From Farrell and Carlson, 1989.)

ment of elevated mixed layers seems to be less extensive. Fundamentally, the preservation of the elevated mixed layer's properties for great distances downstream from the source seems to be dependent on the absence of large-scale baroclinicity. When the elevated mixed layer moves poleward in advance of a mobile trough, it rises rapidly and is subsequently destroyed by lifting at shorter distances from the source than in regions where the atmosphere is relatively barotropic. Lid destruction is associated with the polar front jet, more specifically mid-latitude jet streaks where forced ascent by the synoptic-scale dynamics is strong.

Although the lateral boundaries of capping inversions near their source region tend to remain relatively fixed in space, the elevated mixed layer can migrate eastward and poleward ahead of traveling waves in the westerlies. Thus, it is not uncommon to find an elevated mixed layer from Mexico over the central Great Plains and Ohio Valley (the extreme northeastern corner of Figures 16.6 and 16.7). Over Europe, elevated mixed layers from Spain and North Africa are found intermittently over France and Great Britain. Elevated mixed layers from the Sahara are common at middle levels over the Caribbean, and migrating elevated mixed layers from Iran and Saudi Arabia persist over the Arabian Sea.

During the early summer the main belt of westerlies shifts poleward over the United States (and, indeed, throughout the Northern Hemisphere). Accordingly, the southern plains of the United States experience more barotropic conditions later in the season, and the Mexican air is able to advance to a more northerly latitude (Fig. 16.8).

16.5 The elevated mixed layer front

Figure 16.9, which is a cross section along the chain of radiosonde stations in Figure 16.5, shows three air streams, each representing a different geographical origin. Moist air, labeled M, originates from the Gulf of Mexico, and moves northward into the plane of the figure. There is a baroclinic zone along the western edge of the air stream, which coincides with the surface dry line. The dry line is denoted most by very large gradients of relative or absolute humidity but is reflected in the distribution of potential temperature. It therefore resembles the surface fronts discussed in Chapter 13, except that the cool air corresponds to the moist air. The lid inversion is strongest some distance to the east of the dry line.

Above the capping inversion, the Mexican air (CD in Fig. 16.9) exhibits a layer of relatively slowly changing potential temperature with height. Note that, because of the warmth at the base of the Mexican air, the baroclinicity reverses with height, the air becoming warmest east of the dry line above the base of the Mexican air stream. Table 16.1 shows that dry desert air over western Texas (SP in Fig. 16.9) generally is cooler than the CD air originating over Mexico. Accordingly, this climatological temperature difference is reflected in the baroclinic zones along the lateral boundary of the CD air streams. The baroclinic zone slopes upward toward the west (the cooler air) along the western side of the Mexican (CD) air between 750 and 500 mb. Because of the rapid decrease in temperature with height in the Mexican elevated mixed layer, the temperature gradient reverses again with height near 500 mb, where there is a baroclinic zone tilting upward toward the east (toward the cooler air). A very weak inversion is present at the top of the Mexican air stream.

16.6 Severe local convection and elevated mixed layers

Severe local storms occur where moist air in the planetary boundary layer attains high values of θ_w. This situation can occur after air near the surface is subject to prolonged heating and moistening as the result of surface sensible and latent-heat fluxes. This latently unstable air typically arrives from a uniform source region, such as the Gulf of Mexico, where it may possess a rather high value of θ_w near the surface (Table 16.1). Over the United States, the moist air from the Gulf of Mexico may encounter a capping inversion beneath an elevated mixed layer that arrives from Mexico or the desert regions of the American southwest. Capping of the moist air over moist terrain allows θ_w in the boundary layer to increase further in the absence of cloud. If that boundary-layer air moves toward cooler air aloft, as is likely the case ahead of upper troughs, the potential instability can become very large.

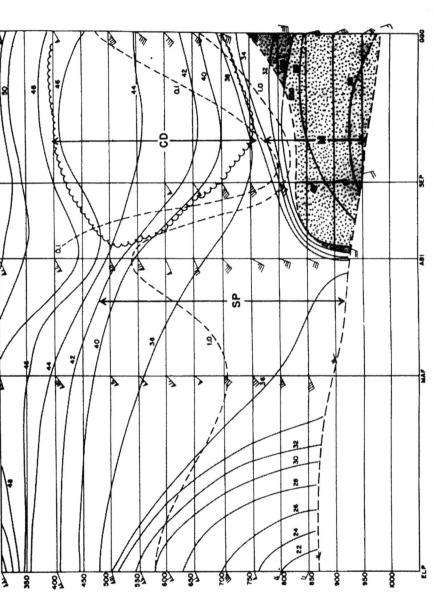

Figure 16.9 Vertical cross section from El Paso, TX (ELP) to Longview, TX (GGG) along the series of radiosonde stations shown in Figure 16.5 for 0000 GMT 11 April 1979. Full curves correspond to isentropes (°C), broken curves to isopleths of mixing ratio r (g kg^{-1}) and stippling to mixing ratios in excess of 10 g kg^{-1}. Scalloping is intended to offset the Mexican desert air (CD), from the moist air (M) below and from the modified subsided polar air (SP) to the west. Winds are shown at the station locations (1 flag = 5 m s^{-1}). The surface dry line was near Abilene (ABI). (From Carlson *et al.*, 1983.)

The rise of θ_w in the boundary layer beneath a lid is an essential element in the outbreak of severe storms in various parts of the world. For example, over western Europe an increase of θ_w near the surface occurs as the result of insolation at the ground for one or two days beneath a lid that arrives from Spain or the Sahara. The potential temperature in the mixing layer over the Spanish plateau is typically about 35–38°C and that of the Saharan air is 42–46°C during summertime. Air arriving over France from the Spanish plateau forms an elevated mixed layer over moist air situated in the lee of the Pyrenees.

Favorable situations for the presence of a lid at middle latitudes occur

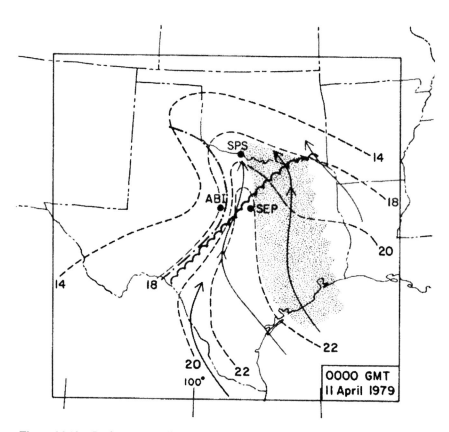

Figure 16.10 Surface streamline analysis for 0000 GMT 11 April 1979, showing isotherms of the wet-bulb potential temperature averaged over the lowest 50 mb (broken curves in °C). The surface dry line position is indicated by a chain curve. The western edge of the lid formed by the Mexican desert air stream is shown by the scalloped border. Light shading is the western edge of a low stratus deck (the location of the first tornadoes began about 1800 GMT west of station SPS). (From Carlson *et al.*, 1983.)

poleward of such arid, elevated source regions in approach of a migrating trough in the westerlies. Severe convection occurs because the lid delays the release of latent instability until rather large values of that instability are achieved. There are documented cases of severe convection and heavy rains occurring in China and along the coast in Australia in association with elevated mixed layers originating over nearby deserts. Mexican air has been found as far north as 40°N on occasions where severe local storms have broken out over the northern Great Plains and, at least on one occasion (31 May 1985), over the northeastern part of the United States. The most famous tornado outbreak over the United States (that of 3–4 April 1974), in which 144 tornadoes were sighted, occurred when an elevated mixed layer arrived over the Ohio Valley from a source over the American southwest. In other instances over North America, elevated mixed layers form by differential advection of a dry, deep mixing layer from the high plains or the southern Rocky Mountains over cooler, moister air to the north and east.

Figure 16.10 shows underrunning taking place just to the south of station SPS during the afternoon and evening of 10 April. During this time, latently unstable air in the planetary boundary layer moved from a location beneath the lid over south Texas at the close of the previous afternoon into the region where the negative area on the sounding was effectively zero. Highest values of θ_w near the ground extended along a tongue across the lid edge during the early evening hours. Underrunning (the component of velocity normal to the edge of the lid in the moist air) was most pronounced during the afternoon. This increase in the low-level component of motion across the lid edge was the result of an increase in the ageostrophic motion towards the west, not because of an increased frictional drag, but in response to an isallobaric pattern in advance of approaching trough and attendant jet streak (Fig. 15.7). Thus the isallobaric wind increased during the day in the direction of the maximum surface pressure falls ahead of the migrating cyclone (Fig. 16.5). During the early morning, these maximum pressure falls moved into Texas and subsequently increased in magnitude over a relatively small region near the outbreak of the severe local convection.

16.7 Large-scale aspects of the elevated mixed layer

The role of ageostrophic motion in creating a vigorous flow in the direction of the maximum pressure falls and the coupling of lower- and upper-level wind fields are discussed in Chapter 15 (see Fig. 15.7). In the case illustrated in Figures 16.5, 16.6, 16.7, 16.9 and 16.10, the large-scale pattern of pressure falls was produced by forcing associated with an upper-tropospheric mobile trough and attendant jet streak. Intensification of the isallobaric minimum with time was produced by several factors, principally surface sensible heating and lower static stability west of the dry front. Underrunning was most pronounced near the axis of the surface wet-bulb potential temperature maximum (Fig. 16.10). Since the western boundary of the lid was aligned parallel with the 700 mb winds, the pattern shown in Figure 16.10 is consistent with warm air advection (veering of the winds with height) and with large-scale ascent along the edge of the lid.

The surface mixing layer (SP in Fig. 16.9) originated in air that descended from west of the trough (the dry air stream of Chapter 12), and crossed the southwestern part of the United States and extreme northern Mexico. In contrast, the Mexican elevated mixed layer (CD) was formed in air that originated east of the trough and at middle levels over the Caribbean Sea where the air is also dry. In moving more slowly than the subsiding air crossing the trough axis, this Mexican air stream was able to remain for two days over northern Mexico, where it was subject to intense heating at the surface. In contrast, the subsiding air stream (SP) represented initially cooler (mid-latitude) air, which remained in contact with the dry terrain for a shorter time than had the Mexican elevated mixed layer.

Movement of the moist air stream is influenced by the blocking effect of the high Mexican plateau and by the horizontal distribution of soil moisture (which is itself a product of the terrain effect). The trade wind moist layer is prevented from ascending over the plateau and it is deflected northward over the southern United States by the plateau. Numerical simulations show that the presence of a topographic barrier contributes to enhancing the poleward component of the motion at low levels; this component is confluent with the dry air stream arriving from the west of the trough, thereby forming the dry front. Like the Mexican elevated mixed layer, the dry front tends to be rooted to the geography in the region immediately north of the Mexican plateau.

During the late spring or early summer, a poleward shift occurs in the westerlies, resulting in a movement of moist air over Mexico from the Gulf of Mexico and from the Pacific Ocean. As a result, precipitation increases over northern Mexico and over the mountainous regions just to the northwest of Mexico during June and July. With the inception of this monsoon type of precipitation pattern (reflecting a global shift toward monsoon rains elsewhere at those latitudes), a corresponding decrease in maximum surface potential temperature occurs over Mexico (reflected in the June values of Table 16.1), a weakening of lids over the southern United States and a decrease in the occurrences of dry front formation over the southern Great Plains of the United States.

A configuration of limiting streamlines for a severe local storm episode is illustrated in Figure 16.11, showing the limiting streamlines for the Mexican and low-level moist air and the flow at high levels. As in the April 1979 severe storm outbreak, there was a region in which moist air emerges from beneath

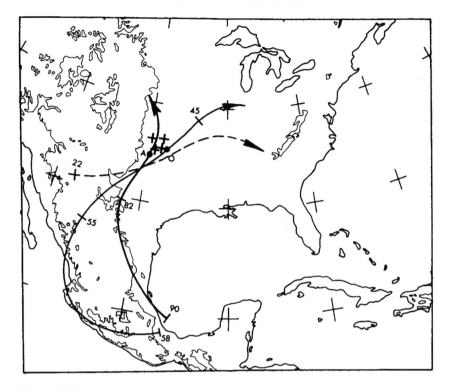

Figure 16.11 Composite relative wind isentropic chart for a tornado outbreak, 4 May 1961. Limiting streamlines are shown for the moist trade wind flow (thick full curve; height of top of layer marked in tens of mb) and the top of the Mexican air (thin full curve; height marked in tens of mb). The broken curve shows the flow in the high troposphere, at about 220 mb. Note that in the vicinity of the severe storms (crosses) the winds veer with height. Stations marked by the full circles are Altus (A) and Oklahoma City (O). (From Carlson and Ludlam, 1968.)

the Mexican elevated mixed layer over a small area between the edge of the moist air and the edge of the Mexican air.

Similar events occurred in the case presented in Figure 16.12. Here, an out-break of severe storms began near the limiting streamline of moist air (the dry front) from a series of growing cumulonimbus cells situated at the locations marked by crosses in the figure. These cells moved eastward across the region, becoming more intense and producing tornadoes, hail and wind damage. Severe convection was almost entirely confined to a small region between the dry front and the edge of the Mexican lid, which is the streamline labeled M. Formation of these storms was triggered by underrunning of the moist air (arrow) and by ascent along the edge of the elevated mixed layer. Dissipation began after the thunderstorm cells reached the lid edge.

469

Soundings near the sites of the severe convection are typified by that at ABI in the cross section of Figure 16.9. Note that there is no restraining inversion but there is a deep layer of high θ_w in the boundary layer. This implied latent instability can be released by large-scale ascent coupled with a small-scale transverse/vertical circulation along the edge of the lid.

The particular arrangement of topography, large-scale forcing and differential heating found over the southern plains of the United States is illustrated in Figure 16.13. The underrunning air stream (M) beneath the Mexican air (CD) leads to the inception of convective cells outside the

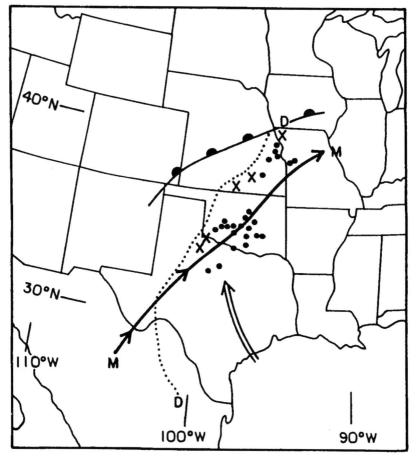

Figure 16.12 Map showing position (on afternoon of 26 May 1962) of the warm front, the surface dry front (dotted curve labeled D) and the limiting streamline of the Mexican air (full curve labeled M). Reports of severe storm incidents between 15 and 23 h CST, 26 May 1962, are indicated by small filled circles. Large crosses refer to locations of initial cumulonimbus groups. (From Carlson and Ludlam, 1968.)

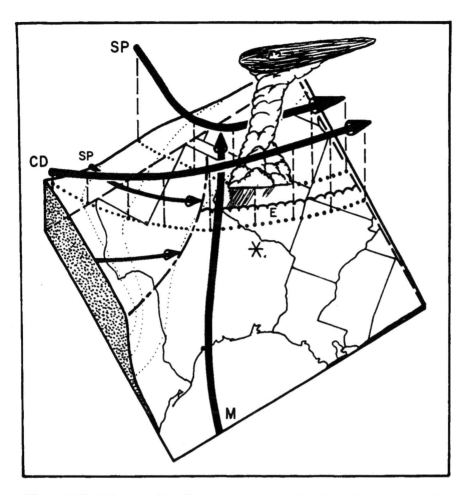

Figure 16.13 Schematic flow diagram, representing in three dimensions the air streams M, CD and SP (see text and Figure 16.9 for further explanation of these symbols). Airflow is shown in perspective against topography of the southern Great Plains of the United States and Mexico. Thin dotted curves denote surface terrain elevation contours and illustrate the gradient of surface elevation north and east of the high plateau of Mexico (shown in cut-away section). The left edge of the moist air stream (M) is shown bounded by the dry line (chain curve); the left edge of the Mexican air stream (labeled CD), which is forming a lid over the moist air, is denoted by the scalloped border. Underrunning (see text) is occurring at the location of the thunderstorm near E, which is a region with no capping inversion, large- and small-scale ascent and strong vertical wind shear. Thunderstorms can also occur underneath the lid at the location of the star, where large-scale ascent coupled with surface heating may be removing the lid. The third air stream (SP) is subsiding polar air which originates west of the trough. (From Carlson *et al.*, 1983.)

boundary of the restraining inversion, whose edge projected on the surface is shown by the dotted curve. Underrunning is promoted by strong ageostrophic motion toward the west. The dry descending air stream (SP) is also heated within the boundary layer over the southwestern part of the United States and it arrives west of the surface position of the dry front (the chain curve). Synoptic-scale ascent occurs over most of the region. Convection is usually initiated outside the lid region, except when the effects of the lid are overcome locally due to the forced ascent by existing convection cells or as the result of local orography or peculiarities of surface heating. As the base of the Mexican air rises toward the north and as daytime heating proceeds under the lid, isolated convection may occur where local heating and orography combine to overcome the restraining effect of the lid (near the star).

According to the omega equation, differential advection of warm air (as, for example, from a heated source region such as Mexico) contributes to ascent. The rise of the inversion along the direction of flow within the elevated mixed layer is consistent with sloping isentropes and with warm air advection. In cross section (e.g. Fig. 16.2), the base of the elevated mixed layer constitutes both a horizontal and a vertical frontal boundary, which slopes upward away from its source. Consider the horizontal segment (the horizontal broken line in Figure 16.2), which intersects the base of the elevated mixed layer and the cooler, moist boundary layer downstream. Clearly, a horizontal temperature gradient exists because the inversion is inclined upward. Motion of the elevated mixed layer toward cooler air constitutes warm air advection on an isobaric surface. It is customary to consider that such ascent is *forced* by the warm air advection. The point here is that the warm air advection can be considered a consequence of the rise of the elevated mixed layer base downstream as much as it constitutes a forcing of that ascent.

Figure 16.14 shows transverse/vertical circulation (in the exit region of the jet) that involves an underrunning across the lateral boundary of the jet and the release of latent instability. Note that the figure implies large-scale ascent up the isentropes in the direction of the left-hand side of the jet (veering of the geostrophic wind with height). Further, the ageostrophic motion implies an acceleration of the low-level winds into the plane of the figure.

In view of the discussions in Chapters 14 and 15, it is not surprising that baroclinic zones along the edges of elevated mixed layers are associated with transverse/vertical circulations; these circulations are suggested in Figure 16.2. Like cloud edges, the lateral edges of elevated mixed layers coincide with a zone of confluence between air streams of differing origins. The sharp definition of these baroclinic zones contrasts with the synoptic-scale distribution of temperature and vertical motion, which exhibit more uniform horizontal gradients at middle levels. Where smaller-scale observations can be made, as for example in the cross section through the edge of the elevated mixed layer in Fig. 16.9, the temperature structure resembles a front.

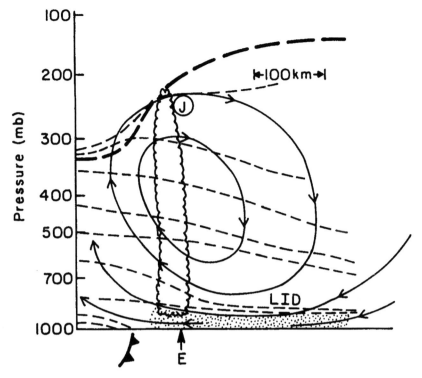

Figure 16.14 Schematic illustration of underrunning produced by transverse/vertical circulation in exit region of a straight jet streak (labeled J). Broken contours are isentropes, thin full curves the streamfunction of the circulation ellipse, the letter E the lid edge, shading the region of high θ_w in the boundary layer, scalloping the severe convection and the surface cold front is shown by conventional symbols. (From Farrell and Carlson, 1989.)

In accordance with thermal wind considerations, one finds a relative wind speed maximum along the western edge of the Mexican elevated mixed layer (e.g. near 500 mb at the station labeled ABI in Figure 16.9), where the horizontal gradient of temperature reverses with height. The reverse temperature gradient below 700 mb in this figure (colder air to the east) corresponds to a rapid decrease of the southerly wind component with height at low levels along the western border of the Mexican air stream.

16.8 Ageostrophic motion along the lid edge

Let $-\partial\theta/\partial p$ serve as a measure of the lid strength and take the total time rate of change of this term to diagnose the change in lid strength with time.

The kinematic vertical front strength equation, analogous to that for front strength ($- \partial\theta/\partial y'$) in equation (13.2), is formed by expanding the total differential ($d(-\partial\theta/\partial p)/dt$):

$$-\frac{d}{dt}\left(\frac{\partial\theta}{\partial p}\right) = -\frac{\partial}{\partial p}\left(\frac{d\theta}{dt}\right) + \frac{\partial u}{\partial p}\left(\frac{\partial\theta}{\partial x'}\right) + \frac{\partial v}{\partial p}\left(\frac{\partial\theta}{\partial y'}\right) + \frac{\partial\omega}{\partial p}\frac{\partial\theta}{\partial p}. \quad (16.1)$$

Writing the total vertical wind shears ($\partial u/\partial p$; $\partial v/\partial p$) in terms of their geostrophic and ageostrophic components, and substituting the thermal wind relationships for the geostrophic vertical wind shear components, results in a cancellation of the vertical geostrophic wind shear terms and leads to the following expression:

$$\frac{d}{dt}\left(-\frac{\partial\theta}{\partial p}\right) = \frac{\partial u_{ag}}{\partial p}\left(\frac{\partial\theta}{\partial x'}\right) + \frac{\partial v_{ag}}{\partial p}\left(\frac{\partial\theta}{\partial y'}\right) + \frac{\partial\omega}{\partial p}\left(\frac{\partial\theta}{\partial p}\right) + \frac{\partial}{\partial p}\left(\frac{d\theta}{dt}\right) \quad (16.2)$$

which states that, in the absence of diabatic heating, *an inversion cannot be created or destroyed following a parcel except through the action of ageostrophic and vertical motion.* More specifically, ageostrophic motion is solely responsible for producing or destroying a lid. (This result is general for any change in static stability following a parcel.) In the case of the lid, the movement of a warm mixed layer over a cool moist layer is due to differential advection of differing air streams. (One may properly refer to the lid as a "differential advection" inversion as opposed to a subsidence or radiation inversion.) Underrunning therefore implies the existence of a transverse/vertical circulation, in which $d(-\partial\theta/\partial p)/dt < 0$. The latter is positive if u_{ag} (the ageostrophic wind component underneath the lid) increases normal to the lid edge with decreasing height and $\partial\theta/\partial x'$ is negative. This arrangement is often the case for lid edges aligned parallel to an upper jet streak.

Underrunning tends to be associated with large-scale ascent and ageostrophic motion along the lid edge. Smaller-scale transverse/vertical circulations may develop as the result of the lid itself. The cross section in Figure 16.15 illustrates the salient aspects of these circulations along the lateral boundary of the lid. Shading represents the elevated mixed layer, below which the inversion is readily apparent. As in the cross section of Figure 16.9, there is a weak mid level jet maximum associated with the horizontal temperature gradient along the left edge of the elevated mixed layer. The wind speed maximum is connected to the principal jet streak at the upper left-hand corner of the diagram. The transverse/vertical circulation associated with the lid is depicted by the full streamlines. This vertical motion is weak (1–2 μb s^{-1}) but it is concentrated in the region along the lid edge. (A similar, but somewhat stronger, circulation typically occurs along the dry front, which constitutes the confluence between dry, warm and moist, cool air streams.)

Two circulation cells are indicated in Figure 16.15a, one indirect (the upper, clockwise cell) and one direct (the lower, counterclockwise cell). Both cells are produced by a combination of shear and confluence. Confluence is associated with the merging of air streams of differing origins, the Mexican elevated mixed layer (CD) and the dry descending air stream (SP) of Figures 16.9 and 16.13. The distribution of u_g and v_g in the cross section is shown in Figure 16.15b. The signs of Q_{yp} are opposite for confluence and shear terms. (In contrast, however, when the kinematic frontal equations are applied to specific humidity rather than temperature, as in equation (13.2), the humidity front is almost entirely produced by confluence.)

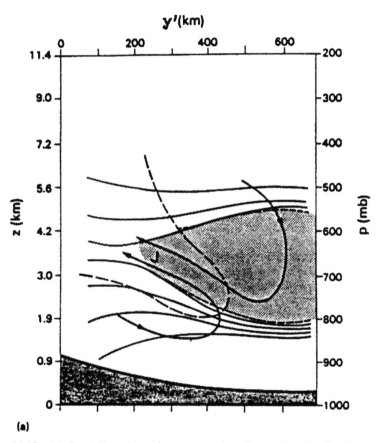

(a)

Figure 16.15 (a) Partially schematic cross section illustrating idealized elevated mixed layer structure and lid edge. Thin full curves are isentropes, and shaded region bounded above and below by broken curves depicts elevated mixed layer. The broken contour represents an isotach depicting a downward extension of higher wind speed and the letter J a local wind speed maximum. Thick full curves are streamlines of the transverse ageostrophic circulation diagnosed in this study.

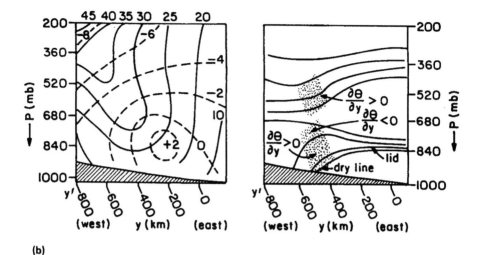

(b)

Figure 16.15 (b) Partially idealized cross sections approximately west to east across western Oklahoma and northern Texas through the dry front and the lid edge at approximately 2100 GMT 9 May 1979, for the case of (a). On the left are isopleths of u_g and v_g (geostrophic components of the wind, respectively, the full and broken isopleths labeled in m s^{-1}). The right-hand figure shows the isentropes, the location of the surface dry front (dry line), the lid and the location of baroclinic zones (shading) associated with the dry line and lid edge. The directions of the horizontal temperature gradients in these baroclinic zones are indicated by the derivatives (positive y' axis to left). (Adapted from Keyser and Carlson, 1984.)

It is worthwhile noting that the transverse/vertical circulation is ascending over a narrow zone along the edge of the elevated mixed layer, where the lid is weak. The small-scale circulation depicted in Figure 16.15a contributes to a lifting of the inversion in this region, and therefore to a release of convection along the lid edge. In summary, the larger-scale circulation associated with a migrating trough and jet streak (represented in Fig. 16.14) is primarily responsible for underrunning, which is aided by the smaller-scale circulations associated with the lid (represented in Fig. 16.15).

16.9 Differential soil moisture and static stability

A key to the intensification of the isallobaric wind over the southern Great Plains in advance of a migrating cyclone is the static stability, which is highly influenced by the pattern of differential surface heating. Figure 16.5 suggests that the isallobaric pattern can become very intense over regions in which a deep, dry adiabatic mixing layer and strong surface sensible heating are not capped by a lid. The formation of deep, isentropic layers accompanied by intense surface sensible heating is closely tied to the climatological distribution of soil moisture. Over the United States, the largest gradient of soil moisture lies west–east over the southern Great Plains (Fig. 16.16).

It is easy to demonstrate using conventional formulae for evaporation that latent heating exceeds sensible heating by a factor of 5–10 over wet surfaces. Over moist terrain, most of the net available radiant energy is used to evaporate liquid water from the ground. Conversely, the sensible heating over dry ground comprises a large fraction of the available radiant energy reaching the surface. The differences in heating between dry and wet ground can lead to horizontal gradients of surface sensible heating, surface tempera-

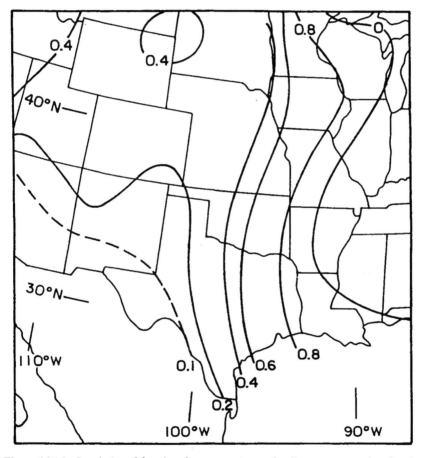

Figure 16.16 Isopleths of fractional water content of soil water saturation for the month of April, as derived from an analysis of climatological data by Fennessey and Sud (1983).

ture, static stability, vertical mixing of momentum and the depth of the surface mixed layer. These gradients can be large between regions that differ greatly in rainfall and substrate moisture. In fact, rainfall is the primary meteorological variable governing soil moisture.

Deep isentropic layers are regions of lowered static stability that foster the intensification of transverse/vertical circulations (and an augmentation of the surface isallobaric pattern) when a jet streak or upper disturbance migrates through the region (see Fig. 15.3). Although it is difficult to assign a true cause and effect to the evolution of a series of atmospheric events, surface heating and its effect on static stability constitute true precursors of development. For example, during the 10 April 1979 thunderstorms outbreak discussed above, the isallobaric pattern intensified during the day due to intense surface heating and low static stability in the mixing layer ahead of an advancing cold front. Intensification of the isallobaric pattern (Fig. 16.5) accompanied the formation of a mesoscale surface low, which moved across Texas during the afternoon and evening hours of 10 April. As the cold front advanced eastward, the migrating isallobaric minimum weakened, moved beneath the lid during the evening, and the meso-cyclone vanished as a separate entity.

Numerical simulations show that the surface heating pattern is critical in the formation of thunderstorms in several different ways. Without intense heating over Mexico there would be no restraining inversion and the effects of the lid and the process of underrunning would not occur. Intense surface heating over western Texas and New Mexico is essential for enhancing the isallobaric pattern in that region, for increasing the northwestward flow out of the Gulf of Mexico and for enhancing the underrunning. Furthermore, both the low-level airflow and the strength of the transverse/ageostrophic circulation associated with the upper-level jet are significantly enhanced by the effects of surface heating. The role of surface sensible heating in changing the synoptic-scale circulation pattern is illustrated in Figure 16.17. The circulation pattern represents the difference near the surface for numerical simulations with and without surface sensible heating. This pattern is due to the horizontal variation in surface moisture availability, which governs the evaporation. Clearly, there is a large increase in cyclonic inflow over the driest terrain when surface heating is included. Wind speeds in the southerly flow over Texas are increased by about 7.5 m s^{-1} and there is increased confluence along the dry front and a change in the location of the dry front.

In summary, the following events take place with the approach of an upper trough and surface cyclone toward a region such as the southern Great Plains of the United States. A southerly flow is initiated over the Gulf of Mexico. At the level of the Mexican plateau, a deep surface mixing layer is formed by the intense surface heating within an air stream that originates over the western Caribbean. Low static stability over the desert southwest

478

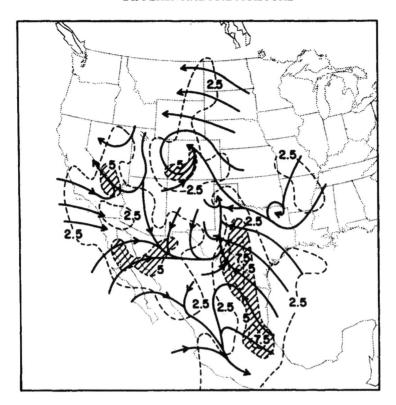

Figure 16.17 Streamlines representing vector difference in near-surface winds between two numerical simulations with and without surface sensible heating, using climatological values of soil moisture availability similar to that shown in Figure 16.16 (sense of arrows is to show wind vectors for heating minus those for no heating). Broken curves are isopleths of wind speed differences (values greater than 5 m s^{-1} shaded). (From Benjamin and Carlson, 1986.)

leads to an enhancement of the pressure falls ahead of the migrating wave. These pressure falls, combined with the effects of diabatic heating and topography, promote an increased poleward component at low levels from the Gulf of Mexico and at mid-levels from Mexico toward the region of geopotential height falls, with the moist air stream being capped by the elevated mixed layer from Mexico. Confluence of the moist air with dry air from the west occurs along the dry front, which may extend a considerable distance from the Mexican border. In response to the approaching trough, ageostrophic winds near the surface increase toward the west, causing underrunning of the moist air along the western border of the Mexican elevated mixed layer. Convergence in the isallobaric (ageostrophic) flow west of the lateral boundary of the lid, enhanced by frontal-type circulations along the lid edge (and along the dry front), promote ascent along and just outside of the lid edge, thereby triggering the release of the latent instability. Forcing of ascent occurs most frequently in the left exit or right entrance regions of a jet streak. The distribution of soil moisture strongly modulates the location of the lid edge and dry front, and is uniquely dependent on the surface climatology of the region.

Problems

16.1 Consider Figure 16.15b, which is a cross section of potential temperature and geostrophic wind components through a dry front and the edge of an elevated mixed layer of Mexican origin. (See also Fig. 16.15a.)

  (a) Show that quasi-geostrophic forcing is responsible for the presence of a transverse/vertical circulation ellipse and a narrow zone of ascent along the lid edge. Use the left-hand rule to determine the sense, location and magnitude of the circulation ellipses, and sketch the streamlines. Explain whether the circulation ellipses are direct or indirect.

  (b) Indicate whether shear or confluence is primarily responsible for each cell. Try to do this using both a mathematical and a graphical argument.

16.2 If the vertical motion is 5 cm s^{-1}, the horizontal motion 20 m s^{-1}, the value of wet bulb potential temperature θ_{wb} in the surface layer 22°C and the potential temperature above the base of the capping inversion is 44°C, how long will it take for the negative area to disappear on the sounding and the result of lifting, and how far downstream will this take place?

Further reading

Benjamin, S. G. and T. N. Carlson 1986. Some effects of surface heating and topography on the regional severe storm environment. Part I: Three-dimensional simulations. *Mon. Wea. Rev.* **114**, 307–29.

Carlson, T. N., R. A. Anthes, M. Schwarz, S. G. Benjamin and D. G. Baldwin 1980. Analysis and prediction of severe storms environment. *Bull. Am. Met. Soc.* **61**, 1018–32.

Carlson, T. N., S. G. Benjamin, G. S. Forbes and Y.-F. Li 1983. Elevated mixed layers in the regional severe storm environment: conceptual model and case studies. *Mon. Wea. Rev.* **111**, 1453–73.

Carlson, T. N. and F. H. Ludlam 1968. Conditions for the occurrence of severe local storms. *Tellus* **20**, 203–26.

Farrell, R. and T. N. Carlson 1989. Evidence for the role of the lid and underrunning in an outbreak of tornadic thunderstorms. *Mon. Wea. Rev.* 117, 857–71.

Fawbush, E. J. and R. C. Miller 1954. The types of air masses in which North American tornadoes form. *Bull. Amer. Meteor. Soc.* 35, 154–65.

Fennessey, M. J. and Y. Sud 1983. *A study of the influence of soil moisture on future precipitation.* NASA Technical Memorandum no. 85042, NASA, Greenbelt, MD.

Keyser, D. and T. N. Carlson 1984. Transverse ageostrophic circulations associated with elevated mixed layers. *Mon. Wea. Rev.* 112, 2465–78.

Lakhtakia, M. and T. T. Warner 1987. A real-data numerical study of the development of precipitation along the edge of an elevated mixed layer. *Mon Wea. Rev.* 115, 156–68.

McCarthy, J. and S. E. Koch 1982. The evolution of an Oklahoma dryline. Part I: A meso- and subsynoptic-scale analysis. *J. Atmos. Sci.* 39, 225–36.

Palmen, E. and C. W. Newton 1969. *Atmospheric circulation systems.* New York: Academic Press.

Appendix: list of symbols

Roman symbols

R_E	radius of Earth
R_t	radius of curvature of trajectory
i, j, k	unit vectors
$u, v, w\ (\omega)$	components of wind speed in x, y, z (or x, y, p) coordinates
Z	geopotential height of constant-pressure surface
s, n	tangential and normal directions in natural coordinate system
R_s	radius of curvature of streamlines
r	radial distance from an origin in curved system; ratio of wind speed at 10 m to geostrophic wind speed at 1000 mb; mixing ratio
g	gravitational constant
f	Coriolis parameter
f_0	spatially averaged Coriolis parameter – independent of latitude
d	finite difference grid increment
A, σ	area of arbitrarily closed curve
$A(p), B(p), N(p), F(p)$	functions in equivalent barotropic or two-parameter (baroclinic) models
A_s	area of advection solenoid
Δp	arbitrary depth (in mb) of an atmospheric column (subscripts denote depth in 100s of mb)
K_0	value of $\partial Z / \partial p$ at the surface
C_d	drag coefficient
F_0	friction term in vorticity equation
L, L_x, L_y	wavelengths
F, F_x, F_y	frictional force and frictional force components per unit mass in x and y directions
U_*	friction velocity
D_p	horizontal divergence on a constant-pressure surface

R_d, R_m, R_v	universal gas constants for dry air, for moist air and for water vapor
L_e	latent heat of vaporization for water
c_p	specific heat at constant pressure
c_v	specific heat at constant volume
q, q_s	specific humidity (subscript s for saturation)
p_0	reference pressure (1000 mb)
Ro; R_T	Rossby number; Thermal Rossby number
T_w	wet-bulb temperature
Ri	Richardson number
M	Montgomery streamfunction
C_*	empirical constant
D, a, b, a', b', a''	constants in linearized omega equation
c	phase velocity
L_R	Rossby radius of deformation
k	kinetic energy of parcel per unit mass
e	potential energy of a parcel per unit mass
$[A]$, $[K]$	available potential energy and kinetic energy per unit surface area
$[A_E]$, $[A_Z]$	available potential energy per unit surface area (eddy, zonal components)
$[K_E]$, $[K_Z]$	kinetic energy per unit surface area (eddy, zonal components)
$[G_E]$, $[G_Z]$	generation rate of available potential energy per unit surface area (eddy, zonal components) due to diabatic heating
$[D_E]$, $[D_Z]$	dissipation rate of kinetic energy per unit surface area (eddy, zonal components) due to friction; (negative numbers)
G	geopotential height of G ("beta wind") field
U, u	zonal wind speed averaged in x or over a latitude band
$(\overset{..}{\bar{}})$	vertical average; time average
()	zonal average
$(\hat{})$	area average; amplitude
$(\bar{})$	vertical (linear or logarithm of pressure) or time average
$()''$	deviation from area average
$()^*$	deviation from zonal average
$()'$	deviation from time average
s_d, s_m, s', s'_m	dry, moist and scaled static stability; static stability increment $(s_d - s_m)$
a_h	zonally averaged meridional thickness gradient

a_z	zonally averaged geopotential meridional height gradient
T, T_v	temperature virtual temperature
H_s	height of smoothed terrain
V	wind speed
ΔZ	thickness
h	1000–500 mb thickness
R	rainfall rate
p^*	depth of disturbance
$(\tilde{\ })$	spatial average
F_1, F_2, F_3	forcing functions
kt	knot
mb	millibar (= 100 Pa)
K	degrees Kelvin
°C	degrees Celsius
m, m	meters, mass
s	second
∂	partial derivative
p	pressure
$\dot{Q}$	rate of addition of heat to a parcel ($= c_p\,\theta$) per unit mass
LCL	lifting condensation level
T_e	equivalent temperature
u_{ag}, v_{ag}	ageostrophic u and v wind components
V_T	thermal wind (generally 1000–500 mb)
$\overline{T_v}$	mean vertical temperature averaged over column
$(\dot{\ })$	d( )/dt
Q_{yp}, Q_{xp}	Q-vector component in yp and xp planes
Q_{yp}^d	diabatic component of Q_{yp}
Z_r	domain-averaged reference value or zonally averaged geopotential height
D_{pg}	horizontal geostrophic divergence
C_f	vorticity drag coefficient
θ_f	freezing adiabat
$\Delta t, T$	time interval
P, P_θ	potential vorticity; potential vorticity on a potential temperature surface; potential energy
h_d, h_m	mixing layer depths
L	distance along latitude circle; wavelength
$J_{xy}(a, b)$	Jacobian in x and y of a and b
W	precipitable water

Greek symbols

ζ	relative vorticity (vertical component)
θ	latitude; potential temperature; angle between axis of dilatation and x axis
$\dot{\theta}$	rate of change of potential temperature with time
ϕ	phase lag (distance)
β	beta parameter $(\partial f/\partial y)$
γ	wind direction angle (positive in clockwise sense); thermal wind constant
ω	vertical motion in p coordinates
τ, τ_x, τ_y	wind stress and components
γ	lapse rate $(-\partial T/\partial z)$
γ_d	adiabatic lapse rate (g/c_p)
γ_m	moist adiabatic lapse rate
∇	total gradient operator
∇_p	horizontal (constant-pressure) gradient operator
α	angle between isotherms and axis of dilatation; angle between thickness gradient vector and geostrophic wind velocity; angle between two vectors
∇_z	gradient operator on constant-height surface
$\theta, \theta_w, \theta_e, \theta_{sw}, \theta_{wb}$	potential temperature, wet-bulb potential temperature, equivalent potential temperature, saturation and saturation wet-bulb potential temperature, value of θ_w near surface
$\omega_d, \omega_f, \omega_m, \omega_{db}$	vertical motion in p coordinates obtained from quasi-geostrophic omega equation without diabatic heating or terrain effects, due to frictional forcing, due to terrain slope and due to diabatic heating
Δ	potential temperature jump
ρ_w	density of liquid water
Ω	angular rotation rate of Earth
ζ_e	Earth's vorticity
∇_p^2	two-dimensional Laplacian operator on pressure surface
μ	coefficient of (eddy) viscosity; convective parameter constant of proportionality
Γ	potential temperature lapse rate $(-\partial\theta/\partial p$ or $\partial\theta/\partial z)$
$\bar{\gamma}$	mean lapse rate over column
ζ_a	absolute vorticity
χ	tendency of the geopotential height $(\partial Z/\partial t)$
ψ	streamfunction
κ	R_d/c_p

Subscripts and superscripts

f	at top of planetary boundary layer
g	geostrophic
z (vertical coordinate)	constant-height surface
p (pressure coordinate)	constant pressure
0, 9, 8, ...	1000, 900, 800, ... mb surfaces
s	surface or sea level
F	frictionally induced vertical motion
M	orographically induced vertical motion
db	diabatic
nd	non-dry adiabatic
ad	dry adiabatic
ma	moist adiabatic
m	following a moist adiabat; at mean level
T	geostrophic thermal wind; usually 1000–500 mb layer
*	at equivalent barotropic level
b, t	bottom, top of column
R	relative; Rossby radius
ag	ageostrophic
gm	geostrophic momentum
3	three-dimensional vector (in x, y, z)
d	dry adiabatic, dynamic forcing
E	earth
a	absolute

Selected references, by subject area

Vorticity and vertical motion

Astling, E. G. 1976. Some aspects of cloud and precipitation features associated with mid-latitude cyclones. *Mon. Wea. Rev.* **104**, 1466–73.

Charney, J. G. 1947. The dynamics of long waves in a baroclinic westerly current. *J. Meteor.* **4**, 135–62.

Charney, J. G. 1948. *On the scale of atmospheric motions.* Geophys. Publ. no. 17.

Charney, J. G. and N. A. Phillips 1953. Numerical integration of the quasi-geostrophic equations for barotropic and simple baroclinic flows. *J. Meteor.* **10**, 71–99.

Dutton, J. A. 1976. *The ceaseless wind.* New York: McGraw-Hill.

Glossary of Meteorology 1959. Rittuschke (ed.). Boston, MA: American Meteorological Society.

Haltiner, G. J. 1971. *Numerical weather prediction.* New York: Wiley.

Holton, J. R. 1979. *An introduction to dynamic meteorology*, 2nd edn (Int. Geophys. Ser., Vol. 23). New York: Academic Press.

McIlveen, R. 1988. *Basic meteorology. A physical outline.* London: Van Nostrand Reinhold.

MacVean, M. K. and I. N. James 1986. On the differences between the life cycles of some baroclinic waves using the primitive and quasi-geostrophic equations on a sphere. *J. Atmos. Sci.* **43**, 741–8.

Ninomiya, K. 1971. Mesoscale modifications of synoptic situations from thunderstorm development as revealed by ATS III and aerological data. *J. Appl. Meteor.* **10**, 1103–21.

Petterssen, S. 1956. *Weather analysis and forecasting II.* New York: McGraw-Hill.

Saucier, W. J. 1955. *Principles and practice of synoptic analysis.* Chicago, IL: University of Chicago Press.

Younkin, R. J., R. A. LaRue and F. Sanders 1965. The objective prediction of clouds and precipitation using vertically integrated moisture and adiabatic vertical motions. *J. Appl. Meteor.* **4**, 3–17.

Zwack, P. and B. Okossi 1986. A new method for solving the quasi-geostrophic omega equation by incorporating surface pressure tendency data. *Mon. Wea. Rev.* **114**, 655–66.

Energetics

Bullock, B. R. and D. R. Johnson 1971. The generation of available potential energy by latent heat release in a mid-latitude cyclone. *Mon. Wea. Rev.* **99**, 1–14.

Colucci, S. J. 1987. Comparative diagnosis of blocking versus nonblocking planetary-scale circulation changes during synoptic-scale cyclogenesis. *J. Atmos. Sci.* **44**, 124–39.

Danard, M. B. 1964. On the influence of released latent heat on cyclone development. *J. Appl. Meteor.* **3**, 27–37.

Danard, M. B. 1966. On the contribution of released latent heat to changes in available potential energy. *J. Appl. Meteor.* **5**, 81–4.

Edmond, H. J. Jr, B. J. Hoskins and M. E. McIntyre 1980. Eliassen–Palm cross sections for the troposphere. *J. Atmos. Sci.* **37**, 2600–16.

Fuelberg, H. E., M. G. Rumenski and E. O'C. Starr 1985. Some aspects of available potential energy in the warm sector of an extratropical cyclone. *Mon. Wea. Rev.* **113**, 1150–65.

Hensen, A. L. and A. Sutuna 1984. A comparison of the spectral energy and entropy budgets of blocking versus non-blocking periods. *Tellus* **36A**, 52–63.

Holopainen, E. and C. Fortelius 1987. High-frequency transient eddies and blocking. *Mon. Wea. Rev.* **44**, 1632–45.

Kenny, S. E. and P. J. Smith 1983. On the release of eddy available potential energy in an extratropical cyclone system. *Mon. Wea. Rev.* **111**, 745–55.

Kohler, T. L. and K.-D. Min 1984. Available potential energy and extratropical cyclone activity during the FGGE year. *Tellus* **36A**, 64–75.

Kung, E. C. 1977. Energy sources in mid-latitude synoptic-scale disturbances. *J. Atmos. Sci.* **34**, 1352–65.

Kung, E. C. and W. Baker 1986. Spectral energetics of the observed and simulated northern hemisphere general circulation during blocking episodes. *J. Atmos. Sci.* **43**, 2792–812.

Lin, S. C. and P. J. Smith 1979. Diabatic heating and generation of available potential energy in a tornado-producing extratropical cyclone. *Mon. Wea. Rev.* **107**, 1169–83.

Lorenz, E. N. 1955. Available potential energy and maintenance of the general circulation. *Tellus* **7**, 157–67.

Lorenz, E. N. 1967. *The nature and theory of the general circulation of the atmosphere.* WMO Publ. no. 218.TP.115.

McGinley, J. A. and J. S. Goerss 1986. Effects of terrain height and blocking initialization on numerical simulation of Alpine lee cyclogenesis. *Mon. Wea. Rev.* **114**, 1578–90.

McIntyre, M. E. 1980. An introduction to the generalized Lagrangian-mean description of wave, mean-flow interaction. *Pure Appl. Geophys.* **118**, 152–76.

Mullen, S. L. 1987. Transient eddy forcing of blocking flows. *J. Atmos. Sci.* **44**, 3–22.

Newell, R. E., J. W. Kidson, D. G. Vincent and G. J. Boer 1974. *The general circulation of the tropical atmosphere*, Vol. II. Cambridge, MA: M.I.T. Press.

Oort, A. H. 1964. On the estimates of the atmospheric energy cycle. *Mon. Wea. Rev.* **82**, 483–93.

Palmen, E. and E. O. Holopainen 1962. Divergence, vertical velocity and conversion between potential and kinetic energy in an extratropical disturbance. *Geophysica (Helsinki)* **8**, 89–113.

Pauley, P. M. and P. J. Smith 1988. Direct and indirect effects of latent heat release on a synoptic-scale wave system. *Mon. Wea. Rev.* **116**, 1209–35.

Plumb, R. A. 1983. A new look at the energy cycle. *J. Atmos. Sci.* **40**, 1669–88.

Robertson, F. R. and P. J. Smith 1983. The impact of model moist processes on the energetics of extratropical cyclones. *Mon. Wea. Rev.* **111**, 723–44.

Salmon, R. and M. G. Hendershott 1976. Large scale air–sea interactions. *Tellus* **28**, 228–42.

Saltzman, B. 1957. Equations governing the energetics of the larger scales of atmospheric turbulence in the domain of wave number. *J. Meteor.* **14**, 513–23.

Saltzman, B. 1970. Large-scale atmospheric energetics in the wave-number domain. *Rev. Geophys. Space Phys.* **8**, 289–302.

Smith, P. J. 1973. The kinetic energy budget over North America during a period of major cyclone development. *Tellus* **28**, 411–23.

Smith, P. J. 1980. The energetics of extratropical cyclones. *Rev. Geophys. Space Phys.* **18**, 378–86.

Tsou, C.-H., P. J. Smith and P. M. Pauley 1987. A comparison of adiabatic and diabatic forcing in an intense extratropical cyclone system. *Mon. Wea. Rev.* **115**, 763–86.

Vincent, D. G., G. B. Pant and H. J. Edmon Jr 1977. Generation of available potential energy of an extratropical cyclone system. *Mon. Wea. Rev.* **105**, 1252–63.

Winston, J. S. and A. F. Krueger 1961. Some aspects of a cycle of available potential energy. *Mon. Wea. Rev.* **89**, 307–18.

Zeng, Q.-L. 1983. The development characteristics of quasi-geostrophic baroclinic disturbances. *Tellus* **35A**, 337–49.

Barotropic motions

Buch, H. 1954. *Hemispheric wind conditions during the year 1950*. Final Report, Part 2, Contract AF 19(122)-153, Dept Meteor., M.I.T., Cambridge, MA, pp. 32–80.

Fjortoft, R. 1952. On a numerical method of integrating the barotropic vorticity equation. *Tellus* **4**, 179–94.

Haltiner, G. J. and F. L. Martin 1957. *Dynamical and physical meteorology*. New York: McGraw-Hill.

Lorenz, E. N. 1972. Barotropic instability of Rossby wave motion. *J. Atmos. Sci.* **29**, 258–64.

McClain, E. P. 1960. Some effects of the western cordillera of North America on cyclone activity. *J. Meteor.* **17**, 104–15.

Randall, W. J. and J. L. Stanford 1985. The observed life cycle of a baroclinic instability. *J. Atmos. Sci.* **42**, 1364–73.

Staff Members, Forecasting Branch, U.S. Dept of Commerce Weather Bureau 1961. *Synoptic meteorology as practiced by the National Meteorological Center*. The NAWAC Manual, Part II.

Some simple models of vertical motion and development

Hoskins, B. J., I. Draghici and H. C. Davies 1978. A new look at the *x*-equation. *Q. J. R. Met. Soc.* **104**, 31–8.

Hoskins, B. J. and M. A. Pedder 1980. The diagnosis of middle latitude synoptic development. *Q. J. R. Met. Soc.* **106**, 707–19.

Reed, R. J. 1963. *Experiments in 1000 mb prognosis*. Nat. Meteor. Center, Wea. Bur., ESSA (NOAA). Tech. Memo no. 26. U.S. Dept of Commerce.

Roebber, P. J. 1984. Statistical analysis and updated climatology of explosive cyclones. *Mon. Wea. Rev.* **112**, 1577–89.

Sutcliffe, R. C. 1947. A contribution to the problem of development. *Q. J. R. Met. Soc.* **73**, 370–83.

Trenberth, K. E. 1978. On the interpretation of the diagnostic quasi-geostrophic omega equation. *Mon. Wea. Rev.* **106**, 131–7.

Some dynamic aspects of the wave/cyclone

Anthes, R. A., Y.-H. Kuo and J. R. Gyakum 1983. Numerical simulations of a case of explosive marine cyclogenesis. *Mon. Wea. Rev.* **111**, 1174–88.

Blackmon, M. L., S. L. Mullen and G. T. Bates 1986. The climatology of blocking events in a perpetual January simulation of a spectral general circulation model. *J. Atmos. Sci.* **43**, 1379–405.

Bosart, L. F. 1981. The President's Day snowstorm of 18–19 February 1979: a subsynoptic scale event. *Mon. Wea. Rev.* **109**, 1542–66.

Buzzi, A. and A. Speranza 1986. A theory of deep cyclogenesis in the lee of the Alps. Part II: Effects of finite topographic slope and height. *J. Atmos. Sci.* **43**, 2826–37.

Buzzi, A. and S. Tibaldi 1978. Cyclogenesis in the lee of the Alps: a case study. *Q. J. R. Met. Soc.* **104**, 271–87.

Chen, S.-J. and L. Dell'Osso 1987. A numerical case study of east Asian coastal cyclogenesis. *Mon. Wea. Rev.* **115**, 477–87.

Chen, T.-C., C.-B. Chang and D. Perkey 1983. Numerical study of an AMTEX '75 oceanic cyclone. *Mon. Wea. Rev.* **111**, 1818–29.

Cho, H.-R. 1986. Comments on Mak's closure for CISK in geostrophic systems. *J. Atmos. Sci.* **43**, 312–16.

Chung, Y.-S., K. D. Hage and E. R. Reinhelt 1976. On lee-side cyclogenesis in the Canadian Rocky Mountains and the East Asian mountains. *Mon. Wea. Rev.* **104**, 879–91.

Colucci, S. J. 1985. Explosive cyclogenesis and large-scale circulation changes: implications for atmospheric blocking. *J. Atmos. Sci.* **42**, 2701–17.

Colucci, S. J. and J. C. Davenport 1987. Rapid surface anticyclogenesis, synoptic climatology and attendant large-scale circulation changes. *Mon. Wea. Rev.* **115**, 822–36.

Danard, M. B. and G. E. Ellenton 1980. Physical influences of east coast cyclogenesis. *Atmosphere–Ocean* **18**, 354–63.

Farrell, B. 1985. Transient growth of damped baroclinic waves. *J. Atmos. Sci.* **42**, 2718–27.

Gyakum, J. R. 1983a. On the evolution of the QE II storm. I: Synoptic aspects. *Mon. Wea. Rev.* **111**, 1137–55.

Gyakum, J. R. 1983b. On the evolution of the QE II storm. II: Dynamic and thermodynamic structure. *Mon. Wea. Rev.* **111**, 1156–73.

Hanson, H. P. and B. Long 1985. Climatology of cyclogenesis over the East China Sea. *Mon. Wea. Rev.* **113**, 697–707.

Hovanec, R. D. and L. H. Horn 1975. Static stability and the 300 mb isotach field in the Colorado cyclogenetic area. *Mon. Wea. Rev.* **103**, 628–38.

Lejenas, H. H. and H. Oakland 1983. Hemispheric blocking as determined from a long series of observational data. *Tellus* **35A**, 350–62.

Mattocks, C. and R. Bleck 1986. Jet streak dynamics and geostrophic adjustment processes during the initial stages of lee-cyclogenesis. *Mon. Wea. Rev.* **114**, 2033–56.

Miller, J. E. 1946. Cyclogenesis in the Atlantic coastal region of the United States. *J. Meteor.* **3**, 31–44.

Nuss, W. A. and R. A. Anthes 1987. A numerical investigation of low-level processes in rapid cyclogenesis. *Mon. Wea. Rev.* **115**, 2728–43.

Pagnotti, V. and L. F. Bosart 1984. Comparative diagnostic case study of east coast secondary cyclogenesis under weak versus strong synoptic-scale forcing. *Mon. Wea. Rev.* **112**, 5–30.

Reed, R. J. and M. D. Albright 1986. A case study of explosive cyclogenesis in the eastern Pacific. *Mon. Wea. Rev.* **114**, 2297–319.

Rogers, E. and L. F. Bosart 1986. An investigation of explosively deepening oceanic cyclones. *Mon. Wea. Rev.* **114**, 702–18.

Sanders, F. 1986a. Explosive cyclogenesis in the west-central North Atlantic Ocean. 1981–1984. Part I: Composite structure and mean behavior. *Mon. Wea. Rev.* **114**, 1781–94.

Sanders, F. 1986b. Explosive cyclogenesis over the west-central North Atlantic Ocean. 1981–1984. Part II: Evaluation of LFM model performance. *Mon. Wea. Rev.* **114**, 2207–18.

Sanders, F. and J. R. Gyakum 1980. Synoptic–dynamic climatology of the "bomb". *Mon. Wea. Rev.* **108**, 1589–606.

Smith, P. J., P. M. Dare and S.-J. Lin 1984. The impact of latent heat release on synoptic-scale vertical motions and the development of an extratropical cyclone system. *Mon. Wea. Rev.* **112**, 2421–30.

Smith, R. B. 1982. Synoptic observations and theory of orographically disturbed wind and pressure. *J. Atmos. Sci.* **39**, 60–70.

Smith, R. B. 1984. A theory of lee cyclogenesis. *J. Atmos. Sci.* **41**, 1159–68.

Smith, R. B. 1986. Further development of a theory of lee cyclogenesis. *J. Atmos. Sci.* **43**, 1583–602.

Sparanza, A., A. Buzzi, A. Trevisan and P. Malguzzi 1985. A theory of deep cyclogenesis in the lee of the Alps. Part I. Modification of baroclinic instability by localized topography. *J. Atmos. Sci.* **42**, 1521–35.

Tibaldi, S. and A. Buzzi 1983. Effects of orography on Mediterranean lee cyclogenesis and its relationship to European blocking. *Tellus* **35A**, 269–86.

Tibaldi, S., A. Buzzi and P. Malguzzi 1980. Orographically induced cyclogenesis: analysis of numerical experiments. *Mon. Wea. Rev.* **108**, 1302–14.

Tosi, E., M. Fantini and A. Trivasan 1983. Numerical experiments on orographic cyclogenesis: relationships between the development of the lee cyclone and the basic flow characteristics. *Mon. Wea. Rev.* **111**, 799–814.

Tracton, M. S. 1973. The role of cumulus convection in the development of extra tropical cyclones. *Mon. Wea. Rev.* **101**, 573–93.

Uccellini, L., D. Keyser, K. F. Brill and C. H. Wash 1985. The President's Day cyclone of 18–19 February 1979: influence of upstream trough amplification and associated tropopause folding on rapid cyclogenesis. *Mon. Wea. Rev.* **113**, 962–88.

Baroclinic development

Bjerknes, J. and H. Solberg 1922. Life cycle of cyclones and the polar front theory of atmospheric circulation. *Geofys. Publ.* **3**, 1–18.

Browning, K. A. 1986. Conceptual models of precipitation systems. *Wea. Forecast.* **1**, 25–41.

Carlson, T. N. 1982. A simple model illustrating baroclinic development. *Bull. Am. Met. Soc.* **63**, 1302–8.

Forbes, G. S. and W. D. Lottes 1985. Classification of mesoscale vortices in polar airstreams and the influence of the large-scale environment on their evolutions. *Tellus* **37A**, 132–55.

Gall, R. 1976a. A comparison of linear baroclinic instability theory with the eddy statistics of a general circulation model. *J. Atmos. Sci.* **33**, 349–73.

Gall, R. 1976b. Structural changes of growing baroclinic waves. *J. Atmos. Sci.* **33**, 374–90.

Gall, R. 1976c. The effects of released latent heat in growing baroclinic waves. *J. Atmos. Sci.* **33**, 1686–701.

Gall, R., R. Blakeslee and R. C. Somerville 1979. Baroclinic instability and the selection of the zonal scale of the transient eddies of middle latitudes. *J. Atmos. Sci.* **36**, 767–84.

Green, J. S. A. 1960. A problem in baroclinic stability. *Q. J. R. Met. Soc.* **86**, 237–51.

Harrold, T. W. and K. A. Browning 1969. The polar low as a baroclinic disturbance. *Q. J. R. Met. Soc.* **95**, 719–30.

Kutzbach, G. 1979. *The thermal theory of cyclones*, (Hist. Monogr. Ser.). Boston, MA: American Meteorological Society.

Lilly, D. 1960. On the theory of disturbances in a conditionally unstable atmosphere. *Mon. Wea. Rev.* **88**, 1–17.

Mansfield, D. A. 1974. Polar lows: the development of baroclinic disturbances in cold air outbreaks. *Q. J. R. Met. Soc.* **100**, 541–9.

Moorthi, S. and A. Arakawa 1985. Baroclinic instability with cumulus heating. *J. Atmos. Sci.* **42**, 2007–31.

Mullen, S. L. 1979. An investigation of small synoptic-scale cyclones in polar air-streams. *Mon. Wea. Rev.* **107**, 1636–47.

Mullen, S. L. 1982. Cyclone development in polar air streams over the wintertime continent. *Mon. Wea. Rev.* **110**, 1664–76.

Petersen, R. A., L. W. Uccellini, A. Moster and D. A. Keyser 1984. Delineating mid- and low-level water vapor patterns in pre-convective environments using VAS moisture channels. *Mon. Wea. Rev.* **112**, 2178–98.

Rasmussen, E. 1979. The polar low as an extratropical CISK disturbance. *Q. J. R. Met. Soc.* **105**, 531–49.

Rasmussen, E. 1981. An investigation of a polar low with a spiral cloud structure. *J. Atmos. Sci.* **38**, 1785–92.

Reed, R. J. 1979. Cyclogenesis in polar air streams. *Mon. Wea. Rev.* **107**, 38–52.

Reiter, E., D. W. Beran, J. D. Mahlman and G. Wooldridge 1965. *Effects of large mountain ranges on atmospheric flow patterns as seen from TIROS satellites.* Rep. no. 2, Project WISP. Atmos. Sci. Pap. no. 69, Colo. State Univ.

Sanders, F. 1971. Analytic solutions of the nonlinear omega and vorticity equations for a structurally simple model of disturbances in the baroclinic westerlies. *Mon. Wea. Rev.* **99**, 393–407.

Sardie, J. M. and T. T. Warner 1983. On the mechanism for the development of polar lows. *J. Atmos. Sci.* **40**, 869–81.

Simmons, A. J. and B. J. Hoskins 1978. The life cycles of some nonlinear baroclinic waves. *J. Atmos. Sci.* **35**, 414–32.

Sinclair, M. R. and R. L. Elsberry 1986. A diagnostic study of baroclinic disturbances in polar air streams. *Mon. Wea. Rev.* **114**, 1957–83.

Staley, D. O. and R. Gall 1977. On the wavelength of maximum baroclinic instability. *J. Atmos. Sci.* **34**, 1679–88.

White, G. H. 1982. An observational study of the northern hemisphere extratropical summertime general circulation. *J. Atmos. Sci.* **39**, 24–40.

Young, M. V., G. A. Monk and K. A. Browning 1987. Interpretation of satellite imagery of a rapidly deepening cyclone. *Q. J. R. Met. Soc.* **113**, 1089–116.

Airflow through mid-latitude synoptic-scale disturbances

Atkinson, B. W. and P. A. Smithson 1978. Mesoscale precipitation areas in a warm frontal wave. *Mon. Wea. Rev.* **106**, 211–22.

Betts, A. K. and J. F. R. McIlveen 1979. The energy formula in a moving reference frame. *Q. J. R. Met. Soc.* **95**, 639–42.

Browning, K. A. 1974. Mesoscale structure of rain systems in the British Isles. *J. Met. Soc. Japan* **50**, 314–27.

Browning, K. A. 1985. *Conceptual models of precipitation systems.* Met. Off. RRL Res. Rep. no. 43.

Browning, K. A., M. E. Hardman, T. W. Harrold and C. W. Pardoe 1973. The structure of rainbands within a midlatitude depression. *Q. J. R. Met. Soc.* **99**, 215–312.

Browning, K. A. and T. W. Harrold 1969. Air motion and precipitation growth in a wave depression. *Q. J. R. Met. Soc.* **95**, 288–309.

Browning, K. A., F. F. Hill and C. W. Pardoe 1974. Structure and mechanism of precipitation and the effect of orography in a wintertime warm sector. *Q. J. R. Met. Soc.* **100**, 309–30.

Browning, K. A. and G. A. Monk 1982. A simple model for the synoptic analysis of cold fronts. *Q. J. R. Met. Soc.* **100**, 435–52.

Browning, K. A. and C. W. Pardoe 1973. Structure of low-level jet stream ahead of midlatitude cold fronts. *Q. J. R. Met. Soc.* **99**, 619–38.

Carlson, T. N. 1980. Airflow through midlatitude cyclones and the comma cloud pattern. *Mon. Wea. Rev.* **108**, 1498–509.

Carlson, T. N. 1981. Speculations on the movement of polluted air to the Arctic. *Atmos. Env.* **15**, 1473–7.

Carr, F. H. and J. P. Millard 1985. A composite study of comma clouds and their association with severe weather over the Great Plains. *Mon. Wea. Rev.* **113**, 349–61.

Danielsen, E. F. 1966. *Research in four-dimensional diagnosis of cyclonic storm cloud systems.* Penn. State Univ., Sci. Rep. no. 2, Project Rep. no. 6698 to Air Force Cambridge Research Laboratories, AF19(628)-4762.

Danielsen, E. F. 1968. Stratospheric–tropospheric exchange based on radioactivity, ozone and potential vorticity. *J. Atmos. Sci.* **25**, 502–18.

Danielsen, E. F. 1974. The relationship between severe weather, major dust storms and rapid large-scale cyclogenesis (II). Subsynoptic extratropical weather systems: observations, analysis and prediction. Notes from a colloquium. Summer 1974, Vol II. NCAR Rep. no. ASP-CO-3-V-2 226-241.

Danielsen, E. F. and R. Bleck 1967. *Research in four-dimensional diagnosis of cyclonic storm cloud systems.* Final Sci. Rep. to Air Force Cambridge Research Laboratories, AF19(628)-4762.

Durran, D. R. and D. Weber 1988. An investigation of the poleward edges of cirrus clouds associated with mid-latitude jet streams. *Mon. Wea. Rev.* **116**, 702–14.

Eliassen, A. and E. Kleinschmidt 1957. *Dynamic meteorology* (Hand. Phys., Vol. 48). Berlin: Springer-Verlag.

Green, J. S. A., F. H. Ludlam and J. F. R McIlveen 1966. Isentropic relative-flow analysis and the parcel theory. *Q. J. R. Met. Soc.* **92**, 210–19.

Grotjahn, R. 1979. Cyclone development along weak thermal fronts. *J. Atmos. Sci.* **36**, 2049–74.

Grotjahn, R. and C.-H. Wang 1989. On the source of air modified by ocean surface fluxes to enhance frontal cyclone development. *Ocean–Air Interact.* 257–88.

Harrold, T. W. 1973. Mechanisms influencing the distribution of precipitation within baroclinic disturbances. *Q. J. R. Met. Soc.* **99**, 232–51.

Harrold, T. W. and P. M. Anston 1974. The structure of precipitation systems – a review. *J. Rech. Atmos.* **8**, 41–57.

Jackson, M. L., D. A. Gillette, E. F. Danielsen, I. H. Blifford, R. A. Bryson and

J. K. Syers 1973. Global dustfall during the Quaternary as related to environments. *Soil Sci.* **116**, 135–45.

Ludlam, F. H. 1980. *Clouds and storms*. University Park, PA: Pennsylvania State University Press.

Mullen, S. 1985. On the maintenance of a blocking anticyclone in a general circulation model. Ph.D. Thesis, University of Washington. NCAR Cooperative Thesis no. 86.

Newton, C. W. 1956. Mechanism of circulation change during a lee cyclogenesis. *J. Meteor.* **13**, 528–39.

Palmen, E. and C. W. Newton 1969. *Atmospheric circulation systems*. New York: Academic Press.

Reiter, E. and J. D. Mahlman 1965. Heavy radioactive fallout in the southern United States, November, 1962. *J. Geophys. Res.* **70**, 4501–20.

Saltzman, B. and C.-M. Tang 1972. Analytical study of the evolution of an amplifying baroclinic wave. *J. Atmos. Sci.* **29**, 427–44.

Saltzman, B. and C.-M. Tang 1972. Analytical study of the evolution of an amplifying baroclinic wave: Part II. Vertical motions and transport properties.. *J. Atmos. Sci.* **32**, 243–59.

Saltzman, B. and C.-M. Tang 1985. The effect of finite-amplitude baroclinic waves on passive low-level, atmospheric constituents, with applications to comma cloud evolution. *Tellus* **37A**, 41–55.

Staley, D. O. 1960. Evaluation of potential vorticity changes near the tropopause and the related vertical motions, vertical advection of vorticity, and transfer of radioactive debris from stratosphere to troposphere. *J. Meteor.* **17**, 591–620.

Weldon, R. 1979. *Cloud patterns and the upper air wind field*. Satellite Training Service Course Notes, Part IV. Air Weather Service Document AWS/TR-79/003.

Blocking and downstream development

Austin, J. F. 1980. The blocking of mid-latitude westerly wind by planetary waves. *Q. J. R. Met. Soc.* **106**, 327–50.

Boyle, J. S. and L. F. Bosart 1983. A cyclone/anticyclone couplet over North America: an example of anticyclone evolution. *Mon. Wea. Rev.* **111**, 1025–45.

Colucci, S. J. 1985. Explosive cyclogenesis and large-scale circulation changes: implications for atmospheric blocking. *J. Atmos. Sci.* **42**, 2701–17.

Cressman, G. P. 1959. On the forecasting of long waves in the upper westerlies. *J. Meteor.* **5**, 54–7.

Dole, R. M. 1983. Persistent anomalies of the extra-tropical Northern Hemisphere winter-time circulation. In *Large-scale dynamical processes in the atmosphere*, 95–109. New York: Academic Press.

Farrell, B. 1985. Transient growth of damped baroclinic waves. *J. Atmos. Sci.* **42**, 2718–27.

Fredericksen, J. S. 1984. The onset of blocking and cyclogenesis in Southern Hemisphere flows: linear theory. *J. Atmos. Sci.* **41**, 1116–31.

Fredricksen, J. S. and K. Puri 1985. Nonlinear instability and error growth in Northern Hemisphere three-dimensional flows: cyclogenesis, onset-of-blocking and mature anomalies. *J. Atmos. Sci.* **42**, 1374–97.

Gall, R. J., R. T. Williams and T. L. Clark 1987. On the minimum scale of surface fronts. *Mon. Wea. Rev.* **44**, 2562–74.

Illari, L. 1984. A diagnostic study of the potential vorticity in a warm blocking anticyclone. *J. Atmos. Sci.* **41**, 3518–26.

Illari, L. and J. C. Marshall 1983. On the interpretation of eddy-fluxes during a blocking episode. *J. Atmos. Sci.* **40**, 2232–42.

Keshishian, L. G. and L. F. Bosart 1987. A case study of extended east coast frontogenesis. *Mon. Wea. Rev.* **115**, 100–12.

Kidson, J. W. 1985. Index cycles in the Northern Hemisphere during the Global Weather Experiment. *Mon. Wea. Rev.* **113**, 607–23.

Kidson, J. W. 1986. Index cycles in the Southern Hemisphere during the Global Weather Experiment. *Mon. Wea. Rev.* **114**, 1654–63.

Legras, B. and M. Ghil 1985. Persistent anomalies, blocking and variations in atmospheric predictability. *J. Atmos. Sci.* **42**, 433–71.

Miles, M. K. 1959. Factors leading to the meridional extension of thermal troughs and some forecasting criteria derived from them. *Met. Mag.* **88**, 193–205.

Namais, J. and P. F. Clapp 1949. Confluence theory of the high tropospheric jet stream. *J. Meteor.* **6**, 330–6.

Rex, D. F. 1950a. Blocking action in the middle troposphere and its effect on regional climate. Part I: An aerological study of blocking action. *Tellus* **2**, 196–211.

Rex, D. F. 1950b. Blocking action in the middle troposphere and its effect on regional climate. Part II: The climatology of blocking action. *Tellus* **2**, 275–301.

Shutts, G. J. 1983. The propagation of eddies in diffluent jet streams: eddy vorticity forcing of blocking flow fields. *Q. J. R. Met. Soc.* **109**, 737–61.

Simmons, A. J. and B. J. Hoskins 1979. The downstream and upstream development of unstable baroclinic waves. *J. Atmos. Sci.* **36**, 1239–54.

Trenberth, K. E. 1986. The signature of a blocking episode on the general circulation in the southern hemisphere. *J. Atmos. Sci.* **43**, 2061–9.

Trenberth, K. E. and K. C. Mo 1985. Blocking in the Southern Hemisphere. *Mon. Wea. Rev.* **113**, 3–21.

Fronts and ageostrophic motion

Ballentine, R. J. 1980. A numerical investigation of New England coastal frontogenesis. *Mon. Wea. Rev.* **108**, 1479–97.

Bergeron, T. 1937. On the physics of fronts. *Bull. Am. Met. Soc.* **18**, 265–75.

Bosart, L. F. 1975. New England coastal frontogenesis. *Q. J. R. Met. Soc.* **101**, 957–78.

Bosart, L. F., C. J. Vaudo and J. H. Helsdon Jr 1972. Coastal frontogenesis. *J. Appl. Meteor.* **11**, 1236–58.

Eliassen, A. 1962. On the vertical circulation in frontal zones. *Geofys. Publ.* **24**, 147–60.

Gidel, L. T. 1978. Simulation of differences and similarities of warm and cold surface frontogenesis. *J. Geophys. Res.* **83**, 915–28.

Hoskins, B. J. 1975. The geostrophic momentum approximation and the semi-geostrophic equations. *J. Atmos. Sci.* **32**, 233–42.

Hoskins, B. J. and F. P. Bretherton 1972. Atmospheric frontogenesis models: mathematical formulation and solution. *J. Atmos. Sci.* **29**, 11–37.

Hoskins, B. J. and W. A. Heckley 1981. Cold and warm fronts in baroclinic waves. *Q. J. R. Met. Soc.* **107**, 79–90.

Hoskins, B. J. and N. V. West 1979. Baroclinic waves and frontogenesis. Part II: Uniform potential vorticity jet flows – cold and warm fronts. *J. Atmos. Sci.* **36**, 1663–80.

Keyser, D. and M. A. Shapiro 1986. A review of the structure and dynamics of upper-level frontal zones. *J. Atmos. Sci.* **114**, 452–99.

Miller, J. E. 1948. On the concept of frontogenesis. *J. Meteor.* **5**, 169–71.

Newton, C. W. 1954. Frontogenesis and frontolysis as a three-dimensional process. *J. Meteor.* **13**, 449–61.

Reed, R. J. and F. Sanders 1953. An investigation of the development of a mid-tropospheric frontal zone and its associated vorticity field. *J. Meteor.* **10**, 338–49.

Sanders, F. 1955. An investigation of the structure and dynamics of an intense surface frontal zone. *J. Meteor.* **12**, 542–52.

Sawyer, J. S. 1956. The vertical circulation at meteorological fronts and its relation to frontogenesis. *Proc. R. Soc.* **A234**, 246–62.

Schubert, W. H. 1985. Semigeostrophic theory. *J. Atmos. Sci.* **42**, 1770–4.

Shapiro, M. A. 1981. Frontogenesis and geostrophically forced secondary circulations in the vicinity of jet stream–frontal zone systems. *J. Atmos. Sci.* **38**, 954–73.

Shapiro, M. A. 1982. *Mesoscale Weather Systems of the Central United States.* Cooperative Institute for Research in Atmospheric Sciences, University of Colorado/National Oceanic and Atmospheric Administration.

Uccellini, L. W. and D. R. Johnson 1979. The coupling of upper and lower tropospheric jet streaks and implications for the development of severe convective storms. *Mon. Wea. Rev.* **107**, 682–701.

Upper-tropospheric fronts and jet streaks

Brill, K. F., L. W. Uccellini, R. P. Burkhart, T. T. Warner and R. A. Anthes 1985. Numerical simulations of a transverse indirect circulation and low-level jet in the exit region of an upper level jet. *J. Atmos. Sci.* **42**, 1306–20.

Hoskins, B. J. 1971. Atmospheric frontogenesis models: some solutions. *Q. J. R. Met. Soc.* **97**, 139–53.

Hoskins, B. J. 1972. Non-Boussinesq effects and further development in a model of upper tropospheric frontogenesis. *Q. J. R. Met. Soc.* **29**, 11–37.

Keyser, D. and M. J. Pecnick 1984. Diagnosis of ageostrophic circulations in a two-dimensional primitive equation model of frontogenesis. *J. Atmos. Sci.* **41**, 1283–305.

Keyser, D. and M. J. Pecnick 1985. A two-dimensional primitive equation model of frontogenesis forced by confluence and horizontal shear. *J. Atmos. Sci.* **42**, 1259–82.

Keyser, D. and M. J. Pecnick 1987. The effect of along-front temperature variation in a two-dimensional primitive equation model of surface frontogenesis. *J. Atmos. Sci.* **44**, 577–604.

Keyser, D., M. J. Pecknick and M. A. Shapiro 1986. Diagnosis of the role of vertical deformation in a two-dimensional primitive equation model of upper-level frontogenesis. *J. Atmos. Sci.* **43**, 839–50.

Kocin, P. J., L. W. Uccellini and R. A. Petersen 1988. Rapid evolution of a jet streak circulation in a pre-convective environment. *Meteor. Atmos. Phys.* **35**, 103–38.

Newton, C. W. and A. Trevisan 1984. Clinogenesis and frontogenesis in jet-stream waves. Part I: Analytical relations to wave structure. *J. Atmos. Sci.* **41**, 2717–34.

Reed, R. J. 1955. A study of a characteristic type of upper-level frontogenesis. *J. Meteor.* **12**, 226–37.

Reed, R. J. and E. F. Danielsen 1959. Fronts in the vicinity of the tropopause. *Arch. Meteor. Geophys. Bioklim.* **A11**, 1–17.

Reeder, M. and R. K. Smith 1987. A study of frontal dynamics with applications to the Australian summertime "Cool Change". *J. Atmos. Sci.* **44**, 687–705.

Reiter, E. R. 1975. Stratospheric–tropospheric exchange processes. *Rev. Geophys. Space Phys.* **13**, 459–74.

Shapiro, M. A. 1970. On the applicability of the geostrophic approximation to upper-level, frontal-scale motions. *J. Atmos. Sci.* **27**, 409–20.

Shapiro, M. A. 1976. The role of turbulent heat flux in the generation of potential vorticity in the vicinity of upper-level jet stream systems. *Mon. Wea. Rev.* **104**, 892–906.

Shapiro, M. A. 1978. Further evidence of the mesoscale and turbulent structure of upper-level jet stream frontal systems. *Mon. Wea. Rev.* **106**, 1100–11.

Shapiro, M. A. 1980. Turbulent mixing within tropopause folds as a mechanism for the exchange of chemical constituents between the stratosphere and troposphere. *J. Atmos. Sci.* **37**, 994–1004.

Uccellini, L. W. 1986. The possible influence of upstream upper-level baroclinic processes on the development of the QE II storm. *Mon. Wea. Rev.* **114**, 1019–27.

Uccellini, L. W., K. F. Brill, R. A. Petersen, D. Keyser, R. Aune, P. J. Kocin and M. des Jardins 1988. A report on the upper-level wind conditions preceding and during the shuttle Challenger (STS 51L) explosion. *Bull. Am. Met. Soc.* 1248–65.

Uccellini, L. W., D. Keyser, K. F. Brill and C. H. Wash 1985. The President's Day cyclone of 18–19 February 1979: influence of upstream trough amplification and associated tropopause folding on rapid cyclogenesis. *Mon. Wea. Rev.* **113**, 941–61.

Uccellini, L. W. and P. J. Kocin 1987. The interaction of jet streak circulations during heavy snow events along the east coast of the United States. *Wea. Forecast.* **1**, 289–308.

Uccellini, L. W., P. J. Kocin, R. A. Petersen, C. H. Wash and K. F. Brill 1984. The President's Day cyclone of 18–19 February 1979: synoptic overview and analysis of the subtropical jet streak influencing the pre-cyclogenetic period. *Mon. Wea. Rev.* **112**, 31–55.

Uccellini, L. W., R. A. Petersen, K. F. Brill, P. J. Kocin and J. J. Tuccillo 1987. Synergistic interactions between an upper-level jet streak and diabatic processes that influence the development of a low-level jet and a secondary coastal cyclone. *Mon. Wea. Rev.* **115**, 2227–61.

Mid-tropospheric fronts and elevated mixed layers

Benjamin, S. G. and T. N. Carlson 1986. Some effects of surface heating and topography on the regional severe storm environment. Part I: Three-dimensional simulations. *Mon. Wea. Rev.* **114**, 307–29.

Carlson, T. N., R. A. Anthes, M. Schwartz, S. G. Benjamin and D. G. Baldwin 1980. Analysis and prediction of severe storms environment. *Bull. Am. Met. Soc.* **61**, 1018–32.

Carlson, T. N., S. G. Benjamin, G. S. Forbes and Y.-F. Li 1983. Elevated mixed layers in the regional severe storm environment: conceptual model and case studies. *Mon. Wea. Rev.* **111**, 1453–73.

Carlson, T. N. and F. H. Ludlam 1968. Conditions for the occurrence of severe local storms. *Tellus* **20**, 203–26.

Fawbush, E. J. and R. C. Miller 1954. The types of air masses in which North American tornadoes form. *Bull. Amer. Meteor. Soc.* **35**, 154–65.

Fennessey, M. J. and Y. Sud 1983. *A study of the influence of soil moisture on future precipitation.* NASA Technical Memorandum no. 85042, NASA, Greenbelt, MD.

Hirschberg, P. A. and J. M. Fritsch 1990. Tropopause undulations and the development of extratropical cyclones: an observational study. *Mon. Wea. Rev.* **118** (in press).

Iskenderian, H. 1988. Three-dimensional airflow and precipitation structure in a non-deepening cyclone. *Wea. Forecast.* **3**, 18–32.

Keyser, D. and T. N. Carlson 1984. Transverse ageostrophic circulations associated with elevated mixed layers. *Mon. Wea. Rev.* **112**, 2465–78.

Lakhtakia, M. and T. T. Warner 1987. A real-date numerical study of the development of precipitation along the edge of an elevated mixed layer. *Mon. Wea. Rev.* **115**, 156–68.

McCarthy, J. and S. E. Koch 1982. The evolution of an Oklahoma dryline. Part I: A meso- and subsynoptic-scale analysis. *J. Atmos. Sci.* **39**, 225–36.

Index

CPSIA information can be obtained at www.ICGtesting.com

262111BV00002B/1/P

9 781878 220301